Industrial and Applied Mathematics

Industrial and Applied Mathematics, a Scopus-indexed book series, publishes high-quality research-level monographs, lecture notes and contributed volumes focusing on areas where mathematics is used in a fundamental way, such as industrial mathematics, bio-mathematics, financial mathematics, applied statistics, operations research and computer science.

Durdimurod K. Durdiev

Inverse Problems for Fractional Diffusion Equations

Springer

Durdimurod K. Durdiev
The Institute of Mathematics
Academy of Sciences
Tashkent, Uzbekistan

Bukhara State University
Bukhara, Uzbekistan

ISSN 2364-6837 ISSN 2364-6845 (electronic)
Industrial and Applied Mathematics
ISBN 978-981-96-5337-9 ISBN 978-981-96-5338-6 (eBook)
https://doi.org/10.1007/978-981-96-5338-6

Mathematics Subject Classification: 35R30, 35A01, 35A08, 35A02, 45D05, 35R11

This Springer imprint is published by the registered company Springer Nature Singapore Pte Ltd.
The registered company address is: 152 Beach Road, #21-01/04 Gateway East, Singapore 189721, Singapore

This book is dedicated to my wife, Maraljan M. Kadirova, whose patience throughout our life together and unwavering support have always inspired me

Preface

The aim of this book is to introduce readers to the rich and fascinating field of inverse problems for time-fractional diffusion equations in modern mathematics.

The problem of determining the density of a diffusing material at interior points of a domain, given the initial and boundary conditions along with a specified source term, is referred to as a *direct (forward) diffusion problem.* Many important real-world applications lead to inverse problems, which involve recovering unknown causes based on observations of their effects. Mathematically, in an inverse problem, the differential (or integro-differential) equation is only partially specified. Some functions that are part of the equation—such as terms on the right-hand side, kernels in integro-differential equations, or initial and boundary conditions—remain unknown. Problems where these unknowns are determined using additional information (referred to as overdetermination) about the solution of the direct problem are called *inverse (backward) problems.* In the case of a diffusion equation, these unknowns may include the reaction and diffusion coefficients, the source term, or the initial and boundary functions. Such problems are ill-posed in the sense of Hadamard.

Direct problems typically study well-posed problems, meaning problems that have solutions which are unique and stable with respect to small changes in the input data within suitable functional spaces. In the case of evolution differential equations, these include *initial problems (Cauchy problems)* and *initial-boundary value problems.* In such problems, the differential equation, along with certain initial and boundary conditions, is specified. Over the past few decades, fractional calculus has garnered significant attention not only from mathematicians and engineers but also from scientists across various fields. Indeed, the tools of fractional calculus have found numerous applications in applied sciences, including nanotechnology, control theory, viscoplastic flow, biology, image processing, and more.

Many applications of fractional calculus have been successful because fractional-order models are often more accurate than integer-order models. This is due to the fact that fractional-order models provide more degrees of freedom compared to their classical counterparts. By accommodating non-integer-order dynamics, fractional calculus serves as a powerful tool for explaining the memory and hereditary properties of various materials and processes. Since the 19th century, the theory of fractional calculus has advanced rapidly across many scientific fields, including fractional geometry, fractional differential equations, and fractional dynamics, owing to its broad range of applications.

The diffusion process is one of the most well-known transport mechanisms in nature. At the microscopic level, it is associated with the random motion of individual particles. The use of the Laplace operator and the first-order derivative in the canonical diffusion model is based on the assumption of Gaussian particle motion. Over the past three decades, extensive research has demonstrated that anomalous diffusion models, where the mean squared displacement grows faster (superdiffusion) or slower (subdiffusion) than in Gaussian processes, can accurately describe experimental data under certain conditions. For instance, anomalous diffusion is frequently observed in memory materials such as viscoelastic substances and in heterogeneous media, including soil, heterogeneous aquifers, and underground fluid flows.

In this book, we discuss various inverse problems for the time-fractional diffusion equation, including inverse coefficient problems (nonlinear problems) and inverse problems involving the determination of the right-hand sides of equations and initial functions (linear problems). The study of inverse problems necessitates a thorough investigation of the corresponding direct problems, which involves deriving representation formulas, establishing a priori estimates, and analyzing the differential properties of the solution, among other aspects. This is particularly crucial for nonlinear problems, where obtaining solvability theorems requires a detailed understanding of how the differential properties of the solution to the direct problem depend on the smoothness of the coefficients and other problem data. As a result, direct problems, such as initial problems (Cauchy problems) and initial-boundary value problems with various boundary conditions, occupy a significant portion of the book.

The content of the book can be described as follows.

Chapter 1 addresses the direct initial-boundary value problem and the coefficient inverse problems for diffusion equations with fractional time derivatives. The overdetermination conditions correspond to imposing the value of the heat flux at the left end of the rod. The usual initial condition (Cauchy condition) and Dirichlet boundary conditions are also specified. The unknowns in the inverse problems are the time-dependent reaction and diffusion coefficients. Theorems on the local existence and global uniqueness are proved, and conditional stability estimates for the solutions to the direct and inverse problems are obtained.

Chapter 2 deals with inverse problems of determining the coefficient of the lower-order term in a multidimensional fractional diffusion equation, where the direct problem is the initial-boundary value problem for the equation with non-local initial and homogeneous Dirichlet conditions. To determine the unknown coefficient, an overdetermination condition in the integral form is imposed on the solution of the direct problem. By applying the fixed-point argument in suitable Sobolev spaces, global theorems on the existence and uniqueness of solutions to the inverse problem are established. Additionally, the inverse problem of simultaneously determining two unknown functions in a one-dimensional fractional diffusion equation with periodic boundary conditions (non-local conditions) is also studied.

Chapter 3 is devoted to inverse problems of determining the time-dependent reaction-diffusion coefficient in the Cauchy problem for time-fractional diffusion equations with the Caputo and Hilfer (generalized Riemann-Liouville) operators. As an overdetermination condition, a single observation at the point $x = 0$ of the fractional diffusion process is specified. New theorems on the unique solvability of these problems are established, along with stability estimates for the solutions.

Chapter 4 discusses inverse source and coefficient problems for first- and half-order time-fractional diffusion equations using the Carleman estimate method. Conditional stability estimates of Hölder and Lipschitz types are established for solutions to these inverse problems.

Chapter 5 addresses the inverse problems (linear problems) of recovering source and initial functions in fractional diffusion equations under non-local boundary and integral overdetermination conditions. The existence theorems and stability estimates for solutions to the direct and inverse problems are derived. Additionally, the inverse source problem for a mixed-type equation, comprising a fractional diffusion equation in the parabolic part and a wave equation in the hyperbolic part of a cylindrical domain, is studied. The existence, uniqueness, and stability results are proved, and examples are provided to illustrate the obtained results.

Chapter 6 examines the inverse problems of determining the convolution kernel in the fractional integro-differential diffusion equation for both bounded and unbounded domains. These problems are reduced to equivalent integral equations of the Volterra type. In the case of a bounded domain, the overdetermination condition corresponds to imposing the value of the heat flux at the endpoint $x = 0$ of the rod. For an unbounded domain, it involves specifying the "temperature" at the point $x = 0$. Using the contraction mapping principle, the local existence and global uniqueness of solutions are proven. Additionally, stability estimates are obtained.

In Chap. 7, the method introduced in Chap. 1 is adapted to the case where two unknowns of the equation are determined simultaneously. The nonlinear inverse problems involving the simultaneous determination of the time-dependent zero-order coefficient and the source function in the time-fractional diffusion equation, as well as the simultaneous determination of the source function and the kernel in the equation with a convolution integral, are considered. In this case, the direct problem is formulated as an initial-boundary value problem, with Dirichlet boundary conditions in the first section and Neumann boundary conditions in the second section. The overdetermination conditions are specified in the form of integral constraints. New theorems regarding the solvability of these problems are obtained.

This book presents various methods for investigating direct and inverse problems for time-fractional diffusion equations. The results included here do not aim to be comprehensive; they represent only a small part of the rapidly developing theory of direct and inverse problems for fractional differential equations. It should also be noted that the literature lists provided at the end of each chapter are not exhaustive but rather reflect the author's preferences.

The author would like to thank Askar Rahmonov, Jonibek Jumaev, Halim Turdiev, Umidjon Durdiev, and Dilshod Durdiev for their valuable discussions on the problems addressed in this book and for their collaboration on several joint papers.

This book is intended for mathematicians and young researchers working on inverse problems, and it may also serve as a resource for doctoral courses.

Bukhara, Tashkent, Uzbekistan
September 2024

Durdimurod K. Durdiev

Contents

Chapter 1
Coefficient Determination Problems with Local Conditions

The time-fractional diffusion equations have received much attention to the most of the research communities due to their applications in various fields of science and engineering, such as physics, chemistry, biology, control, viscoelasticity, electrical, and mechanical engineering (see [1–4] and references therein).

This chapter will consider two inverse problems of determining the time-dependent reaction coefficient and one problem of finding the time-dependent diffusion coefficient. In the first two problems, the direct problems are initial-boundary value problems with Dirichlet boundary conditions for the first case and periodic conditions for the second case. The overdetermination condition for both inverse problems corresponds to the assignment of the heat flux value at the left end of the rod. Theorems on the local existence and global uniqueness of inverse problems are obtained, as well as estimates of the conditional stability of the solution are constructed. In the third inverse problem, the problem of determining the time-dependent diffusion coefficient is studied. For the direct problem, the existence, uniqueness, and some regularity properties of a weak solution of the fractional diffusion equation with a given diffusion coefficient are proven. For the inverse problem, in order to determine the unknown coefficient, an operator is introduced and its monotonicity is shown. Using this property, the uniqueness of the determination of this coefficient is established.

The unique solvability theorems are proved and the conditional stability estimates for the solutions to the direct and the inverse problems are obtained.

D. K. Durdiev, *Inverse Problems for Fractional Diffusion Equations*,
Industrial and Applied Mathematics, https://doi.org/10.1007/978-981-96-5338-6_1

1.1 Zero-Order Coefficient Determination Problem

Nowadays, significant attention has been given to the study of initial and initial-boundary value problems for fractional differential equations, due to their widespread applications in engineering, physics, chemistry, biology, and other fields. Various applications can be found in [5] and references therein.

The direct problems for fractional diffusion equations, such as initial or boundary value problems, have been extensively studied in [6–11] and related works. A general representation of solutions for the fractional diffusion equation, along with the construction of Green's functions for first, second, and mixed boundary value problems, is explored in [12]. Additionally, [13] investigates the uniqueness and existence of solutions for initial-boundary value problems in generalized time-fractional diffusion equations over open bounded domains. The uniqueness of the solution is established using a maximum principle for these equations.

Inverse problems for classical integro-differential heat and wave propagation equations have been widely studied. Most of the existing literature focuses on nonlinear inverse problems with various overdetermination conditions (see, e.g., [14–20] and references therein). These studies discuss unique solvability, stability estimates, and numerical approaches for solving such problems. However, inverse problems for fractional differential heat equations remain less explored. The literature primarily addresses linear source determination problems and nonlinear coefficient inverse problems for the Cauchy problem with different overdetermination conditions (see, e.g., [21–32]). The key results in these works include existence and uniqueness theorems, as well as stability estimates for solving the reaction coefficient in time-fractional diffusion equations.

The linear heat equation involving a fractional derivative in time with a nonlocal boundary condition was examined in [33]. This study focuses on determining a source term independent of the spatial variable and analyzing the temperature distribution for a problem with an integral-type overdetermination condition. Furthermore, it establishes the existence and uniqueness of the solution, along with its continuous dependence on the given data.

1.1.1 Introduction and Setting Up the Problem

The time-fractional diffusion equation is deduced by replacing the classical time derivative with a time-fractional derivative. The direct problems: initial-value problems and initial-boundary value problems for the time-fractional diffusion equation have been studied extensively in recent years (see [5] and references therein).

Let $T > 0$ be a fixed number and $D_T := \{(x, t) : 0 < x < l,\ 0 < t \leq T\}$. Consider the time-fractional diffusion equation

$$\partial_t^\alpha u - u_{xx} + q(t)u(x, t) = f(x, t), \quad (x, t) \in D_T, \tag{1.1}$$

with the initial condition

$$u(x,0) = \varphi(x), \quad x \in [0,l], \tag{1.2}$$

the boundary conditions

$$u|_{x=0} = u|_{x=l} = 0, \quad \varphi(0) = \varphi(l) = 0, \quad t \in [0,T], \tag{1.3}$$

where ∂_t^α is the Caputo fractional derivative of order $0 < \alpha < 1$ of the integrable function u in the time variable:

$$\partial_t^\alpha u(x,t) = I_t^{1-\alpha} \frac{\partial}{\partial t} u(x,t) = \frac{1}{\Gamma(1-\alpha)} \int_0^t (t-\tau)^{-\alpha} \frac{\partial}{\partial \tau} u(x,\tau) d\tau, \tag{1.4}$$

$$(I_t^\alpha u)(x,t) = \frac{1}{\Gamma(\alpha)} \int_0^t (t-\tau)^{\alpha-1} u(x,\tau) d\tau \tag{1.5}$$

is the Riemann-Liouville fractional integral of the function $u(x,t)$ with respect to t, $\Gamma(\cdot)$ is Euler's Gamma function; $q(t), t > 0$ is the reaction coefficient, $f(x,t)$ is known source term, $\varphi(x)$ is the initial temperature.

The Caputo fractional derivative for $0 < \alpha < 1$ with the Riemann-Liouville fractional derivative

$$\frac{\partial^\alpha}{\partial t^\alpha} u(x,t) = \frac{\partial}{\partial t} I_t^{1-\alpha} u(x,t) = \frac{1}{\Gamma(1-\alpha)} \frac{\partial}{\partial t} \int_0^t (t-\tau)^{-\alpha} u(x,\tau) d\tau \tag{1.6}$$

is connected by the following relation:

$$\partial_t^\alpha u(x,t) = \frac{\partial^\alpha}{\partial t^\alpha} u(x,t) - \frac{u(x,0)}{\Gamma(1-\alpha) t^\alpha}. \tag{1.7}$$

In general, the γth-order Riemann-Liouville and Caputo fractional derivatives in time variable are defined by

$$\frac{\partial^\gamma}{\partial t^\gamma} u(x,t) = \frac{1}{\Gamma(n-\gamma)} \frac{\partial^n}{\partial t^n} \int_0^t (t-\tau)^{n-1-\gamma} u(x,\tau) d\tau,$$

and

$$\partial_t^\gamma u(x,t) = \frac{1}{\Gamma(n-\gamma)} \int_0^t (t-\tau)^{n-1-\gamma} \frac{\partial^n u(x,\tau)}{\partial \tau^n} d\tau,$$

respectively, for $n-1 < \gamma < n$, $n \in \mathbb{N}$ (see, e.g., [5], pp. 69–99).

We pose the inverse problem as follows: find the function $q(t)$ in (1.1), if the solution of the problem (1.1)–(1.3) satisfies

$$u_x(x,t)|_{x=0} = h(t), \quad t \in [0,T], \tag{1.8}$$

$h(t)$ is given function of $t \in [0, T]$.

The additional condition (1.8) corresponds to a given heat flow at the left end of a rod of length l.

Definition 1.1 A function $u(x,t)$ is called a classical solution to the initial-boundary problem (1.1)–(1.3) if:

(1) $u(x,t)$ is twice continuously differentiable in x for each $t > 0$;
(2) for each $x \in (0,l)$ function $u(x,t)$ is continuous in t on $[0, T]$, and its fractional integral (1.5) is continuously differentiable in t for $t > 0$;
(3) $u(x,t)$ satisfies (1.1)–(1.3).

The class of functions satisfying conditions (1), (2) of Definition 1.1 denote by $C^{2,\alpha}(D_T)$.

By $C^{m,k}(D_T)$ we denote the class of functions that are m times continuously differentiable with respect to x and k times continuously differentiable with respect to t in the domain D_T.

Let $K_\alpha(t)$ be the kernel of the Riemann-Liouville fractional integral, i.e., $K_\alpha(t) = \frac{1}{\Gamma(\alpha)} t^{\alpha-1}$.

Lemma 1.1 *The relation* $\lim\limits_{\alpha\to 0+0} K_\alpha(t) = \delta(t)$ *is valid in the sense of generalized functions,* $\delta(t)$ *is the Dirac function.*

Proof (see [34]). Let $f(x)$ be a function locally integrable everywhere except at x_o where it has a nonintegrable singularity (for instance, $f(x) = 1/x$ on the line). Then in general the integral

$$\int\limits_{\mathbb{R}} f(x)\phi(x)dx, \tag{1.9}$$

where $\phi(x)$ is a test function, will diverge. But this integral will converge if $\phi(x)$ vanishes in a neighborhood of x_0. We may now ask whether it is possible to use this result to redefine a functional, that is to construct an generalized function $f(x)$ such that for all test functions $\phi(x)$ vanishing in a neighborhood of x_0, the functional has the value given by (1.9). Any such functional f is called a regularization of the divergent integral of (1.9) ([34], pp. 45–78).

Proposition If there exists an integer $m > 0$ such that $f(x)x^m$ is locally integrable, the integral of (1.9) can be regularized.

For this case we can construct the regularization f in accordance with the equation

$$(f,\phi) = \int\limits_{\mathbb{R}} f(x)\left\{\phi(x) - \left[\phi(0) + \phi'(0)x + \cdots + \phi^{(m)}(0)\frac{x^m}{m!}\right]\theta(1-x)\right\}dx, \tag{1.10}$$

where we subtract enough terms of Taylor's series to leave a remainder of order greater than x^m. The function $\theta(1-x)$ is Heaviside's function: $\theta(1-x) = 1$ for $x < 1$ and $\theta(1-x) = 0$ for $x \geq 1$.

Clearly the integral in (1.10) converges for all test functions $\phi(x)$ and is a continuous linear functional. If, further, $\phi(x)$ vanishes in a neighborhood of the origin, (1.10) becomes

$$(f, \phi) = \int_{\mathbb{R}} f(x)\phi(x)dx$$

so that except at the origin f coincides with $f(x)$. For this case, therefore, the regularization exists.

Another approach to constructing a regularization of a functional is the analytical continuation method. The analytic continuation method is the following: let $f_\lambda(x)$ be a locally integrable function (of x) when λ is in some domain Λ of the complex plane, but not in general otherwise. Further, for $\lambda \in \Lambda$ let (f_λ, ϕ) be analytic for every test function $\phi(x)$, and assume that it can be extended analytically to a wider domain Λ_1, independent of $\phi(x)$. Then, with the function $f_{\lambda_0}(x)$ for $\lambda_0 \in \Lambda - \Lambda_1$ we may associate the functional $\left(f_{\lambda_0}, \phi\right)$ obtained by analytic continuation of (f_λ, ϕ) out of λ . In other words we shall write

$$\int_{\mathbb{R}} f_{\lambda_0}(x)\phi(x)dx = \text{anal. cont.}_{\lambda \to \lambda_0} \int_{\mathbb{R}} f_\lambda(x)\phi(x)dx.$$

For instance, to define the generalized function $x^{-3/2}$, we shall consider function

$$x_+^\lambda = \begin{cases} x^\lambda & \text{for } x > 0, \\ 0 & \text{for } x \leq 0. \end{cases}$$

For $\Re e\lambda > -1$ this is the regular functional given by

$$\left(x_+^\lambda, \phi\right) = \int_0^\infty x^\lambda \phi(x)dx \tag{1.11}$$

Now (1.11) is a function which is obviously analytic in λ, for its derivative with respect to λ is

$$\int_0^\infty x^\lambda \ln x \phi(x)dx.$$

Let us rewrite the right-hand side of (1.11) in the form

$$\int_0^{\infty} x^{\lambda}\phi(x)dx = \int_0^{1} x^{\lambda}\left[\phi(x)-\phi(0)\right]dx + \int_1^{\infty} x^{\lambda}\phi(x)dx + \frac{\phi(0)}{\lambda+1}. \tag{1.12}$$

The first term here is defined for $\Re e\lambda > -2$, the second for all λ, and the third for $\lambda \neq -1$. Thus the functional defined in (1.11) can be analytically continued to $\Re e\lambda > -2$, $\lambda \neq -1$. In particular, for $\lambda = -3/2$ we have

$$\left(x_+^{-3/2}, \phi\right) = \int_0^{1} x^{-3/2}\left[\phi(x)-\phi(0)\right]dx + \int_1^{\infty} x^{-3/2}\phi(x)dx - 2\phi(0). \tag{1.13}$$

The right-hand side of (1.13) agrees with that of (1.9), if we choose $m = 1$ and since in general

$$\int_1^{\infty} x^{\lambda}dx = -\frac{1}{\lambda+1}.$$

Consider now (1.12) for $\lambda = \alpha - 1 \in (-1, 0)$ regarding the function $\frac{x_+^{\alpha-1}}{\Gamma(\alpha)}$:

$$\left(\frac{x_+^{\alpha-1}}{\Gamma(\alpha)}, \phi\right) = \int_0^{\infty} \frac{x^{\alpha-1}}{\Gamma(\alpha)}\phi(x)dx$$

$$= \alpha \int_0^{1} \frac{x^{\alpha}}{\Gamma(1+\alpha)} \frac{\phi(x)-\phi(0)}{x}dx + \alpha \int_1^{\infty} \frac{x^{\alpha-1}}{\Gamma(1+\alpha)}\phi(x)dx + \frac{\phi(0)}{\Gamma(1+\alpha)}. \tag{1.14}$$

In (1.14) we used the equality $\Gamma(\alpha) = \Gamma(1+\alpha)/\alpha$. On the right side of (1.14), the integrals are finite for any test function $\phi(x)$. Then, from (1.14) we get $\lim\limits_{\alpha\to 0+0} \left(K_\alpha(x), \phi(x)\right) = \phi(0)$, that means

$$\lim_{\alpha\to 0+0} \frac{x_+^{\alpha-1}}{\Gamma(\alpha)} = \delta(x),$$

in the generalized sense.

Based on Lemma 1.1 we have $\lim\limits_{\alpha\to 1-0} \partial_\alpha^t u(x,t) = \frac{\partial u(x,t)}{\partial t}$, $\frac{\partial}{\partial t}$ is the classical partial derivative. Now, in the future, we can also put $\alpha = 1$ and consider $\alpha \in (0,\ 1]$.

1.1.2 Derivation of the Equivalent Problem

Let $u(x, t)$ be a classical solution to the problem (1.1)–(1.3) and f, φ, h be enough smooth functions. We carry out the next converting of the inverse problem (1.1)–(1.8). Denote for this purpose the second derivative of $u(x, t)$ with respect to x, by $\vartheta(x, t)$, i.e., $\vartheta(x, t) := u_{xx}(x, t)$. Differentiating (1.1) and (1.2) twice in x, we get

$$\partial_t^\alpha v - v_{xx} + q(t)v(x, t) = f_{xx}(x, t), \quad (x, t) \in D_T, \tag{1.15}$$

$$\vartheta(x, 0) = \varphi''(x), \quad x \in [0, l]. \tag{1.16}$$

To obtain boundary conditions for the function $v(x, t)$, we note that the second term in (1.1) is $v(x, t)$. Setting $x = 0$, $x = l$ and using equality (1.3), we obtain

$$v(0, t) = -f(0, t), \ v(l, t) = -f(l, t) \ x \in [0, l].$$

Suppose $f(0, t) = f(l, t) = 0$. Then we have the following boundary condition

$$v(0, t) = v(l, t) = 0. \tag{1.17}$$

To obtain an additional condition for the function $\vartheta(x, t)$, we differentiate equation (1.1) with respect to x and using equality $u_{xx}(x, t) = \vartheta(x, t)$ and additional condition (1.8), we get

$$v_x(0, t) = q(t)h(t) + \partial_t^\alpha h(t) - f_x(0, t). \tag{1.18}$$

Thus, if the problem (1.1)–(1.8) has a solution $(u, \ q)$, then the problem (1.15)–(1.18) also has a solution $(v, \ q)$ with the same q, and $v(x, t) = u_{xx}(x, t)$.

Conversely, let $(v, \ q)$ satisfies the equations (1.15)–(1.18). First of all, we will show that conditions (1.3) are fulfilled. Since $u_{xx}(x, t) = v(x, t)$, from (1.1) we have

$$v(x, t) = \partial_t^\alpha u + q(t)u(x, t) - f(x, t), \quad (x, t) \in D_{Tl}. \tag{1.19}$$

Denoting $u|_{x=0} = u|_{x=l} =: y(t)$ we prove $y(t) = 0$. By $f(0, t) = f(l, t) = 0$, setting $x = 0$ and $x = l$ in (1.19), from (1.17) and (1.19) one obtains

$$\partial_t^\alpha y(t) + q(t)y(t) = 0, \quad t \in (0, T), \ y(0) = 0. \tag{1.20}$$

As follows from (1.7) the fractional Caputo derivative $\partial_t^\alpha y(t)$ for $y(0) = 0$ coincides with the fractional Riemann-Liouville derivative $\frac{\partial^\alpha}{\partial t^\alpha} y(t) = \frac{d}{dt} I_t^{1-\alpha} y(t)$. Due to this, the problem (1.20) is equivalent to the homogeneous Abel's integral equation $y(t) = -I_t^\alpha \left[q(t)y(t)\right]$, which in accordance with the general theory of integral equations for $q(t) \in C[0, T]$ has only the solution $y(t) = 0$.

Further, from $u_{xx}(x,t) = v(x,t)$ we have

$$u(x,t) = u_x(0,t)x + \int_0^x (x-\xi)v(\xi,t)d\xi.$$

When the matching condition $\varphi'(0) = h(0)$ is valid, it follows

$$u(x,0) = h(0)x + \int_0^x (x-\xi)v(\xi,0)d\xi = \varphi'(0)x + \int_0^x (x-\xi)\varphi''(\xi)d\xi = \varphi(x),$$

i.e., the condition (1.2) is satisfied. Further, the equations (1.15) and (1.18) imply

$$\partial_t^\alpha u - u_{xx} + q(t)u(x,t) - f(x,t) = \int_0^x (x-\xi)\Big[\partial_t^\alpha v(\xi,t) - v_{\xi\xi}(\xi,t) + q(t)v(\xi,t)\Big]d\xi$$

$$+ \partial_t^\alpha h(t)\cdot x - \big[q(t)h(t) + \partial_t^\alpha h(t) - f_x(0,t)\big]x - f(0,t) + q(t)h(t)\cdot x.$$

Since $\partial_t^\alpha v(\xi,t) - v_{\xi\xi}(\xi,t) + q(t)v(\xi,t) = f_{\xi\xi}(\xi,t)$, from the previous equalities, we get $\partial_t^\alpha u - u_{xx} + q(t)u - f = 0$, that is equation (1.1).

The next subsection is devoted to study of the direct problem (1.15)–(1.17).

1.1.3 The Direct Problem: Existence and Uniqueness

Definition 1.2 We will say that the function $\phi(x) \in \Phi^k[0,l]$, k is an even number, if the following conditions are met:

(1) $\phi^i(x) \in C[0,l],\ i = 0, 1, ..., k$ and $\phi^{k+1}(x) \in L_2(0,l)$;

(2) $\phi(0) = \phi(l) = \phi''(0) = \phi''(l) = \cdots = \phi^{(k)}(0) = \phi^{(k)}(l) = 0$.

We also use for functions depending on both variables x and t the class $C^m\left([0,T];\ \Phi^k[0,l]\right)$ $m-$ time continuously differentiable in $t \in [0,T]$ with values in $\Phi^k[0,l]$ of functions; $C^0\left([0,T];\ \Phi^k[0,l]\right) = C\left([0,T];\ \Phi^k[0,l]\right)$.

We assume that the data of the problem (1.1)–(1.3), (1.8) satisfy the following conditions:

(A1) $\varphi(x) \in \Phi^4[0,l]$, $f(x,t) \in C\left([0,T],\ \Phi^4[0,l]\right)$,

(A2) $h(t) \in C^1[0,T]$ and $|h(t)| \geq h_0 = const > 0$, h_0 is a given number, $\varphi'(0) = h(0)$, $\varphi'''(0) = q(0)h(0) + \partial_t^\alpha h(t) - f_x(0,0)$.

First we show that the equivalent direct problem (1.15)–(1.17) has a unique solution. We will seek the solution of this problem in the form

$$\vartheta(x,t)=\sum_{n=1}^{\infty}\vartheta_n(t)\sin(\lambda_n x),\ \ \lambda_n=\frac{\pi n}{l},\ \ \vartheta_n(t)=\sqrt{\frac{2}{l}}\int_0^l \vartheta(x,t)\sin(\lambda_n x)dx. \tag{1.21}$$

In view of equalities (1.21), from (1.15) we obtain the following equation:

$$\left(\partial_t^\alpha \vartheta_n\right)(t)+\lambda_n^2\vartheta_n(t)=F_n(t;\vartheta,q,f), \tag{1.22}$$

where

$$F_n(t;\vartheta,q,f):=f_n(t)-q(t)\vartheta_n(t),\ \ f_n(t)=\sqrt{\frac{2}{l}}\int_0^l f_{xx}(x,t)\sin(\lambda_n x)dx.$$

The initial condition (1.16) gives

$$\vartheta_n(0)=\sqrt{\frac{2}{l}}\int\limits_0^l \vartheta(x,0)\sin(\lambda_n x)dx$$

$$=\sqrt{\frac{2}{l}}\int\limits_0^l \varphi''(x)\sin(\lambda_n x)dx=:\varphi_n. \tag{1.23}$$

To investigate the problem (1.22), (1.23) we use the following theorem (see [5], Theorem 5.15):

Theorem 1.1 *Let* $n-1<\alpha\le n,\ n\in\mathbb{N},\ \ \lambda\in\mathbb{R}$ *and let* $f(x)$ *be a given real function defined on* $\mathbb{R}_+:=\{x\in\mathbb{R}:\ x>0\}$. *Then the equation*

$$\partial_t^\alpha y(x)-\lambda y(x)=f(x),\ \ x\in\mathbb{R}_+ \tag{1.24}$$

is solvable, and its general solution is given by

$$y(x)=\int\limits_0^x (x-t)^{\alpha-1}E_{\alpha,\alpha}\left(\lambda(x-t)^\alpha\right)f(t)dt$$

$$+\sum_{j=0}^{l-1}c_j x^j E_{\alpha,j+1}\left(\lambda t^\alpha\right),$$

where $c_j,\ j=0,...,n-1$, *are arbitrary real constants.*

In particular, the general solution to the equation (1.24) *with* $0<\alpha\le 1$ *and* $1<\alpha\le 2$ *have the forms*

$$y(x) = \int_0^x (x-t)^{\alpha-1} E_{\alpha,\alpha}\left(\lambda(x-t)^{\alpha}\right) f(t)dt + c_1 E_{\alpha}\left(\lambda t^{\alpha}\right) \tag{1.25}$$

and

$$y(x) = \int_0^x (x-t)^{\alpha-1} E_{\alpha,\alpha}\left(\lambda(x-t)^{\alpha}\right) f(t)dt$$

$$+ c_1 E_{\alpha}\left(\lambda t^{\alpha}\right) + c_2 x E_{\alpha,2}\left(\lambda t^{\alpha}\right),$$

respectively, where c_1 *and* c_2 *are arbitrary real constants,* $E_{\alpha,\beta}(z)$ *is the Mittag-Leffler function defined by the following series (see, for example, [5] pp. 40–45):*

$$E_{\alpha,\beta}(z) = \sum_{k=0}^{\infty} \frac{z^k}{\Gamma(\alpha k + \beta)} \tag{1.26}$$

for $\alpha, \beta, z \in \mathbb{C}$ *with* $\Re(\alpha) > 0$, $\Re(\alpha)$ *denotes the real part of the complex number* α *and*

$$E_{\alpha,1}(z) =: E_{\alpha}(z) = \sum_{k=0}^{\infty} \frac{z^k}{\Gamma(\alpha k + 1)}. \tag{1.27}$$

In particular, when $\alpha = 1$ *and* $\alpha = 2$*, we have*

$$E_1(z) = e^z \quad \text{and} \quad E_2(z) = \cosh\left(\sqrt{z}\right).$$

Applying this theorem to the initial-value problem (1.22), (1.23), we note this problem is equivalent in the space $C[0, T]$ to the following Volterra integral equation of the second kind

$$\vartheta_n(t) = \varphi_n E_{\alpha}(-\lambda_n^2 t^{\alpha}) + \int_0^t (t-\tau)^{\alpha-1} E_{\alpha,\alpha}\left(-\lambda_n^2 (t-\tau)^{\alpha}\right) f_n(\tau)d\tau$$

$$+ \int_0^t (t-\tau)^{\alpha-1} E_{\alpha,\alpha}\left(-\lambda_n^2 (t-\tau)^{\alpha}\right) q(\tau)\vartheta_n(\tau)d\tau. \tag{1.28}$$

Further, to evaluate expressions with the Mittag-Leffler function, we use the following three statements (see [5] pp. 40–45). The proof of these assertions comes from the definition of the Caputo fractional derivative and differentiation of the Mittag-Leffler function.

Proposition 1.1 *Let* $0 < \alpha < 2$ *and* $\beta \in \mathbb{R}$ *be arbitrary. We suppose that* κ *is such that* $\pi\alpha/2 < \kappa < \min\{\pi, \pi\alpha\}$*. Then there exists a constant* $C = C(\alpha, \beta, \kappa) > 0$ *such that*

$$\left|E_{\alpha,\beta}(z)\right| \le \frac{C}{1+|z|}, \quad \kappa \le |arg(z)| \le \pi.$$

Proposition 1.2 *For* $0 < \alpha < 1$, $t > 0$, *we have* $0 < E_\alpha(-t) < 1$. *Moreover,* $E_\alpha(-t)$ *is completely monotonic, that is*

$$(-1)^n \frac{d^n}{dt^n} E_\alpha(-t) \geq 0, \quad for\, n \in \mathbb{N}.$$

Proposition 1.3 *For* $0 < \alpha < 1$, $\eta > 0$, *we have* $0 \leq E_{\alpha,\alpha}(-\eta) \leq \frac{1}{\Gamma(\alpha)}$. *Moreover,* $E_{\alpha,\alpha}(-\eta)$ *is a monotonic decreasing function with* $\eta > 0$.

Using Propositions 1.1–1.3 and equation (1.28), we estimate $v_n(t)$. As result, for $t \in [0, T]$, we get the following integral inequality:

$$|\vartheta_n(t)| \leq |\varphi_n| + \frac{t^\alpha \|f_n\|}{\alpha\Gamma(\alpha)} + \frac{||q||}{\Gamma(\alpha)} \int_0^t (t-\tau)^{\alpha-1} |\vartheta_n(\tau)| d\tau, \quad n \in \mathbb{N}, \tag{1.29}$$

where $\|f_n\| = \max_{t\in[0,T]} |f_n(t)|$, $\|q\| = \max_{t\in[0,T]} |q(t)|$.

The following results concern some properties of ∂_t^α, which are crucial for the future development. The proofs of the lemmas below can be found in ([5], pp. 90–99).

Lemma 1.2 *For* $0 < \alpha < 1$, $u(t)$, $\partial_t^\alpha u(t) \in C[0, T]$, *we have*

$$\partial_t^\alpha \left(I_t^\alpha\right) u(t) = I_t^\alpha \left(\partial_t^\alpha\right) u(t) = u(t) - u(0), \ t \in [0, T].$$

Lemma 1.3 *Fix* $0 < \alpha < 1$ *and given* $u(t) \in C[0, T]$ *with* $\partial_t^\alpha u \in C[0, T]$. *If* $u(t)$ *attains its maximum over the interval* $[0, T]$ *at the point* $t = t_0 \in (0, T]$, *then* $\partial_{t_0}^\alpha u \geqslant 0$.

Next, we also need the following assertion on a Gronwall-type inequality with a weakly singular kernel (see [35], p. 188).

Lemma 1.4 *Let* $m(t) \in C[t_0, T]$ $(t_0 \in \mathbb{R}_+ = [0, \infty), T \leq +\infty)$ *and suppose that*

$$m(t) \leq m_0 + \frac{L}{\Gamma(\gamma)} \int_{t_0}^t (t-s)^{\gamma-1} m(s) ds, \ \ t \in [t_0, T].$$

Then we have

$$m(t) \leq m_0 E_{\gamma,1}(L(t-t_0)^\gamma), \ \ t \in [t_0, T],$$

where m_0 *and* L *are nonnegative constants,* $\gamma \in (0, 1)$.

Applying Lemmas 1.4–(1.29), we have the following estimate for $|\vartheta_n(t)|$:

$$|\vartheta_n(t)| \leq \left(|\varphi_n| + \frac{t^\alpha \|f_n\|}{\alpha\Gamma(\alpha)}\right) E_\alpha(||q|| t^\alpha). \tag{1.30}$$

Using equality (1.22), from (1.30), we obtain an estimate for $|\partial_t^\alpha \vartheta_n(t)|$:

$$|\partial_t^\alpha \vartheta_n(t)| \leq (\lambda_n^2 + \|q\|)\left(|\varphi_n| + \frac{t^\alpha \|f_n\|}{\alpha\Gamma(\alpha)}\right)E_\alpha(\|q\|t^\alpha) + \|f_n\|.$$

From here, based also on Propositions 1.1–1.3, we have the following lemma:

Lemma 1.5 *For any $t \in [0, T]$ and $n \in \mathbb{N}$, the following estimates are valid:*

$$|\vartheta_n(t)| \leq \overline{C}_1\Big(|\varphi_n| + \|f_n\|\Big), \quad |\partial_t^\alpha \vartheta_n(t)| \leq \overline{C}_2\Big(n^2|\varphi_n| + n^2\|f_n\|\Big),$$

where $\overline{C}_i$, $i = 1, 2$, are positive constants, depending only on α, T and $\|q\|$.

Formally, from (1.21) the term-by-term differentiation we compose the series

$$\partial_t^\alpha \vartheta(x,t) = \sum_{n=1}^{\infty} \partial_t^\alpha \vartheta_n(t)\sin(\lambda_n x), \quad \vartheta_{xx}(x,t) = \sum_{n=1}^{\infty} \lambda_n^2 \vartheta_n(t)\sin(\lambda_n x). \tag{1.31}$$

Based on Lemma 1.5, we note the series (1.21), (1.31) for any $(x,t) \in D_T$ are estimated by

$$\Big\{|v|, \ |\partial_t^\alpha v|, \ |v_{xx}|\Big\} \leq \overline{C}_3 \sum_{n=1}^{\infty}\Big(n^2|\varphi_n| + n^2\|f_n\|\Big), \tag{1.32}$$

where the constant $\overline{C}_3 > 0$ depends only on α, T and $\|q\|$.

The following assertion is true:

Lemma 1.6 *If the conditions (A1) hold, then there are equalities*

$$\varphi_n = \frac{1}{\lambda_n^3}\widetilde{\varphi}_n, \quad f_n(t) = \frac{1}{\lambda_n^3}\widetilde{f}_n(t), \tag{1.33}$$

where

$$\widetilde{\varphi}_n = \sqrt{\frac{2}{l}}\int_0^l \varphi^{(5)}(x)\cos(\lambda_n x)dx, \quad \widetilde{f}_n(t) = \sqrt{\frac{2}{l}}\int_0^l f^{(5)}_{xxxxx}(x,t)\cos(\lambda_n x)dx,$$

with the following estimates:

$$\sum_{n=1}^{\infty} \widetilde{\varphi}_n^2 \leq \|\varphi^{(5)}\|^2_{L_2(0,l)}, \quad \sum_{n=1}^{\infty} \widetilde{f}_n^2(t) \leq \|f^{(5)}_{xxxxx}\|^2_{L_2(0,l)\times C[0,T]}. \tag{1.34}$$

If the functions $\varphi(x)$ and $f(x,t)$ satisfy the conditions (A1), then due to representations (1.33) and (1.34) series (1.21), (1.31) converge uniformly in the domain

D_T. In fact, the sum of these series obeys the condition (1.32). The numerical series on the right side of (1.32) converge. This is shown, for example, for the first series using (1.33) and the Cauchy-Bunyakovsky inequality as

$$\sum_{n=1}^{\infty} n^2 |\varphi_n| \leq \left(\frac{l}{\pi}\right)^3 \sum_{n=1}^{\infty} \frac{1}{n} |\widetilde{\varphi}_n| \leq \left(\frac{l}{\pi}\right)^3 \left(\sum_{n=1}^{\infty} \frac{1}{n^2}\right)^{1/2} \cdot \left(\sum_{n=1}^{\infty} \widetilde{\varphi}_n^2\right)^{1/2}.$$

The first series on the right-hand side of the last relations is convergent, and the second series converges according to the first inequality (1.34). The convergence of the second series in (1.32) is identical to that proved. Thus, the series (1.21), (1.31) converge absolutely and uniformly in D_T, defining a function $v(x, t)$ that satisfies (1.15)–(1.17) in the usual sense.

Using the above results, we obtain the following assertion:

Lemma 1.7 *Let $q(t) \in C[0, T]$, conditions (A1) are satisfied, then there exists a unique solution of the direct problem* (1.15)–(1.17) *such that $\vartheta(x, t) \in C^{2,\alpha}(D_T)$.*

Remark 1.1 In the proof of Lemma 1.7, we have used Proposition 1.2 and the estimate $\left|E_{\alpha,\beta}(z)\right| \leq C$, following from Proposition 1.3. In fact, this ensures the inclusion $\vartheta(x, t) \in C^{2,\alpha}\left(\overline{D}_T\right)$ holds.

Next, we derive an estimate for the norm of the difference between the solution of the original integral equation (1.28) and the solution of this equation with perturbed functions $\widetilde{q}, \widetilde{\varphi}_n, \widetilde{f}_n$. Let $\widetilde{\vartheta}_n(t)$ be solution of the integral equation (1.28) corresponding to the functions $\widetilde{q}, \widetilde{\varphi}_n, \widetilde{f}_n$, i.e.,

$$\widetilde{\vartheta}_n(t) = \widetilde{\varphi}_n E_\alpha(-\lambda_n^2 t^\alpha) + \int_0^t (t-\tau)^{\alpha-1} E_{\alpha,\alpha}\left(-\lambda_n^2 (t-\tau)^\alpha\right) \widetilde{f}_n(\tau) d\tau$$

$$+ \int_0^t (t-\tau)^{\alpha-1} E_{\alpha,\alpha}\left(-\lambda_n^2 (t-\tau)^\alpha\right) \widetilde{q}(\tau) \widetilde{\vartheta}_n(\tau) d\tau. \tag{1.35}$$

Composing the difference $\vartheta - \widetilde{\vartheta}$ with the help of the equations (1.28), (1.35) and introducing the notations $\vartheta - \widetilde{\vartheta} = \overline{\vartheta}_n, q - \widetilde{q} = \overline{q}, f_n - \widetilde{f}_n = \overline{f}_n$, we obtain the integral equation

$$\overline{\vartheta}_n(t) = \overline{\varphi}_n E_\alpha(-\lambda_n^2 t^\alpha) + \int_0^t (t-\tau)^{\alpha-1} E_{\alpha,\alpha}\left(-\lambda_n^2 (t-\tau)^\alpha\right) \overline{f}_n(\tau) d\tau +$$

$$+ \int_0^t (t-\tau)^{\alpha-1} E_{\alpha,\alpha}\left(-\lambda_n^2 (t-\tau)^\alpha\right) \overline{q}(\tau) \vartheta_n(\tau) d\tau$$

$$+ \int_0^t (t-\tau)^{\alpha-1} E_{\alpha,\alpha}\left(-\lambda_n^2 (t-\tau)^\alpha\right) \widetilde{q}(\tau) \overline{\vartheta}_n(\tau) d\tau. \tag{1.36}$$

from which, it is derived the following linear integral inequality for $|\overline{\vartheta}_n(t)|$:

$$|\overline{\vartheta}_n(t)| \le |\overline{\varphi}_n| + \frac{t^\alpha \|\overline{f}_n\|}{\alpha\Gamma(\alpha)} + \frac{\|\overline{q}\| t^\alpha}{\alpha\Gamma(\alpha)}\Big(|\varphi_n| + \frac{t^\alpha \| f_n \|}{\alpha\Gamma(\alpha)}\Big) E_\alpha(\|q\| t^\alpha)$$

$$+ \frac{\|\widetilde{q}\|}{\Gamma(\alpha)} \int_0^t (t-\tau)^{\alpha-1} |\overline{\vartheta}_n(\tau)| d\tau.$$

Using Lemma 1.5, from last inequality, we arrive at the estimate:

$$|\overline{\vartheta}_n(t)| \le \Big\{|\overline{\varphi}_n| + \frac{t^\alpha \|\overline{f}_n\|}{\alpha\Gamma(\alpha)} + \frac{\|\overline{q}\| t^\alpha}{\alpha\Gamma(\alpha)}\Big(|\varphi_n| + \frac{t^\alpha \| f_n \|}{\alpha\Gamma(\alpha)}\Big) E_\alpha(\|q\| t^\alpha)\Big\}$$

$$\times E_\alpha(\|\widetilde{q}\| t^\alpha), \quad t \in [0, T], \ n \in \mathbb{N}. \tag{1.37}$$

Indeed, the expression (1.37) is a stability estimate for Fourier coefficients of the solution to the problem (1.15)–(1.17). The uniqueness of the solution to the problem (1.15)–(1.17) follows from (1.21) and (1.37).

1.1.4 Inverse Problem of Finding the Reaction Coefficient

In this section, we study the inverse problem as the problem of determining of functions $q(t)$ from relations (1.15)–(1.18), using the contraction mapping principle.

Firstly, by differentiating (1.21) with respect to x, we get the following equality:

$$\vartheta_x(x, t) = \sum_{n=1}^{\infty} \lambda_n \vartheta_n(t) \cos(\lambda_n x). \tag{1.38}$$

In view of Remark 1.1, setting in (1.38) $x = 0$ and using overdetermination condition (1.18), after simple converting, we obtain the integral equation for determining $q(t)$:

$$q(t) = q_0(t) - \frac{1}{h(t)} \sum_{n=1}^{\infty} \lambda_n \vartheta_n(t; q), \tag{1.39}$$

where

$$q_0(t) = \frac{1}{h(t)}\Big[f_x(0, t) - \partial^\alpha h(t)\Big],$$

and $\vartheta_n(t; q)$ means that the solution of integral equation (1.28) depends on $q(t)$.

Here we use the contraction mapping principle. The idea is to write the integral equation (1.39) for unknown function $q(t)$ as a nonlinear operator equation, and prove that this operator is a contraction mapping operator. Then, the existence and uniqueness of solution to the inverse problem follows immediately.

We recall that some operator F is a contraction mapping operator in a closed set Ω which is a subset of a Banach space if it satisfies the following two properties:

(1) if $y \in \Omega$, then $Fx \in \Omega$ (i.e., F maps Ω into itself);

(2) if $y, z \in \Omega$, then $\|Fy - Fz\| \leq \rho \, \|y - z\|$ with some $\rho \in (0, 1)$ (constant ρ does not depend on y and z).

We introduce an operator B, defining it by the right-hand side of (1.39)

$$B[q](t) = q_0(t) - \frac{1}{h(t)} \sum_{n=1}^{\infty} \lambda_n \vartheta_n(t; q). \tag{1.40}$$

Then, the equation (1.39) is written in more convenient form

$$q(t) = B[q](t). \tag{1.41}$$

Let $q_{00} := \max_{t\in[0,T]} |q_0(t)|$. Fix a sufficiently large number $\rho > 0$ and consider the ball in the space $C[0, T]$

$$\Psi^T(q_0, \rho) := \{q(t) : q(t) \in C[0, T], \|q - q_0\|_{C[0,T]} \leq \rho\}.$$

Theorem 1.2 *Let (A1), (A2) are satisfied. Then there exists a number $T^* \in (0, T)$, such that there exists a unique solution $q(t) \in C[0, T^*]$ of the inverse problem* (1.15)–(1.18).

Proof Let us first prove that for an enough small $T > 0$ the operator B maps the ball $\Psi^T(q_0, \rho)$ into itself. Indeed, for any function $q(t) \in C[0, T]$, the function $B[q](t)$ calculated using formula (1.40) will be continuous. Moreover, estimating the norm of the differences, we find

$$\|B[q](t) - q_0(t)\| \leq \frac{1}{h_0} E_{\alpha,1}(\|q\|T^\alpha) \sum_{n=1}^{\infty} \lambda_n \left(|\varphi_n| + \frac{T^\alpha \|f_n\|}{\alpha\Gamma(\alpha)} \right).$$

Here we have used the estimate (1.30). In view of Lemmas 1.5 and 1.6, last series is convergent. Note that the function occurring on the right-hand side in this inequality is monotone increasing with T, and the fact that the function $q(t)$ belongs to the ball $\Psi^T(q_0, \rho)$ implies the inequality

$$\|q\| \leq \rho + \|q_0\|. \tag{1.42}$$

Therefore, we only strengthen the inequality if we replace $\|q\|$ in this inequality with the expression $\rho + \|q_0\|$. Performing these replacements, we obtain the estimate

$$\|B[q](t) - q_0(t)\| \leq \frac{1}{h_0} E_{\alpha,1}\left((\rho + \|q_0\|)T^\alpha\right) \sum_{n=1}^{\infty} \lambda_n \left(|\varphi_n| + \frac{T^\alpha \|f_n\|}{\alpha\Gamma(\alpha)} \right).$$

Let $\rho > m_1(0) = (1/h_0)\sum_{n=1}^{\infty} \lambda_n |\varphi_n|$ and T_1 be a smallest positive root of the equation

$$m_1(T) = \frac{1}{h_0} E_{\alpha,1}\left((\rho + \|q_0\|)T^\alpha\right) \sum_{n=1}^{\infty} \lambda_n \left(|\varphi_n| + \frac{T^\alpha \|f_n\|}{\alpha\Gamma(\alpha)} \right) = \rho.$$

Then for $T \in [0, T_1]$ we have $B[q](t) \in \Psi^T(q_0, \rho)$.

Now consider two functions $q(t)$ and $\widetilde{q}(t)$ belonging to the ball $\Psi^T(q_0, \rho)$ and estimate the distance between $B[q](t)$ and $B[\widetilde{q}](t)$ in the space $C[0, T]$. The function $\widetilde{\vartheta}_n(t)$ corresponding to $\widetilde{q}(t)$ satisfies the integral equation (1.35) with the functions $\varphi_n = \widetilde{\varphi}_n$ and $f_n = \widetilde{f}_n$. Composing the difference $B[q](t) - B[\widetilde{q}](t)$ with the help of equations (1.28), (1.35) and then estimating its norm, we obtain

$$\|B[q](t) - B[\widetilde{q}](t)\| \leq \frac{1}{h_0} \sum_{n=1}^{\infty} \lambda_n \|\overline{\vartheta}_n\|.$$

Using inequality (1.30) with $\varphi_n = \widetilde{\varphi}_n$ and $f_n = \widetilde{f}_n$, we continue the previous inequality in following form:

$$\|B[q](t) - B[\widetilde{q}](t)\|$$
$$\leq \frac{1}{h_0} E_{\alpha,1}(\|\widetilde{q}\|T^\alpha) E_{\alpha,1}(\|q\|T^\alpha) \sum_{n=1}^{\infty} \frac{\lambda_n \|\overline{q}\| T^\alpha}{\alpha\Gamma(\alpha)} \left(|\varphi_n| + \frac{T^\alpha \|f_n\|}{\alpha\Gamma(\alpha)} \right). \quad (1.43)$$

The functions $q(t)$ and $\widetilde{q}(t)$ belong to the ball $\Psi^T(q_0, \rho)$, and hence for each of these functions one has inequality (1.42). Note that the function on the right-hand side in inequality (1.43) at the factor $\|\overline{q}\|$ is monotone increasing with $\|q\|$, $\|\widetilde{q}\|$, and T. Consequently, replacing $\|q\|$ and $\|\widetilde{q}\|$ in inequality (1.43) with $\rho + \|q_0\|$ will only strengthen the inequality. Thus, we have

$$\|B[q](t) - B[\widetilde{q}](t)\|$$
$$\leq \frac{1}{h_0} \left(E_{\alpha,1}((\rho + \|q_0\|)T^\alpha) \right)^2 \sum_{n=1}^{\infty} \frac{\lambda_n T^\alpha}{\alpha\Gamma(\alpha)} \left(|\varphi_n| + \frac{T^\alpha \|f_n\|}{\alpha\Gamma(\alpha)} \right) \|\overline{q}\| \leq m_2(T)\|\overline{q}\|.$$

Let T_2 be a smallest positive root of the equation

$$m_2(T) = \frac{1}{h_0} \left(E_{\alpha,1}((\rho + \|q_0\|)T^\alpha) \right)^2 \sum_{n=1}^{\infty} \frac{\lambda_n T^\alpha}{\alpha\Gamma(\alpha)} \left(|\varphi_n| + \frac{T^\alpha \|f_n\|}{\alpha\Gamma(\alpha)} \right) = 1.$$

Then, for $T \in [0, T_2)$ we have that the distance between the functions $B[q](t)$ and $B[\widetilde{q}](t)$ in $C[0, T]$ is not greater than the distance between the functions $\|q\|$ and $\|\widetilde{q}\|$.

Remark 1.2 The equations $m_1(T) = \rho, \quad \rho > m_1(0) = (1/h_0)\sum_{n=1}^{\infty} \lambda_n |\varphi_n|$ and $m_2(T) = 1, \quad m_2(0) = 0$, always have unique solutions, since $m_i(T),\ i = 1, 2$, are increasing functions of $T > 0$.

If we choose $T^* < \min(T_1, T_2)$, then the operator B is a contraction in the ball $\Psi^{T^*}(q_0, \rho)$. However, in accordance with the Banach theorem, the operator B has unique fixed point in the ball $\Psi^{T^*}(q_0, \rho)$; i.e., there exists a unique solution of equation (1.41). Theorem 1.2 is proven. □

Let T, l be positive fixed numbers. Consider the set $K(\chi_0)(\chi_0 > 0$ is some fixed number) of the given functions (φ, f, h), for which all conditions (A1), (A2) are fulfilled and

$$\max\left\{\|\varphi\|_{C^5[0,T]},\ \|h\|_{C^1[0,T]},\ \|h^{-1}\|_{C[0,T]},\ \|f\|_{C^{5,0}(D_T)}\right\} \le \chi_0.$$

We denote by $Q(\chi_1)$ the set of functions $q(t)$, satisfying the following condition $\|q\|_{C[0,T]} \le \chi_1,\ \ \chi_1 > 0$.

Theorem 1.3 *Let* $(\varphi, f, h) \in K(\chi_0)$, $\left(\widetilde{\varphi}, \widetilde{f}, \widetilde{h}\right) \in K(\chi_0)$ *and* $q \in Q(\chi_1)$, $\widetilde{q} \in Q(\chi_1)$. *Then, for solution of the inverse problem* (1.15)–(1.18) *the following stability estimate holds:*

$$\|q - \tilde{q}\|_{C[0,T]} \le d\Big(\|\varphi - \tilde{\varphi}\|_{C^5[0,l]} + \|f - \tilde{f}\|_{C^{5,0}(D_T)} + \|h - \tilde{h}\|_{C^1[0,T]}\Big), \tag{1.44}$$

where the constant d depends only on $T,\ l,\ \chi_0,\ \chi_1$.

Proof Using (1.39) we write down the equations for $\widetilde{q}(t)$ and compose the difference $q(t) - \widetilde{q}(t)$. Then, after estimating the resulting relation and using (1.30), (1.37), we obtain the following estimates:

$$\begin{aligned}|q(t) - \widetilde{q}(t)| \le d_0\Big(\|\overline{\varphi}\|_{C^5[0,l]} + \|\overline{f}\|_{C^{5,0}(D_T)} + \|\overline{h}\|_{C^1[0,T]}\Big)\\ + d_1\int_0^t (t-\tau)^{\alpha-1}|q(\tau) - \widetilde{q}(\tau)|d\tau, t \in [0, T],\end{aligned} \tag{1.45}$$

where d_0 d_1 depends only on $\chi_0, \chi_1, T, \alpha$ and $\overline{\varphi}(x) = \varphi(x) - \widetilde{\varphi}(x)$, $\overline{f}(x, t) = f(x, t) - \widetilde{f}(x, t)$, $\overline{h}(t) = h(t) - \widetilde{h}(t)$.

Applying a Gronwall-type inequality (see Lemma 1.4) to (1.45) with a weakly singular kernel, we have the following estimate:

$$|q(t) - \widetilde{q}(t)| \le d_0\Big(\|\overline{\varphi}\|_{C^4[0,l]} + \|\overline{f}\|_{C^{4,0}(\overline{D}_T)} + \|\overline{h}\|_{C^1[0,T]}\Big) \times E_{\alpha,1}\Big(d_1\Gamma(\alpha)t^{\alpha}\Big), \ t \in [0, T]. \tag{1.46}$$

This inequality implies the estimate (1.44), if we set $d = d_0 E_{\alpha,1}\Big(d_1\Gamma(\alpha)T^{\alpha}\Big)$. □

The result below shows that the uniqueness of the solution of inverse problem (1.15)–(1.18) is valid for arbitrary $T > 0$. This fact is a consequence of the above-proved stability theorem.

Theorem 1.4 *Let the functions φ, f, h and $\widetilde{\varphi}$, $\widetilde{f}$, $\widetilde{h}$ have the same meaning as in Theorem 1.3 and conditions (A1), (A2). Moreover, if $\varphi = \widetilde{\varphi}$, $f = \widetilde{f}$, $h = \widetilde{h}$ for $t \in [0, T]$, then $q(t) = \widetilde{q}(t)$, $t \in [0, T]$.*

Example 1.1 Consider inverse problem (1.1)–(1.3), (1.17) with the following input data:

$$\varphi(x) = \Gamma(2-\alpha)\sin(x), \quad f(x,t) = \left(t^{1-\alpha} + \Gamma(2-\alpha)(1+t) + 1\right)\sin(x),$$

$$h(t) = \Gamma(2-\alpha)(1+t)$$

for $x \in (0,\ l = 2\pi)$ and $t \in (0,\ T = 1)$. The exact solution is given by

$$u(x,t) = \Gamma(2-\alpha)(1+t)\sin(x), \quad q(t) = \frac{1}{\Gamma(2-\alpha)(1+t)}.$$

The above analysis remains valid if the boundary conditions given by (1.3) are replaced by the Neumann conditions: $u_x\big|_{x=0} = u_x\big|_{x=l} = 0, \quad t \in [0, T]$.

The next section is devoted to the analysis of the inverse problem (1.1)–(1.3), (1.17) in the case of periodic boundary conditions instead of (1.3).

1.2 An Inverse Problem with Periodic Boundary and Local Overdetermination Conditions

Periodic boundary conditions (PBCs) are commonly applied in molecular dynamics, dislocation dynamics and materials modeling to eliminate the existence of surface and avoid huge amount of molecules or large size of simulation box. In (PBCs), an infinite lattice system is formed simply by repeating the simulation box throughout space. When a molecule leaves the box, one of its images will enter through the opposite face with exactly the same way and direction. The molecules in the simulation box will conserve and the system can be thought of as having no surface.

In the present section, time-fractional diffusion equation is used with initial, periodic boundary conditions for the determination of the zero coefficient. The existence of the solution to inverse problem is proved by Schauder's theorem.

We consider the inverse problem (1.1)–(1.3), (1.17) for the periodic boundary conditions instead of (1.3)

$$u(0,t) = u(l,t), \quad u_x(0,t) = u_x(l,t), \quad t \in [0,T], \tag{1.47}$$

$$\varphi(0) = \varphi(l), \quad \varphi'(0) = \varphi'(l).$$

The problem of determining a function $u(x,t)$, $(x,t) \in D_T$, that satisfies (1.1), (1.2), (1.47) with given functions $q(t)$, $f(x,t)$ and $\varphi(x)$ will be called the direct problem.

Let $u(x,t)$ be a classical solution to the problem (1.1), (1.2), (1.47) and $f(x,t)$, $\varphi(x)$, $h(t)$ be enough smooth functions.

First we transform the inverse problem (1.1), (1.2), (1.47), (1.17). Denote, for this purpose as in Sect. 1.1 the second derivative of $u(x,t)$ with respect to x, by $\vartheta(x,t)$, $\vartheta(x,t) := u_{xx}(x,t)$. Differentiating (1.1) and (1.2) twice in x, we get (1.15), (1.16)

To obtain boundary conditions for the new function $\vartheta(x,t)$, we note that the second term in (1.1) is $\vartheta(x,t)$. Suppose $f(0,t) = f(l,t)$, $f_x(0,t) = f_x(l,t)$. Then, we have the following boundary conditions

$$\vartheta(0,t) = \vartheta(l,t), \quad \vartheta_x(0,t) = \vartheta_x(l,t). \tag{1.48}$$

The overdetermination condition for the function $\vartheta(x,t)$ coincides with (1.18) Under fulfilling the matching condition $\varphi'(0) = h(0)$, it is easy to deduce from (1.15), (1.16), (1.48) and (1.18) the equations (1.1), (1.2), (1.47) and (1.17) (see Sect. 1.1.2).

We introduce the spaces $C^{2,\alpha}(D_T) := \left\{ v(x,t) : v,\, v_x,\, v_{xx},\, \partial_t^\alpha \vartheta \in C(D_T) \right\}$ and $C^{1,0}(\overline{D}_T) := \left\{ v(x,t) : v,\, v_x, \in C(\overline{D}_T) \right\}$.

Definition 1.3 The functions $\{u(x,t),\ q(t)\}$ from the class $C^{2,\alpha}(D_T) \cap C^{1,0}(\overline{D}_T) \times C[0,T]$ is said to be a classical solution of problem (1.1), (1.2), (1.47) and (1.17), if the functions $u(x,t)$, $q(t)$ satisfy the following conditions:

(1) the function $u(x,t)$ and its derivatives $\partial_t^\alpha u(x,t)$, $u_{xx}(x,t)$ are continuous in the domain D_T;
(2) the function $q(t)$ is continuous on the interval $[0,T]$;
(3) equation (1.1) and conditions (1.2), (1.47) and (1.17) are satisfied in the classical sense.

Throughout this section the functions φ, f and h are assumed to satisfy the following conditions:

(B1) $\varphi(x) \in \Phi^4[0, l]$;

(B2) $f(x,t) \in C^{4,1}_{x,t}(\overline{D}_T)$; $f^{(5)}_{xxxxx}(x,t) \in C([0,T], L_2(0,l))$, $f(0,t) = f(l,t)$, $f_x(0,t) = f_x(l,t)$, $f_{xx}(0,t) = f_{xx}(l,t)$, $f_{xxx}(0,t) = f_{xxx}(l,t)$, $f_{xxxx}(0,t) = f_{xxxx}(l,t)$;

(B3) $h(t) \in C^2[0,T]$ and $|h(t)| \geq h_0 = const > 0, h_0$ is a given number, $\varphi'(0) = h(0), \varphi'''(0) = q(0)h(0) + \partial_t^\alpha h(t) - f_x(0,0)$, where the class of functions. The class of functions $\Phi^k[0,l]$ was introduced in Sect. 1.1.3.

1.2.1 The Initial-Boundary Value Problem with Periodic Conditions

The use of the Fourier method for solving problem (1.15), (1.16), (1.48) leads to the spectral problem for the operator given by the differential expression and boundary conditions

$$X_n''(x) + \lambda^2 X_n(x) = 0, \quad x \in (0,l),$$

$$X_n(0) = X_n(l), \quad X_n'(0) = X_n'(l), \quad n \in \mathbb{N}. \tag{1.49}$$

It is known [36] that the system of eigenfunctions

$$1, \ \cos\lambda_1 x, \ \sin\lambda_1 x, \ \cos\lambda_2 x, \ \sin\lambda_2 x, ..., \cos\lambda_n x, \ \sin\lambda_n x, ..., \tag{1.50}$$

where $\lambda_n = \frac{2\pi n}{l}$, $n \in \mathbb{N}$, is the solution to spectral problem (1.49) and orthogonal basis of $L_2(0,l)$. Therefore we shall seek the classical solution $u(x,t)$ of the problem (1.15), (1.16), (1.48) in the form

$$\vartheta(x,t) = \sum_{n=0}^{\infty} \vartheta_{1n}(t)\cos\lambda_n x + \sum_{n=1}^{\infty} \vartheta_{2n}(t)\sin\lambda_n x, \tag{1.51}$$

where

$$\vartheta_{10}(t) = \frac{1}{\sqrt{l}}\int_0^l \vartheta(x,t)dx, \quad \vartheta_{1n}(t) = \sqrt{\frac{2}{l}}\int_0^l \vartheta(x,t)\cos\lambda_n x dx,$$

$$\vartheta_{2n}(t) = \sqrt{\frac{2}{l}}\int_0^l \vartheta(x,t)\sin\lambda_n x dx.$$

Then, applying the formal scheme of the Fourier method to determine the unknown coefficients $\vartheta_{10}(t)$ and $\vartheta_{in}(t)$ $(i = 1,2;\ n \in \mathbb{N})$ of function $\vartheta(x,t)$ from (1.15) and (1.16), we have

$$\partial_t^\alpha \vartheta_{10}(t) = -q(t)\vartheta_{10}(t) + f_{10}(t), \tag{1.52}$$

$$\vartheta_{10}(t)|_{t=0} = \varphi_{10}, \tag{1.53}$$

$$\partial_t^\alpha \vartheta_{in}(t) + \lambda_n^2 \vartheta_{in}(t) = -q(t)\vartheta_{in}(t) + f_{in}(t), \tag{1.54}$$

$$\vartheta_{in}(t)|_{t=0} = \varphi_{in}, \quad i = 1, 2, \quad n \in \mathbb{N}, \tag{1.55}$$

where

$$f_{10}(t) = \frac{1}{\sqrt{l}} \int_0^l f_{xx}(x,t)dx, \quad f_{1n}(t) = \sqrt{\frac{2}{l}} \int_0^l f_{xx}(x,t) \cos \lambda_n x dx,$$

$$f_{2n}(t) = \sqrt{\frac{2}{l}} \int_0^l f_{xx}(x,t) \sin \lambda_n x dx,$$

$$\varphi_{10} = \frac{1}{\sqrt{l}} \int_0^l \varphi''(x)dx, \quad \varphi_{1n} = \sqrt{\frac{2}{l}} \int_0^l \varphi''(x) \cos \lambda_n x dx,$$

$$\varphi_{2n} = \sqrt{\frac{2}{l}} \int_0^l \varphi''(x) \sin \lambda_n x dx.$$

According to Theorem 1.1 the solutions of problems (1.52), (1.53) and (1.54), (1.55) satisfy the following integral equations

$$\vartheta_{10}(t) = \varphi_{10} + \frac{1}{\Gamma(\alpha)} \int_0^t (t-\tau)^{\alpha-1}(f_{10}(\tau) - q(\tau)\vartheta_{10}(\tau))d\tau \tag{1.56}$$

and

$$\vartheta_{in}(t) = \varphi_{in} E_\alpha(-\lambda^2 t^\alpha)$$

$$+ \int_0^t (t-\tau)^{\alpha-1} E_{\alpha,\alpha}(-\lambda^2(t-\tau)^\alpha)(f_{in}(\tau) - q(\tau)\vartheta_{in}(\tau))d\tau. \tag{1.57}$$

These equations yield

$$|\vartheta_{10}(t)| \le |\varphi_{10}| + \frac{t^\alpha}{\Gamma(\alpha+1)} \|f_{10}\| + \frac{\|q\|}{\Gamma(\alpha)} \int_0^t |\vartheta_{10}(\tau)|(t-\tau)^{\alpha-1} d\tau,$$

$$|\vartheta_{in}(t)| \le |\varphi_{in}| + \frac{t^\alpha}{\Gamma(\alpha+1)} \|f_{in}\| +$$

$$+ \frac{\|q\|}{\Gamma(\alpha)} \int_0^t (t-\tau)^{\alpha-1} E_{\alpha,\alpha}(-\lambda^2(t-\tau)^\alpha)|\vartheta_{in}(\tau)|d\tau, \quad i = 1, 2, \quad n \in \mathbb{N},$$

where $\|q\| = \max_{t\in[0,T]} |q(t)|$. Applying Gronwall's type inequality (see Lemma 1.4) to last relations leads to the following estimates:

$$|\vartheta_{10}(t)| \le \left(|\varphi_{10}| + \frac{t^\alpha}{\Gamma(\alpha+1)}\|f_{10}\|\right) E_\alpha(\|q\|t^\alpha), \tag{1.58}$$

$$|\vartheta_{in}(t)| \le \left(|\varphi_{in}| + \frac{t^\alpha}{\Gamma(\alpha+1)}\|f_{in}\|\right) E_\alpha(\|q\|t^\alpha). \tag{1.59}$$

Using equalities (1.52), (1.54) and (1.58), (1.59) we obtain estimates for $\partial_t^\alpha \vartheta_{10}(t)$, $\partial_t^\alpha \vartheta_{in}(t)$:

$$|\partial_t^\alpha \vartheta_{10}(t)| \le \|q\| \left(|\varphi_{10}| + \frac{t^\alpha}{\Gamma(\alpha+1)}\|f_{10}\|\right) E_\alpha\left(\|q\|t^\alpha\right) + \|f_{10}\|,$$

$$|\partial_t^\alpha \vartheta_{in}(t)| \le (\lambda^2 + \|q\|) \left(|\varphi_{in}| + \frac{t^\alpha}{\Gamma(\alpha+1)}\|f_{in}\|\right) E_\alpha\left(\|q\|t^\alpha\right) + \|f_{in}\|.$$

Thus we have proved the following lemma:

Lemma 1.8 *For any $t \in [0, T]$ the following estimates are valid:*

$$|\vartheta_{10}(t)| \le \left(|\varphi_{10}| + \frac{T^\alpha}{\Gamma(\alpha+1)}\|f_{10}\|\right) E_\alpha\left(\|q\|T^\alpha\right),$$

$$|\vartheta_{in}(t)| \le \left(|\varphi_{in}| + \frac{T^\alpha}{\Gamma(\alpha+1)}\|f_{in}\|\right) E_\alpha\left(\|q\|T^\alpha\right),\ i = 1, 2,$$

$$|\partial_t^\alpha \vartheta_{10}(t)| \le \|q\| \left(|\varphi_{10}| + \frac{T^\alpha}{\Gamma(\alpha+1)}\|f_{10}\|\right) E_\alpha\left(\|q\|T^\alpha\right) + \|f_{10}\|,$$

$$|\partial_t^\alpha \vartheta_{in}(t)| \le (\lambda^2 + \|q\|) \left(|\varphi_{in}| + \frac{T^\alpha}{\Gamma(\alpha+1)}\|f_{in}\|\right) E_\alpha\left(\|q\|T^\alpha\right) + \|f_{in}\|,\ i = 1, 2.$$

Formally, from (1.51) by the term differentiation we obtain the series

$$\partial_t^\alpha \vartheta(x,t) = \sum_{n=0}^{\infty} \partial_t^\alpha \vartheta_{1n}(t) \cos \lambda_n x + \sum_{n=1}^{\infty} \partial_t^\alpha \vartheta_{2n}(t) \sin \lambda_n x, \tag{1.60}$$

$$\vartheta_{xx}(x,t) = -\sum_{n=0}^{\infty} \lambda_n^2 \vartheta_{1n}(t) \cos \lambda_n x - \sum_{n=1}^{\infty} \lambda_n^2 \vartheta_{2n}(t) \sin \lambda_n x. \tag{1.61}$$

In view of Lemma 1.8, if following series converges, then the series (1.51), (1.60) and (1.61) will converge for any $(x,t) \in D_T$:

$$C_4 \sum_{n=1}^{\infty} (\lambda_n^2 |\varphi_{in}| + \lambda_n^2 \| f_{in} \|), \quad i = 1, 2, \tag{1.62}$$

where the constant C_4 depends only on $T, \alpha, \|q\|$.

We state the following auxiliary lemma:

Lemma 1.9 *If the conditions (B1), (B2) are valid, then there are equalities*

$$\varphi_{in} = \frac{1}{\lambda_n^3} \varphi_{in}^{(3)}, \quad f_{in}(t) = \frac{1}{\lambda_n^3} f_{in}^{(3)}, \ i = 1, 2, \tag{1.63}$$

where

$$\varphi_{1n}^{(3)} = \sqrt{\frac{2}{l}} \int_0^l \varphi^{(5)}(x) \sin \lambda_n x dx, \quad \varphi_{2n}^{(3)} = \sqrt{\frac{2}{l}} \int_0^l \varphi^{(5)}(x) \cos \lambda_n x dx,$$

$$f_{1n}^{(3)}(t) = \sqrt{\frac{2}{l}} \int_0^l f_{xxxxx}^{(5)}(x,t) \sin \lambda_n x dx, \quad f_{2n}^{(3)}(t) = \sqrt{\frac{2}{l}} \int_0^l f_{xxxxx}^{(5)}(x,t) \cos \lambda_n x dx$$

with the following estimates:

$$\sum_{n=1}^{\infty} |\varphi_{in}^{(3)}|^2 \le \|\varphi^{(5)}\|_{L_2(0,1)}^2, \quad \sum_{n=1}^{\infty} |f_{in}^{(3)}(t)|^2 \le \|f^{(5)}\|_{L_2(0,l)\times C[0,T]}^2, \ i = 1, 2. \tag{1.64}$$

If the functions $\varphi(x),\ f(x,t)$ satisfy the conditions of Lemma 1.9, then due to representations (1.63) and (1.64) series (1.51), (1.60) and (1.61) converge uniformly in the domain D_T, therefore, function $\vartheta(x,t)$ satisfies relations (1.15), (1.16), (1.47).

Using the above results, we obtained the following assertion.

Theorem 1.5 *Let $q(t) \in C[0,T]$, and suppose that the conditions (B1) and (B2) are satisfied. Then, there exists a unique solution to the direct problem* (1.15), (1.16), (1.47), *such that* $\vartheta(x,t) \in C^{2,\alpha}(D_T) \cap C^{1,0}(\overline{D}_T)$.

Let us derive an estimate for the norm of the difference between the solution of the original integral equations (1.56), (1.57) and the solution of this equation with perturbed functions $\widetilde{q},\ \widetilde{\varphi}_{in},\ \widetilde{f}_{in}$. Let $\widetilde{\vartheta}_{in}(t)$, $(i := 0, 1, 2)$ be solutions of the integral equation (1.56), (1.57) corresponding to the functions $\widetilde{a},\ \widetilde{\varphi}_{in},\ \widetilde{f}_{in}$, i.e.,

$$\widetilde{\vartheta}_{10}(t) = \widetilde{\varphi}_{10} + \frac{1}{\Gamma(\alpha)} \int_0^t (t-\tau)^{\alpha-1} (\widetilde{f}_{10}(\tau) - \widetilde{q}(\tau)\widetilde{\vartheta}_{10}(\tau)) d\tau, \tag{1.65}$$

$$\widetilde{\vartheta}_{in}(t) = \widetilde{\varphi}_{in} E_\alpha(-\lambda^2 t^\alpha)$$

$$+ \int_0^t (t-\tau)^{\alpha-1} E_{\alpha,\alpha}(-\lambda^2 (t-\tau)^\alpha)(\widetilde{q}(\tau)\widetilde{\vartheta}_{in}(\tau) + \widetilde{f}_{in}(\tau))d\tau. \tag{1.66}$$

Composing the difference $\vartheta_{in} - \widetilde{\vartheta}_{in}$ with the help of the equations (1.56), (1.57), (1.65), (1.66) and introducing the notations $\vartheta_{in} - \widetilde{\vartheta}_{in} = \overline{\vartheta}_{in}$, $q - \widetilde{q} = \overline{q}$, $f_{in} - \widetilde{f}_{in} = \overline{f}_{in}$, we obtain the integral equation

$$\overline{\vartheta}_{10}(t) = \overline{\varphi}_{10} + \frac{1}{\Gamma(\alpha)} \int_0^t (t-\tau)^{\alpha-1} \overline{f}_{10}(\tau)d\tau -$$

$$- \frac{1}{\Gamma(\alpha)} \int_0^t (t-\tau)^{\alpha-1} (\overline{a}(\tau)\vartheta_{10}(\tau) + \widetilde{q}(\tau)\overline{\vartheta}_{10}(\tau))d\tau;$$

$$\overline{\vartheta}_{in}(t) = \overline{\varphi}_n E_{\alpha,1}(-\lambda_n^2 t^\alpha) + \int_0^t (t-\tau)^{\alpha-1} E_{\alpha,\alpha}\left(-\lambda_n^2 (t-\tau)^\alpha\right) \overline{f}_{in}(\tau)d\tau -$$

$$- \int_0^t (t-\tau)^{\alpha-1} E_{\alpha,\alpha}\left(-\lambda_n^2 (t-\tau)^\alpha\right) \overline{q}(\tau)\vartheta_{in}(\tau)d\tau -$$

$$- \int_0^t (t-\tau)^{\alpha-1} E_{\alpha,\alpha}\left(-\lambda_n^2 (t-\tau)^\alpha\right) \widetilde{q}(\tau)\overline{\vartheta}_{in}(\tau)d\tau, \tag{1.67}$$

from which, are derived the following linear integral inequalities for $|\overline{\vartheta}_{10}(t)|$, $|\overline{\vartheta}_{in}(t)|$:

$$|\overline{\vartheta}_{10}(t)| \leq |\overline{\varphi}_{10}| + \frac{t^\alpha \|\overline{f}_{10}\|}{\Gamma(\alpha+1)} +$$

$$+ \frac{\|\overline{q}\| t^\alpha}{\Gamma(\alpha+1)} \left(|\varphi_{10}| + \frac{t^\alpha}{\Gamma(\alpha+1)} \|f_{10}\|\right) E_\alpha(\|q\| t^\alpha) + \frac{\|\widetilde{q}\|}{\Gamma(\alpha)} \int_0^t (t-\tau)^{\alpha-1} |\overline{\vartheta}_{10}(\tau)| d\tau;$$

$$|\overline{\vartheta}_{in}(t)| \leq |\overline{\varphi}_{in}| + \frac{t^\alpha \|\overline{f}_{in}\|}{\Gamma(\alpha+1)} +$$

$$+ \frac{\|\overline{q}\| t^\alpha}{\Gamma(\alpha+1)} \left(|\varphi_{in}| + \frac{t^\alpha}{\Gamma(\alpha+1)} \|f_{in}\|\right) E_\alpha(\|q\| t^\alpha) + \frac{\|\widetilde{q}\|}{\Gamma(\alpha)} \int_0^t (t-\tau)^{\alpha-1} |\overline{\vartheta}_{in}(\tau)| d\tau.$$

Using Lemma 1.4 from last inequality, we arrive at the estimate:

$$|\overline{\vartheta}_{10}(t)| \leq \Big\{ |\overline{\varphi}_{10}| + \frac{t^\alpha \|\overline{f}_{10}\|}{\Gamma(\alpha+1)} +$$

$$+ \frac{\|\overline{q}\| t^\alpha}{\Gamma(\alpha+1)} \left(|\varphi_{10}| + \frac{t^\alpha}{\Gamma(\alpha+1)} \|f_{10}\|\right) E_\alpha(\|q\| t^\alpha) \Big\} E_\alpha(\|\widetilde{q}\| t^\alpha). \tag{1.68}$$

$$|\overline{\vartheta}_{in}(t)| \leq \Big\{ |\overline{\varphi}_{in}| + \frac{t^{\alpha}\|\overline{f}_{in}\|}{\Gamma(\alpha+1)} +$$

$$+ \frac{\|\overline{q}\|t^{\alpha}}{\Gamma(\alpha+1)}\Big(|\varphi_{in}| + \frac{t^{\alpha}}{\Gamma(\alpha+1)}\|f_{in}\|\Big)E_{\alpha}(\|q\|t^{\alpha})\Big\}E_{\alpha}(\|\widetilde{q}\|t^{\alpha}). \tag{1.69}$$

In the following section, we study the inverse problem, which involves determining the functions $q(t)$ from the relations (1.15), (1.16), (1.48), and (1.18) using Schauder's fixed point theorem.

1.2.2 Unique Solvability of the Inverse Problem

Firstly, by differentiating (1.51) with respect to x, we get the following equality

$$\vartheta_x(x,t) = -\sum_{n=0}^{\infty}\lambda_n\vartheta_{1n}(t)\sin\lambda_n x + \sum_{n=1}^{\infty}\lambda_n\vartheta_{2n}(t)\cos\lambda_n x, \tag{1.70}$$

Setting in (1.70) $x = 0$ and in view of the additional condition (1.8), we get the following integral equation for determining $q(t)$:

$$q(t) = q_0(t) - \frac{1}{h(t)}\sum_{n=1}^{\infty}\lambda_n\vartheta_{2n}(t;q), \tag{1.71}$$

where

$$q_0(t) = \frac{1}{h(t)}\Big[f_x(0,t) - \partial_t^{\alpha}h(t)\Big],$$

and $\vartheta_{2n}(t;\,)$ means that the solution of integral equation (1.57) depends on $q(t)$.

Note that in (1.70) $\vartheta_x(x,t)$ is well defined for $x = 0,\ t \in [0,T]$ (see Remark 1.1). In particular, we have

$$q(0) = q_0(0) + \frac{1}{h(0)}\varphi'''(0).$$

The main result of this section is presented as follows:

Theorem 1.6 *Let (B1)–(B3) be satisfied. Then, there exists a number $T^* \in (0,T)$, such that there exists a unique solution $q(t)$ of the inverse problem* (1.15), (1.16), (1.48) *and* (1.18).

Proof We consider the operator equation

$$q = \Lambda q, \tag{1.72}$$

where Λ is defined by the right-hand side of (1.71):

$$\Lambda q(t) = q_0(t) - \frac{1}{h(t)} \sum_{n=1}^{\infty} \lambda_{2n} \vartheta_{2n}(t; q). \tag{1.73}$$

Consider the functional space of functions $q \in [0, T]$ with the norm $\|q\| = \max_{t\in[0,T]} |q(t)|$.

In this subsection, we use the following theorem (see [37]):

Theorem 1.7 (Generalized mean value theorem) *Suppose that* $g \in C[a, b]$ *and* ${}_a\partial_t^\alpha g \in C[a, b]$ *for Let* $0 < \alpha \le 1$, *then we have*

$$g(t) = g(a) + \frac{1}{\Gamma(\alpha)} {}_a\partial_t^\alpha g(\tau)(t - a)^\alpha \tag{1.74}$$

with $a \le \tau \le t$ *for all* $t \in (a, b]$. *Here* ${}_a\partial_t^\alpha$ *is the Caputo derivative with the beginning at point a defined by*

$${}_a\partial_t^\alpha g(t) = \frac{1}{\Gamma(1-\alpha)} \int_a^t (t-\tau)^{-\alpha} \frac{d}{d\tau} g(\tau) d\tau. \tag{1.75}$$

Let $S_d(0) = \{q : \|q\| \le d\}$ and d be some positive quite large number.

Theorem (Schauder principle)(see [38], p. 538). *Let the operator* P *maps a closed bounded convex set* $S_d(0)$ *of a Banach space* X *into itself. Then if* P *is completely continuous on* $S_d(0)$, *then it has a fixed point on* $S_d(0)$.

To show the existence of the solution $q(t)$ to the equation (1.72), we use Shauder's fixed point theorem. To do this, we first find the conditions under which the operator Λ maps the set $S_d(0)$ into itself.

Proof First, we establish the uniform boundedness of the operator Λ. To this end, we show that there exist $T^* \in (0, T]$ and $\rho \in (0, d]$ such that the inequality $\|\Lambda q\| \le \rho$ is hold, where $\|\Lambda q\| = \max_{t\in[0,T]} |\Lambda q(t)|$. For $a \in S_d(0)$ and $t \in [0, T]$, by Lemma 1.8 we have

$$|\Lambda q| \le q_{00} + \frac{1}{h_0} \sum_{n=1}^{\infty} \lambda_n \left(|\varphi_{2n}| + \frac{T^\alpha}{\Gamma(\alpha+1)} |f_{2n}| \right) E_\alpha(dT^\alpha) =: \gamma(T), \tag{1.76}$$

where $q_{00} = \max_{t\in[0,T]} |q(t)|$. Note that $\gamma(T)$ is the positive definite and increasing function of T, and $\gamma(0) = \gamma_0 := q_{00} + \frac{1}{h_0} \sum_{n=1}^{\infty} \lambda_n |\varphi_{2n}|$. Therefore the equation $\gamma(T) = \rho$ has unique positive root if $\rho > \gamma_0$. Then for the sufficiently small T, then operator Λ is uniformly bounded.

Proof of compactness of the operator Λ to individual special cases is not always simple. For sets, situated in some given special metric spaces, special criteria for compactness can be given which are more suitable for practical application. In analysis one of the more important metric spaces is the space $C[a, b]$. For subsets of this space the most important and frequently employed criterion of compactness is given by Arzela's theorem. In order to formulate this theorem it is necessary to first introduce the following concepts.

A family $\chi(x)$ of functions defined on a closed interval is said to be *uniformly bounded* if there exists a number M such that $|\chi(x)| < M$ for all x and for all χ belonging to the given family.

A family of functions is said to be *equicontinuous* if for every $\epsilon > 0$ there is a $\delta > 0$ such that $|\chi(x_1) - \chi(x_2)| < 0$ for all x_1, x_2 such that $|x_1 - x_2| < \delta$ and for all χ in the given family.

Arzela's Theorem (see [38], p. 718; [39], p. 54). A necessary and sufficient condition that a family of continuous functions defined on the closed interval $[a, b]$ be compact (completely continuous) in $C[a, b]$ is that this family be uniformly bounded and equicontinuous.

The following statement holds.

Lemma 1.10 *Suppose (B1)–(B3) conditions are satisfied. Then, for all T, the operator Λ is uniformly bounded and equicontinuous.*

Let $(t_1,\ t_2) \in (0, T)$, $t_1 \neq t_2$ and conditions (B1)–(B3) are met. Then, ${}_{t_1}\partial_t^\alpha q_0(t) \in C[0, T]$ (see [5], pp. 93, 94). In view of Theorem 1.7 we estimate the expression $|\Lambda q(t_1) - \Lambda q(t_2)|$ as follows

$$|\Lambda q(t_1) - \Lambda q(t_2)| \le |q_0(t_1) - q_0(t_2)| + \sum_{n=1}^{\infty} \lambda_n \left| \frac{\vartheta_{2n}(t_1; q)}{h(t_1)} - \frac{\vartheta_{2n}(t_2; q)}{h(t_2)} \right|$$

$$\le \frac{1}{\Gamma(\alpha)} \left| {}_{t_1}\partial_t^\alpha q_0(t) \right| |t_1 - t_2|^\alpha + \sum_{n=1}^{\infty} \frac{\lambda_n}{|h(t_2)|} |\vartheta_{2n}(t_1; q) - \vartheta_{2n}(t_2; q)|$$

$$+ \sum_{n=1}^{\infty} \frac{\lambda_n}{|h(t_2)h(t_1)|} |\vartheta_{2n}(t_2; q)| \, |h(t_1) - h(t_2)|$$

$$\le \frac{1}{\Gamma(\alpha)} \left| {}_{t_1}\partial_t^\alpha q_0(t) \right| |t_1 - t_2|^\alpha + \sum_{n=1}^{\infty} \frac{\lambda_n}{h_0} \frac{1}{\Gamma(\alpha)} \left| {}_{t_1}\partial_t^\alpha \vartheta_{2n}(t; q) \right| |t_1 - t_2|^\alpha$$

$$+ \sum_{n=1}^{\infty} \frac{\lambda_n}{h_0^2} \frac{1}{\Gamma(\alpha)} |\vartheta_{2n}(t_2; q)| \left| {}_{t_1}\partial_t^\alpha h(t) \right| |t_1 - t_2|^\alpha.$$

The definition of the Caputo derivative yields

$${}_{t_1}\partial_t^\alpha \vartheta_{2n}(t; q) = \frac{1}{\Gamma(1-\alpha)} \int_{t_1}^{t} (t - \tau)^{-\alpha} \frac{d}{d\tau} \vartheta_{2n}(\tau; q) d\tau$$

$$= \frac{1}{\Gamma(1-\alpha)} \int_0^t (t-\tau)^{-\alpha} \frac{d}{d\tau}\vartheta_{2n}(\tau; q)d\tau - \frac{1}{\Gamma(1-\alpha)} \int_0^{t_1} (t-\tau)^{-\alpha} \frac{d}{d\tau}\vartheta_{2n}(\tau; q)d\tau$$

$$= \partial_t^{\alpha}\vartheta_{2n}(t; q) - \partial_{t_1}^{\alpha}\vartheta_{2n}(t; q).$$

Now in view of (1.54) we express $\partial_t^{\alpha}\vartheta_{2n}(t; q)$ and $\partial_{t_1}^{\alpha}\vartheta_{2n}(t; q)$ for $t_1 > 0$ as follows

$$\partial_t^{\alpha}\vartheta_{2n}(t; q) + (\lambda_n^2 + q(t))\vartheta_{2n}(t, q) = f_{2n}(t),\ for\ \ t \in (t_1, t_2),$$

$${}_{t_1}\partial_t^{\alpha}\vartheta_{2n}(t; q) + (\lambda_n^2 + q(t_1))\vartheta_{2n}(t_1, q) = f_{2n}(t_1),\ for\ \ t_1 \in (0, T].$$

Subtracting these equations from each other we get

$${}_{t_1}\partial_t^{\alpha}\vartheta_{2n}(t; q) = f_{2n}(t) - (\lambda_n^2 + q(t))\vartheta_{2n}(t; q) - (f_{2n}(t_1) - (\lambda_n^2 + q(t_1))\vartheta_{2n}(t_1; q)).$$

Then in view of (1.61), we have for all $t_1 > 0$ and $t \in (t_1,\ t_2)$ that

$$\left|{}_{t_1}\partial_t^{\alpha}\vartheta_{2n}(t; q)\right|$$
$$\leq 2|f_{2n}(t)| + 2|(\lambda_n^2 + \|q\|)\left(|\varphi_{2n}| + \frac{T^{\alpha}}{\Gamma(\alpha+1)}\|f_{2n}\|\right) E_{\alpha}(\|q\|T^{\alpha}).$$

Using the above estimate, we get

$$\|\Lambda q(t_1) - \Lambda q(t_2)\| \leq \frac{1}{\Gamma(\alpha)}\Bigg[\left\|{}_{t_1}\partial_t^{\alpha}q_0(t)\right\|_{C[0,T]}$$

$$+2\sum_{n=1}^{\infty}\frac{\lambda_n}{h_0}\left(|f_{2n}| + (\lambda_n^2 + d)\left(|\varphi_{2n}| + \frac{T^{\alpha}}{\Gamma(\alpha+1)}\|f_{2n}\|\right) E_{\alpha}\left(dT^{\alpha}\right)\right) +$$

$$+\sum_{n=1}^{\infty}\frac{\lambda_n}{h_0^2}\left(|\varphi_{2n}| + \frac{T^{\alpha}}{\Gamma(\alpha+1)}\|f_{2n}\|\right) E_{\alpha}(dT^{\alpha})\left\|{}_{t_1}\partial_t^{\alpha}h(t)\right\|_{C[0,T]}\Bigg]|t_1 - t_2|^{\alpha}.$$

Due to the fulfillment of the conditions of Theorem 1.6, the series on the right side of this estimate are uniformly convergent.

Fix an arbitrary $\epsilon > 0$. From the above estimate it follows that there exists $\delta = \delta(\epsilon)$ such that for any $t_1,\ t_2 \in [0, T]$ satisfying condition $|t_1 - t_2| < \delta$ the inequality

$$|\Lambda q(t_1) - \Lambda q(t_2)| \leq \epsilon$$

is true. Therefore, the set $\Lambda\ (S_d(0))$ is equicontinuous.

Let us prove the uniqueness of this solution.

Lemma 1.11 *If the operator equation (1.72) has a solution for all $q \in S_d(0)$, then it is unique in the class $C[0, T]$.*

Proof Let there be two solutions q, $\tilde{q}$ of the equation (1.72) with the same data φ, f, h. We denote their differences by $\overline{q}(t) = q - \tilde{q}$. For this $\overline{q}(t)$, we obtain the integral equation

$$\overline{q}(t) = -\frac{1}{h(t)} \sum_{n=1}^{\infty} \lambda_{2n} \overline{\vartheta}_{2n}(t; q), \tag{1.77}$$

where $\overline{\vartheta}_{2n}(t; q) = \vartheta_{2n}(t; q) - \tilde{\vartheta}_{2n}(t; \tilde{q})$. For $\overline{\vartheta}_{2n}(t; q)$ from (1.69) it follows estimate

$$|\overline{\vartheta}_{2n}(t; q)| \leq \frac{\|\overline{q}\| T^\alpha}{\Gamma(\alpha+1)} \Big(|\varphi_{2n}| + \frac{T^\alpha}{\Gamma(\alpha+1)} \| f_{2n} \| \Big) E_\alpha(\|q\| T^\alpha) E_\alpha(\|\widetilde{q}\| T^\alpha).$$

From here and taking into account (1.77) we find $\overline{q}(t) = 0$, i.e., $q(t) = \tilde{q}(t)$ for $t \in [0, T]$.

Lemmas 1.10 and 1.11 complete the proof of Theorem 1.6.

Let T and l be positive fixed constants. Consider the set $\Omega(\chi_0)$, where $\chi_0 > 0$ is a fixed constant, consisting of functions (φ, f, h) that satisfy the conditions (B1)-(B3) and additionally the following requirements:

$\varphi^{(5)}(x) \in C[0, 1]$, $f^{(5)}_{xxxxx}(x, t) \in C^1([0, T], C(0, l))$,

$$\max\left\{ \|\varphi\|_{C^5[0,T]}, \|h\|_{C^1[0,T]}, \|h^{-1}\|_{C[0,T]}, \|f\|_{C^{5,1}(\overline{D}_T)} \right\} \leq \chi_0.$$

We denote by $Q(\chi_1)$ the set of functions $q(t)$, that for some $T > 0$, $l > 0$ satisfy the condition $\|q\|_{C[0,T]} \leq \chi_1$, for a fixed number $\chi_1 > 0$.

Theorem 1.8 *Let $(\varphi, f, h) \in \Omega(\chi_0)$, $(\widetilde{\varphi}, \widetilde{f}, \widetilde{h}) \in \Omega(\chi_0)$ and $q \in Q(\chi_1)$, $\widetilde{q} \in Q(\chi_1)$. Then, for solution of the inverse problem* (1.15), (1.16), (1.48) *and* (1.18) *the following stability estimate holds:*

$$\|q - \tilde{q}\|_{C[0,T]} \leq d\Big(\|\varphi - \tilde{\varphi}\|_{C^5[0,l]} + \|f - \tilde{f}\|_{C^{5,1}(\overline{D}_T)} + \|h - \tilde{h}\|_{C^1[0,T]} \Big), \tag{1.78}$$

where the constant d depends only on T, l, χ_0, χ_1.

Proof First, using (1.71) we write down the equations for $\widetilde{q}(t)$ and compose the difference $q(t) - \widetilde{q}(t)$. Then after evaluating this expression and using estimates (1.59), (1.69), we obtain following estimate:

$$\begin{aligned} &\|q(t) - \widetilde{q}(t)\| \\ &\leq d_0\Big(\|\overline{f}\| + \|\overline{\varphi}\| + \|\overline{h}\| \Big) + d_1 \int_0^t (t-\tau)^{\alpha-1} \|q(\tau) - \widetilde{q}(\tau)\| d\tau, \ t \in [0, T], \end{aligned} \tag{1.79}$$

where d_0 d_1 depends only on χ_0, χ_1, T, α, $\Gamma(\alpha)$ and $\overline{\varphi}(x) = \varphi(x) - \widetilde{\varphi}(x)$, $\overline{f}(x,t) = f(x,t) - \widetilde{f}(x,t)$, $\overline{h}(t) = h(t) - \widetilde{h}(t)$.

From (1.79), we get the estimate

$$\|q(t) - \widetilde{q}(t)\| \le d_0\Big(\|\overline{f}\| + \|\overline{\varphi}\| + \|\overline{h}\|\Big)E_{\alpha,1}\Big(d_1\Gamma(\alpha)t^{\alpha}\Big),\ t \in [0, T],$$

This inequality implies the estimate (1.78), if we set
$d = d_0 E_{\alpha,1}\Big(d_1\Gamma(\alpha)T^{\alpha}\Big)$.

From Theorem 1.8 follows also the next assertion on uniqueness in whole for solution to the inverse problem.

Theorem 1.9 *Let the functions φ, f, h and $\widetilde{\varphi}$, $\widetilde{f}$, $\widetilde{h}$ have the same meaning as in Theorem 1.8 and conditions (B1)–(B3). Moreover, if $\varphi = \widetilde{\varphi}$, $f = \widetilde{f}$, $h = \widetilde{h}$ for $t \in [0, T]$, then $q(t) = \widetilde{q}(t)$, $t \in [0, T]$.*

1.3 Diffusion Coefficient Identification Problem

We consider the fractional diffusion equation $\partial_t^{\alpha} u = a(t)u_{xx}$, where the time-dependent diffusion coefficient $a(t)$ is unknown. First, for the direct problem, we establish the existence, uniqueness, and regularity properties of the weak solution given a known function $a(t)$. Next, for the inverse problem, we introduce an operator to recover $a(t)$ and demonstrate its monotonicity, which allows us to establish the uniqueness of $a(t)$.

Over the past three decades, extensive research has shown that fractional models effectively describe anomalous diffusion processes, particularly subdiffusion, where the expected mean square displacement $\langle x^2 \rangle$ grows more slowly than in a Gaussian process over long timescales. At the microscopic level, subdiffusion can be modeled by the continuous-time random walk [40], whose macroscopic counterpart is a time-fractional differential equation. Anomalous diffusion and fractional derivatives have important applications in various fields, including thermal diffusion in fractal-structured media [41], non-Fickian diffusion in geological formations [42], and underground environmental studies [43]. Additionally, the fractional time derivative ∂_t^{α} is widely used to describe the viscoelastic behavior of materials [44–46].

1.3.1 Formulation of the Diffusion Coefficient Determination Problem

We consider the following fractional diffusion equation model:

$$\begin{cases} \partial_t^\alpha u(x,t) - a(t)u_{xx}(x,t) = 0, & 0 < t \leqslant T, \quad 0 < x < 1; \\ u(x,0) = 0, & 0 < x < 1; \\ u(1,t) = 0, \quad u(0,t) = f(t), & 0 < t \leqslant T, \quad f(0) = 0. \end{cases} \tag{1.80}$$

Here, $0 < \alpha < 1$ and T is a fixed time. This model represents a diffusion process with a time-dependent diffusion coefficient $a(t)$. Given $f(t)$ and $a(t)$, the goal of the direct problem is to determine the temperature distribution in a rod.

We focus on the inverse problem of identifying $a(t)$ using additional boundary data. Specifically, we impose the condition:

$$-a(t)u_x(0,t) = g(t), 0 < t \leqslant T,$$

which corresponds to prescribing the heat flux at the boundary $x = 0$. The inverse problem thus consists of determining $a(t)$ from the known functions $f(t)$ and $g(t)$.

This model describes "anomalous diffusion", where the time evolution is governed by a fractional Caputo derivative, ∂_t^α. The function $a(t)$ is assumed to be positive and continuous. To highlight the dependence of the solution on $a(t)$, we denote it as $u(x,t;a)$.

For model (1.80), we observe that as $\alpha \to 1$, the equation reduces to the classical diffusion equation with an unknown coefficient $a(t)$. The classical parabolic inverse problem was first rigorously solved by Jones in [47, 48]. Jones applied a variational transformation to derive an integral equation, which interestingly was of fractional order. He further demonstrated that the unknown coefficient $a(t)$ is the unique fixed point of this integral equation and used monotonicity properties to prove existence, uniqueness, and stability of the solution. For the generalized fractional case where $\alpha \in (0,2)$, Lopushanskyi and Lopushanska [49] employed Green's functions to express the solution $u(x,t)$ and formulated an operator equation for $a(t)$. Their approach guarantees the existence and uniqueness of $(u(x,t), a(t))$. This method is particularly valuable as it can be extended to other time-fractional differential equations.

This section has two main objectives regarding equation (1.80): analyzing the weak solution for a fixed $a(t)$ and recovering $a(t)$ from $-a(t)u_x(0,t) = g(t)$. We prefer this formulation over using $u_x(0,t)$ directly, as the heat flux $-a(t)u_x(0,t) = g(t)$ is typically measured in practical scenarios. For the direct problem, we establish the existence and uniqueness of the weak solution and derive regularity properties. For the inverse problem, we define an operator whose fixed point corresponds to $a(t)$, proving its monotonicity and uniqueness and deriving a reconstruction algorithm.

The structure of this section is as follows: In Sect. 1.3.2, we introduce a modified model derived from (1.80) using a variational transformation and present preliminary results from fractional calculus. In Sect. 1.3.3, we address the direct problem, providing the spectral representation of the weak solution and proving its existence, uniqueness, and regularity. Section 1.3.4 focuses on the inverse problem, defining an operator K whose fixed point corresponds to $a(t)$ and proving its key properties, including monotonicity and uniqueness, followed by a reconstruction method. The results of this section are detailed in [50] (see also [51]).

Modified Model

To simplify the boundary conditions of (1.80), we introduce the transformation $v(x,t) = u(x,t) + (x-1)f(t)$, leading to the following equivalent model:

$$\begin{cases} \partial_t^\alpha v(x,t) - a(t)v_{xx}(x,t) \\ \qquad = (x-1)\partial_t^\alpha f(t), \quad 0 < t \leqslant T, \quad 0 < x < 1; \\ v(0,t) = v(1,t) = 0, \quad 0 < t \leqslant T, \\ v(x,0) = 0, \quad 0 < x < 1 \end{cases} \tag{1.81}$$

with the associated data:

$$-a(t)\left(v_x(0,t) - f(t)\right) = g(t), 0 < t \leqslant T.$$

This formulation ensures uniform Dirichlet boundary conditions while preserving equivalence with (1.80). Following [7], we analyze the direct problem for (1.81) using spectral decomposition. In the subsequent subsections, we outline the necessary results for this analysis.

1.3.2 Weak Solution of the Initial-Boundary Value Problem

Definition 1.4 We call $v(x,t;a)$ is a weak solution to model (1.81) in $L^2[0,1]$ corresponding with coefficient $a(t)$ if $v(\cdot,t;a) \in H_0^1[0,1]$ for $t \in (0,T]$ and for any $\psi \in H^2[0,1] \cap H_0^1[0,1]$,

$$\Big(\partial_t^\alpha v(x,t;a), \psi(x)\Big) - \Big(a(t)v_{xx}(x,t;a), \psi(x)\Big)$$
$$= \Big((x-1), \psi(x)\Big)\partial_t^\alpha f(t), \quad t \in (0,T];$$
$$\Big(v(x,0;a), \ \psi(x)\Big) = 0,$$

where $(\cdot,\cdot)$ is the inner product in $L^2[0,1]$.

In this subsection, we consider the weak solution $v(x,t;a)$ of model (1.81) under Definition 1.4. Specifically, we will establish the spectral representation, existence, uniqueness, and some regularity results for $v(x, t; a)$. To this end, we state some assumptions on $f(t)$ and the coefficient $a(t)$.

Assumption 1.1 (Direct Problem) Suppose $a(t)$ and $f(t)$ satisfy the following assumptions:

(a) $f(t)$ and $\partial_t^\alpha f(t)$ are continuous and nonnegative on $[0, T]$ with $f(0) = 0$;

(b) $a(t) \in C^+[0, T] := \{\psi \in C[0, T] : \psi(t) > 0, t \in [0, T]\}$.

We suppose Assumption 1.1 be valid in this section. In order to illustrate Assumption 1.1, the following remark is set up.

Remark 1.3 Assumptions 1.1 (a) and 1.1 (b) are crucial to deduce the regularity results of the weak solution $v(x, t; a)$, which can be seen in the future study. Also, without Assumption 1.1 (a) (see [5], Theorem 3.25) can not be used, which leads to the fails of the existence and uniqueness of $v(x, t; a)$.

We prefer the condition $\partial_t^\alpha f(t)$ is continuous and nonnegative on $[0, T]$ rather that $f'(t) \in C^+[0, T]$ since the latter one is stronger that the former. It is not hard to show the former condition can be derived from $f'(t) \in C^+[0, T]$.

Spectral representation of the weak solution

For the operator $-Lv = -v_{xx}$ with homogeneous Dirichlet boundary condition, $\{(n^2\pi^2, \sin n\pi x) : n \in \mathbb{N}\}$ is the eigensystem of $-L$ and $\{\sin n\pi x : n \in \mathbb{N}\}$ forms an orthogonal basis of $L^2[0, 1]$.

Multiplying $\sin n\pi x$ on the both sides of model (1.81) and integrating it on x over $[0, 1]$ give

$$\partial_t^\alpha (v(x, t; a), \sin n\pi x) - a(t)\,(v_{xx}(x, t; a), \sin n\pi x) \\ = ((x - 1), \sin n\pi x)\, \partial_t^\alpha f(t).$$

The zero boundary condition for $v(x, t; a)$ yields that

$$-\,(v_{xx}(x, t; a), \sin n\pi x) = n^2\pi^2 (v(x, t; a), \sin n\pi x),$$

which leads to

$$\begin{aligned}&\partial_t^\alpha (v(x, t; a), \sin n\pi x) + a(t) n^2\pi^2 (v(x, t; a), \sin n\pi x) \\ &= ((x - 1), \sin n\pi x)\partial_t^\alpha f(t), \quad n \in \mathbb{N}.\end{aligned}$$

Set $v_n(t; a) = 2(v(x, t; a), \sin n\pi x)$ because of $(\sin n\pi x, \sin n\pi x) = \frac{1}{2}$. The completeness of $\{\sin n\pi x : n \in \mathbb{N}\}$ in $L^2[0, 1]$ allows us to write $v(x, t; a)$ as

$$v(x, t; a) = \sum_{n=1}^{\infty} v_n(t; a) \sin n\pi x.$$

Also, recall that $(x-1, \sin n\pi x) = -\frac{1}{n\pi}$, then the following fractional ordinary differential equation can be deduced

$$\partial_t^\alpha v_n(t;a) = -n^2\pi^2 a(t) v_n(t;a) - \frac{2}{n\pi}\partial_t^\alpha f(t), \quad n \in \mathbb{N}.$$

The initial condition $v(x,0;a)=0$ and the linear independence of $\{\sin n\pi x : n \in \mathbb{N}\}$ yield that $v_n(0;a)=0,\ n \in \mathbb{N}$.

To sum up, the *spectral representation* of $v(x,t;a)$ is

$$v(x,t;a) = \sum_{n=1}^{\infty} v_n(t;a) \sin n\pi x, \tag{1.82}$$

where $v_n(t;a)$ satisfies

$$\partial_t^\alpha v_n(t;a) + n^2\pi^2 a(t) v_n(t;a) = -\frac{2}{n\pi}\partial_t^\alpha f(t),\ v_n(0;a) = 0,\ n \in \mathbb{N}. \tag{1.83}$$

Existence, uniqueness and some properties of the weak solution

In this subsection, we show the existence and uniqueness of the weak solution (1.82). At first, we state the following lemma (see [5], Theorem 3.25).

Lemma 1.12 *For the Cauchy-type problem*

$$\partial_t^\alpha y(t) = F(y,t), \quad y(0) = c_0,$$

if for any continuous $y(t)$, $F(y,t) \in C[0,T]$, there is $A > 0$ which is independent on $y \in C[0,T]$ and $t \in [0,T]$ such that $|F(t,y_1) - F(t,y_2)| \leqslant A|y_1 - y_2|$, then there exists a unique solution $y(t) \in C[0,T]$ for the Cauchy-type problem which also satisfies $\partial_t^\alpha y(t) \in C[0,T]$.

The following lemma follows from Lemma 1.12.

Lemma 1.13 *Suppose Assumption 1.1 holds. For model (1.81), there exists a unique weak solution $v(x,t;a)$ represented by (1.82) with $v_n(t;a)$, $\partial_t^\alpha v_n(t;a) \in C[0,T]$ for each $n \in \mathbb{N}$.*

Proof Fix $n \in \mathbb{N}$. Due to Assumption 1.1, the right side of (1.83) satisfies the conditions of Lemma 1.12. Hence, for any $n \in \mathbb{N}$, there exists a unique continuous solution $v_n(t;a)$ of (1.83) with $\partial_t^\alpha v_n(t;a) \in C[0,T]$. With the spectral representation (1.82), the existence and uniqueness of each $v_n(t;a)$ lead to the existence and uniqueness of the weak solution $v(x,t;a)$. This completes the proof.

Furthermore, there are two lemmas about $v_n(t;a)$ which play important roles for deriving the regularity of $v(x,t;a)$.

Lemma 1.14 *For any $a(t) \in C^+[0, T]$, there holds $v_n(t; a) \leqslant 0$ on $[0, T]$ for all $n \in \mathbb{N}$.*

Proof For each $a(t) \in C^+[0, T]$ and $n \in \mathbb{N}$, Lemma 1.13 yields that $v_n(t; a) \in C[0, T]$, which leads to $v_n(t; a)$ attains its maximum over $[0, T]$ at $t = t_0 \in [0, T]$.

If $t_0 = 0$, with the fact that $v_n(0, a) = 0$ which comes from Lemma 1.13, then we have 0 is the maximum value of $v_n(t; a)$ on $[0, T]$, which is the desired result. If $t_0 \in (0, T]$, Lemma 1.13 gives that $v_n(t; a),\ \partial_t^\alpha v_n(t; a) \in C[0, T]$, which ensure we can apply Lemma 1.3 to $v_n(t; a)$. With Lemma 1.3 and the definition of t_0, we have $\partial_{t_0}^\alpha v_n \geqslant 0$. Inserting $\partial_{t_0}^\alpha v_n \geqslant 0$ into (1.83) and recalling Assumption 1.1 lead to $n^2\pi^2 a\,(t_0)\, v_n\,(t_0; a) = -\partial_{t_0}^\alpha v_n - \frac{2}{n\pi}\partial_{t_0}^\alpha f \leqslant 0$, i.e., $v_n\,(t_0; a) \leqslant 0$ and complete the proof.

Lemma 1.15 *Given $a_1, a_2 \in C^+[0, T]$ with $a_1 \leqslant a_2$ on $[0, T]$, we have*

$$v_n\,(t; a_1) \leqslant v_n\,(t; a_2) \leqslant 0,\ \ t \in [0, T],\quad n \in \mathbb{N}.$$

Proof Equation (1.83) yields for each $n \in \mathbb{N}$,

$$\begin{aligned}
\partial_t^\alpha v_n\,(t; a_1) &= -n^2\pi^2 a_1(t) v_n\,(t; a_1) - \frac{2}{n\pi}\partial_t^\alpha f(t),\\
\partial_t^\alpha v_n\,(t; a_2) &= -n^2\pi^2 a_2(t) v_n\,(t; a_2) - \frac{2}{n\pi}\partial_t^\alpha f(t),
\end{aligned}$$

i.e.,

$$\partial_t^\alpha v_n\,(t; a_1) + n^2\pi^2 a_1(t) v_n\,(t; a_1) = \partial_t^\alpha v_n\,(t; a_2) + n^2\pi^2 a_2(t) v_n\,(t; a_2)\,. \tag{1.84}$$

Let $w(t) = v_n\,(t; a_1) - v_n\,(t; a_2)$. Lemma 1.13 ensures $w(t),\ \partial_t^\alpha w(t) \in C[0, T]$ and $w(0) = 0$. From (1.84), we deduce that

$$\partial_t^\alpha w(t) + n^2\pi^2 a_1(t) w(t) = n^2\pi^2\,(a_2(t) - a_1(t))\, v_n\,(t; a_2) \leqslant 0,\ t \in [0, T], \tag{1.85}$$

where the last equality follows from Lemma 1.14 and the condition $a_1 \leqslant a_2$. $w(t) \in C[0, T]$ yields that $w(t)$ attains its maximum over $[0, T]$ at some $t = t_0 \in [0, T]$. If $t_0 = 0$, we have $w(t) \leqslant w(0) = 0$. For the case $t_0 \in (0, T]$, applying Lemma 1.3 to $w(t)$ yields that $\partial_{t_0}^\alpha w \geqslant 0$, which together with (1.85) implies $n^2\pi^2 a_1\,(t_0)\, w\,(t_0) \leqslant -\partial_{t_0}^\alpha w \leqslant 0$. With the fact $a_1 \in C^+[0, T]$, we deduce $w\,(t_0) \leqslant 0$ and $w(t) \leqslant w\,(t_0) \leqslant 0$. The desired result follows from $w \leqslant 0$ and Lemma 1.14.

Regularity Results

Here, we deal with the regularity properties of $v(x, t; a)$. Through the next two lemmas, $\|v(x, t; a)\|_{C\left([0,T]; H^2[0,1]\right)}$ is bounded by $\|f\|_{C[0,T]}$ and $\left\|\partial_t^\alpha f\right\|_{C[0,T]}$.

Lemma 1.16 *Suppose Assumption 1.1 holds, then*

$$\|v(x,t;a)\|_{C([0,T];L^2[0,1])} \leqslant C\|f\|_{C[0,T]}.$$

Proof Pick $t \in [0,T]$, take the fractional integral I_t^α on both sides of (1.83) and use Lemma 1.2, then

$$v_n(t;a) + n^2\pi^2 I_t^\alpha\left[a(t)v_n(t;a)\right] = -\frac{2}{n\pi} f(t),\, t \in [0,T],\ n \in \mathbb{N}.$$

Lemma 1.14 together with Assumption 1.2 and the definition of I_t^α gives that $v_n(t;a)$ and $I_t^\alpha\left[a(t)v_n(t;a)\right]$ are both nonpositive on $[0,T]$, which implies

$$|v_n(t;a)| \leqslant \left|\frac{2}{n\pi} f(t)\right| \quad \text{on} \quad [0,T], \quad n \in \mathbb{N}. \tag{1.86}$$

Now, fix $t \in [0,T]$, calculate the L^2 norm of $v(x,t;a)$ with respect to x, then

$$\|v(x,t;a)\|^2_{L^2[0,1]} = \left\|\sum_{n=1}^{\infty} v_n(t;a)\sin n\pi x\right\|^2_{L^2[0,1]} = \frac{1}{2}\sum_{n=1}^{\infty}\left(v_n(t;a)\right)^2, \tag{1.87}$$

where the last equality is due to the orthogonality of $\{\sin n\pi x : n \in \mathbb{N}\}$. Equation (1.86) and (1.87) yield

$$\|v(x,t;a)\|^2_{L^2[0,1]} \leqslant \frac{1}{2}(f(t))^2 \sum_{n=1}^{\infty} \frac{4}{n^2\pi^2} = C(f(t))^2 \leqslant C\|f\|^2_{C[0,T]},$$

i.e., $\|v(x,t;a)\|_{L^2[0,1]} \leqslant C\|f\|_{C[0,T]}$ for each $t \in [0,T]$, which leads to the expected result.

Lemma 1.17 *With Assumption 1.1, we have*

$$\|v(x,t;a)\|_{C([0,T];\ H^2[0,1])} \leqslant C_a\left(\|f\|_{C[0,T]} + \left\|\partial_t^\alpha f\right\|_{C[0,T]}\right),$$

where $C_a > 0$ only depends on the coefficient $a(t)$.

Proof Assumption 1.1 states $a \in C^+[0,T]$, which implies there exists constants $q_a, Q_a > 0$ such that

$$0 < q_a < a(t) < Q_a \quad on \quad [0,T]. \tag{1.88}$$

We denote these constants by q_a, Q_a due to their dependence on $a(t)$. Then by Lemma 1.15, $v_n(t;q_a) \leqslant v_n(t;a) \leqslant 0$, which leads to

$$|v_n(t;q_a)| \geqslant |v_n(t;a)|\,, \quad t \in [0,T],\ n \in \mathbb{N}, \tag{1.89}$$

and $v_n(t; q_a)$ is the solution of

$$\partial_t^\alpha v_n(t; q_a) = -n^2\pi^2 q_a v_n(t; q_a) - \frac{2}{n\pi}\partial_t^\alpha f(t).$$

For the above fractional ordinary differential equation, [7] gives

$$v_n(t; q_a) = -\frac{2}{n\pi}\int_0^t \partial_\tau^\alpha f(\tau)(t-\tau)^{\alpha-1} E_{\alpha,\alpha}\left(-n^2\pi^2 q_a (t-\tau)^\alpha\right) \mathrm{d}\tau, \tag{1.90}$$

which allows us to deduce the regularity of $v_{xx}(x, t; a)$ immediately. Fix $t \in [0, T]$. The representation (1.82) yields that

$$||v(x,t;a)||^2_{H^2[0,1]} = \left\| \sum_{n=1}^{\infty} n^2\pi^2 v_n(t;a)\sin n\pi x \right\|^2_{L^2[0,1]} = \frac{1}{2}\sum_{n=1}^{\infty} n^4\pi^4 (v_n(t;a))^2, \tag{1.91}$$

which implies the following estimate together with (1.89) and (1.90)

$$\begin{aligned}
\|v(x,t;a)\|^2_{H^2[0,1]} &\leqslant \frac{1}{2}\sum_{n=1}^{\infty}\left[\int_0^t \frac{2}{n\pi}\partial_t^\alpha f(\tau) n^2\pi^2 (t-\tau)^{\alpha-1} E_{\alpha,\alpha}\left(-n^2\pi^2 q_a (t-\tau)^\alpha\right)\mathrm{d}\tau\right]^2 \\
&\leqslant \frac{1}{2}\sum_{n=1}^{\infty}\left[\int_0^t \| \frac{2}{n\pi}\partial_t^\alpha f(t)\Big|_{C[0,T]} \right. \\
&\quad \times \left| n^2\pi^2 (t-\tau)^{\alpha-1} E_{\alpha,\alpha}\left(-n^2\pi^2 q_a (t-\tau)^\alpha\right)\right| \mathrm{d}\tau\Big]^2 \\
&= \frac{1}{2}\sum_{n=1}^{\infty}\left\|\frac{2}{n\pi}\partial_t^\alpha f(t)\right\|^2_{C[0,T]}\left[\int_0^t \left|n^2\pi^2\tau^{\alpha-1}E_{\alpha,\alpha}\left(-n^2\pi^2 q_a\tau^\alpha\right)\right|\mathrm{d}\tau\right]^2.
\end{aligned} \tag{1.92}$$

The term $\int_0^t \left|n^2\pi^2\tau^{\alpha-1}E_{\alpha,\alpha}\left(-n^2\pi^2 q_a\tau^\alpha\right)\right|\mathrm{d}\tau$ is bounded by Propositions 1.2 and 1.3 (see Sect. 1.1.3)

$$\begin{aligned}
0 &\leqslant \int_0^t \left|n^2\pi^2\tau^{\alpha-1}E_{\alpha,\alpha}\left(-n^2\pi^2 q_a\tau^\alpha\right)\right|\mathrm{d}\tau = \int_0^t n^2\pi^2\tau^{\alpha-1}E_{\alpha,\alpha}\left(-n^2\pi^2 q_a\tau^\alpha\right)\mathrm{d}\tau \\
&= -\frac{1}{q_a}\int_0^t \frac{\mathrm{d}}{\mathrm{d}\tau}E_{\alpha,1}\left(-n^2\pi^2 q_a\tau^\alpha\right)\mathrm{d}\tau = \frac{1}{q_a}\left(1 - E_{\alpha,1}\left(-n^2\pi^2 q_a t^\alpha\right)\right) \leqslant q_a^{-1},
\end{aligned} \tag{1.93}$$

where the last inequality follows from Proposition 1.3. Inserting (1.93) into (1.92) gives

$$\begin{aligned}
\|v(x,t;a)\|^2_{H^2[0,1]} &\leqslant \frac{1}{2}\sum_{n=1}^{\infty}\frac{4}{n^2\pi^2}\left\|\partial_t^\alpha f(t)\right\|^2_{C[0,T]} q_a^{-2} \\
&= \frac{1}{2}q_a^{-2}\left(\sum_{n=1}^{\infty}\frac{4}{n^2\pi^2}\right)\left\|\partial_t^\alpha f\right\|^2_{C[0,T]},
\end{aligned}$$

i.e.,

$$\|v(x,t;a)\|_{H^2[0,1]} \leqslant C_a \left\|\partial_t^\alpha f(t)\right\|_{C[0,T]}, \tag{1.94}$$

where $C_a > 0$ depends on $a(t)$ only.

For $\|v(x,t;a)\|_{H^1[0,1]}$, the spectral representation (1.82) gives

$$\|v(x,t;a)\|_{H^1[0,1]}^2 = \left\|\sum_{n=1}^{\infty} n\pi v_n(t;a)\cos n\pi x\right\|_{L^2[0,1]}^2$$

$$= \frac{1}{2}\sum_{n=1}^{\infty} n^2\pi^2 (v_n(t;a))^2 \leqslant |v(x,t;a)|_{H^2[0,1]}^2,$$

where the last inequality is ensured by (1.91). Then (1.94) yields

$$\|v(x,t;a)\|_{H^1[0,1]} \leqslant C_a \left\|\partial_t^\alpha f\right\|_{C[0,T]}. \tag{1.95}$$

With Lemma 1.16, (1.94) and (1.95), we have

$$\|v(x,t;a)\|_{H^2[0,1]} \leqslant C_a \left(\|f\|_{C[0,T]} + \left\|\partial_t^\alpha f\right\|_{C[0,T]}\right), \quad t \in [0,T],$$

which gives the expected conclusion. The regularity of $\partial_t^\alpha v(x,t;a)$ is derived from Lemma 1.17 directly.

Corollary 1.1 *Suppose Assumption 1.1 be valid, then we have*

$$\left\|\partial_t^\alpha v(x,t;a)\right\|_{C([0,T];\,L^2[0,1])} \leqslant C_a \left\|\partial_t^\alpha f\right\|_{C[0,T]},$$

where C_a is a positive constant which only depends on the coefficient $a(t)$.

Proof Calculate $\partial_t^\alpha v(x,t;a)$ directly from (1.82) and (1.83)

$$\partial_t^\alpha v(x,t;a) = \sum_{n=1}^{\infty}\left[-n^2\pi^2 a(t) v_n(t;a) - \frac{2}{n\pi}\partial_t^\alpha f(t)\right]\sin n\pi x.$$

Fix an arbitrary $t \in [0,T]$, the above equality yields that

$$\left\|\partial_t^\alpha v(x,t;a)\right\|_{L^2[0,1]}^2 = \frac{1}{2}\sum_{n=1}^{\infty}\left[-n^2\pi^2 a(t) v_n(t;a) - \frac{2}{n\pi}\partial_t^\alpha f(t)\right]^2$$

$$\leqslant \sum_{n=1}^{\infty} n^4\pi^4 (a(t))^2 (v_n(t;a))^2 + \sum_{n=1}^{\infty}\frac{4}{n^2\pi^2}\left|\partial_t^\alpha f(t)\right|^2$$

$$\leqslant \sum_{n=1}^{\infty} n^4\pi^4 (a(t))^2 (v_n(t;a))^2 + C\left\|\partial_t^\alpha f\right\|_{C[0,T]}^2.$$

The bound of $\sum_{n=1}^{\infty} n^4\pi^4(a(t))^2\,(v_n(t;a))^2$ follows from (1.88), (1.91) and (1.94)

$$\sum_{n=1}^{\infty} n^4\pi^4(a(t))^2\,(v_n(t;a))^2 \leqslant \sum_{n=1}^{\infty} n^4\pi^4 Q_a^2\,(v_n(t;a))^2 = 2Q_a^2|v(x,t;a)|_{H^2[0,1]}^2$$

$$\leqslant 2Q_a^2C_a^2\,\left\|\partial_t^\alpha f\right\|_{C[0,T]}^2.$$

Consequently, we have $\left\|\partial_t^\alpha v(x,t;a)\right\|_{L^2[0,1]}^2 \leqslant \left(2Q_a^2C_a^2 + C\right)\left\|\partial_t^\alpha f\right\|_{C[0,T]}^2$, i.e.,

$$\left\|\partial_t^\alpha v(x,t;a)\right\|_{L^2[0,1]} \leqslant C_a\left\|\partial_t^\alpha f\right\|_{C[0,T]},\quad t\in[0,T],$$

which yields the desired result.

Remark 1.4 Lemma 1.17 and Corollary 1.1 guarantee that model (1.81) is well defined in the space $L^2[0,1]$.

Main Theorem for the Direct Problem

In this subsection, we state the main theorem for the direct problem of model (1.81), which follows from Lemmas 1.13 and 1.17 and Corollary 1.1.

Theorem 1.10 *Suppose Assumption 1.1 holds, then there exists a unique weak solution $v(x,t;a)$ for model (1.81) with the spectral representation (1.82) and the following regularity estimates:*

$$\|v(x,t;a)\|_{C([0,T];H^2[0,1])} \leqslant C_a\left(\|f\|_{C[0,T]} + \left\|\partial_t^\alpha f\right\|_{C[0,T]}\right)$$

$$\left\|\partial_t^\alpha v(x,t;a)\right\|_{C([0,T];L^2[0,1])} \leqslant C_a\left\|\partial_t^\alpha f\right\|_{C[0,T]},$$

where $C_a > 0$ only depends on the coefficient $a(t)$.

1.3.3 Problem of Recovering the Diffusion Coefficient

In this subsection, we consider the inverse problem for model (1.81): using the data $-a(t)\,(v_x(0,t;a) - f(t)) = g(t)$ to reconstruct the coefficient $a(t)$. We only consider the case $a(t)\in C^+(0,T]$, which is regarded as the admissible set of $a(t)$.

Operator K and its well-definedness

In order to reconstruct $a(t)$, we introduces an operator $K:\Omega\to C^+(0,T]$ as

$$K\psi(t) := \frac{g(t)}{f(t) - v_x(0,t;\psi)} = \frac{g(t)}{f(t) - \sum_{n=1}^{\infty} n\pi v_n(t;\psi)},$$

where the class Ω of functions $\psi(t)$ is defined as

$$\Omega := \left\{\psi \in C^+(0,T] : 0 < \psi(t) \leqslant g/f, t \in (0,T]\right\}.$$

By the boundary condition $-a(t)\,(v_x(0,t;a) - f(t)) = g(t)$, we can see the coefficient $a(t)$ is one of the fixed points of operator K. In order to analyze this K, we state some restrictions for the data f and g.

Assumption 1.2 Suppose f and g satisfy the following restrictions:
(a) $f(t) \in C[0,T] \cap C^+(0,T]$ with $f(0) = 0$ and $\partial_t^\alpha f(t) \in C[0,T]$ is nonnegative on $[0,T]$;
(b) $g(t) \in C[0,T] \cap C^+(0,T]$ and $g(0) = 0$.

The following two remarks explain the reasonableness and reason of Assumption 1.2 respectively.

Remark 1.5 The function $f(t)$ is input data so that we can control it, which means Assumption 1.2 (a) can be achieved.

For Assumption 1.2 (b), given $a(t) \in C^+(0,T]$, Lemma 1.13 ensures the continuity of $v_n(t;a)$ for each $n \in \mathbb{N}$. This result and the continuity of $f(t)$ and $a(t)$ imply the continuity of $g(t)$. Also, Lemma 1.14 yields that $-v_x(0,t;a) = -\sum_{n=1}^\infty n\pi\, v_n(t;a) \geqslant 0$ on $(0,T]$, which together with $a > 0$ and Assumption 1.2 (a) leads to $g > 0$ on $(0,T]$. At last, the zero initial condition and $f(0) = 0$ give $v_x(0,0) - f(0) = 0$, which implies $g(0) = 0$. The above arguments guarantee the reasonableness of Assumption 1.2 (b).

Remark 1.6 Assumption 1.2 (a) and (b) guarantee the well-definedness of the space Ω in the sense that the upper bound g/f is strictly larger than the lower bound 0 on the interval $(0,T]$. Also, Assumption 1.2 (a) will be used to show the monotonicity of K and the uniqueness of its fixed points.

Furthermore, it is clear that $v_x(0,0) - f(0) = g(0) = 0$ due to the zero initial condition and Assumption 1.2. That means we can not reconstruct $a(t)$ at $t = 0$, but only over the interval $(0,T]$, which can be seen from the definition of Ω.

For operator K, Theorem 1.10 indicates that given $a(t) \in \Omega$, there exists a unique Ka. Also, Assumption 1.2 and the result $-\sum_{n=1}^\infty n\pi\, v_n(t;a) \geqslant 0$ derived from Lemma 1.14 yield the dominator $f(t) - \sum_{n=1}^\infty n\pi\, v_n(t;a)$ of Ka is strictly positive on $(0,T]$, which indicates there is no singularities for Ka on $(0,T]$. In addition, the positivity and continuity of the dominator and numerator of Ka ensure $Ka \in C^+(0,T]$. Therefore, the operator K is well defined and maps Ω to $C^+(0,T]$.

Monotonicity

We show the monotonicity of K.

Theorem 1.11 *With Assumption 1.2, the operator K is monotone, i.e., given a_1, $a_2 \in \Omega$ with $a_1 \leqslant a_2$ on $(0,T]$, we have $Ka_1(t) \leqslant Ka_2(t)$, $t \in (0,T]$.*

Proof Lemma 1.15 gives $v_n(t; a_1) \leqslant v_n(t; a_2) \leqslant 0$ on $(0, T], n \in \mathbb{N}$, which together with Assumption 1.2 yields $f(t) - \sum_{n=1}^{\infty} n\pi v_n(t; a_1) \geqslant f(t) - \sum_{n=1}^{\infty} n\pi v_n(t; a_2) > 0$. Consequently, with Assumption 1.2 (b),

$$\frac{g(t)}{f(t) - \sum_{n=1}^{\infty} n\pi v_n(t; a_1)} \leqslant \frac{g(t)}{f(t) - \sum_{n=1}^{\infty} n\pi v_n(t; a_2)} \quad \text{on } (0, T],$$

which implies the monotonicity.

The following lemma concerns the upper bound g/f of Ω.

Lemma 1.18 *The following inclusion is true:* $K(g/f) \in \Omega$.

Proof Lemmas 1.13 and 1.14 give that $\sum_{n=1}^{\infty} n\pi v_n(t; g/f)$ is nonpositive and continuous; while Assumption 1.2 yields $f, g \in C^{+}(0, T]$. These results ensure the continuity of $K(g/f)$ and

$$0 < K(g/f) = \frac{g(t)}{f(t) - \sum_{r=1}^{\alpha} n\pi v_n(t, g/f)} \leqslant g/f$$

on $(0, T]$, which lead to the expected result.

Uniqueness

Here, we deal with the uniqueness of fixed points of K. To this end, we introduce a successive iteration as $\bar{a}_0 = g/f, \bar{a}_{n+1} = K\bar{a}_n, \ n \in \mathbb{N}$. The reasonableness of the initial guess g/f is guaranteed by Lemma 1.18 and this iteration generates a series $\{\bar{a}_n : n \in \mathbb{N}\}$. The next two lemmas state the above series will converge to a fixed point of K if it exists.

Lemma 1.19 *Suppose Assumption 1.2 holds. If* $a_1, a_2 \in \Omega$ *are two fixed points of* K *with* $a_1 \leqslant a_2$, *then* $a_1 \equiv a_2$ *on* $(0, T]$.

Proof If $a \in \Omega$ is a fixed point of operator K, then

$$\sum_{n=1}^{\infty} n\pi a(t) v_n(t; a) = -g(t) + a(t) f(t), \ t \in (0, T].$$

Taking I_t^{α} on both sides of the above equality and recalling its linearity yield

$$\sum_{n=1}^{\infty} n\pi I_t^{\alpha} [a(t) v_n(t; a)] = -I_t^{\alpha} g(t) + I_t^{\alpha}[a(t) f(t)], \ t \in (0, T]. \tag{1.96}$$

Furthermore, the following equality follows from (1.83) and Lemma 1.2

$$I_t^{\alpha} [a(t) v_n(t; a)] = -\frac{2}{n^3\pi^3} f(t) - \frac{1}{n^2\pi^2} v_n(t; a),$$

which together with (1.96) leads to

$$I_t^\alpha[a(t)f(t)]+\sum_{n=1}^{\infty}\frac{1}{n\pi}v_n(t;a)=-f(t)\sum_{n=1}^{\infty}\frac{2}{n^2\pi^2}+I_t^\alpha g(t),\ t\in(0,T]. \quad (1.97)$$

Now pick two fixed points a_1, a_2 of K with $a_1 \leqslant a_2$. For fixed points a_1, a_2, (1.97) holds respectively. Combining them together yields

$$I_t^\alpha\left[a_1(t)f(t)\right]+\sum_{n=1}^{\infty}\frac{1}{n\pi}v_n\left(t;a_1\right)=I_t^\alpha\left[a_2(t)f(t)\right]+\sum_{n=1}^{\infty}\frac{1}{n\pi}v_n\left(t;a_2\right). \quad (1.98)$$

Lemma 1.15 and $a_1 \leqslant a_2$ give $v_n\left(t;a_1\right) \leqslant v_n\left(t;a_2\right),\ n\in\mathbb{N}$, i.e.,

$$\sum_{n=1}^{\infty}\frac{1}{n\pi}v_n\left(t;a_1\right)\leqslant\sum_{n=1}^{\infty}\frac{1}{n\pi}v_n\left(t;a_2\right),\ t\in(0,T],$$

which together with (1.98) leads to $I_t^\alpha\left[\left(a_1(t)-a_2(t)\right)f(t)\right]\geqslant 0,\ \ t\in(0,T]$. Moreover, recall that $f>0$, $a_1\leqslant a_2$ and the definition of I_t^α, we have $I_t^\alpha\left[\left(a_1(t)-a_2(t)\right)f(t)\right]\leqslant 0,\quad t\in(0,T]$. Therefore, holds $0\leqslant I_t^\alpha$ $\left[\left(a_1(t)-a_2(t)\right)f(t)\right]\leqslant 0$, i.e., $I_t^\alpha\left[\left(a_1(t)-a_2(t)\right)f(t)\right]=0$ on $(0,T]$. The definition of I_t^α (see (1.5)) and the non-positivity of $\left(a_1-a_2\right)f$ imply $\left(a_1(\tau)-a_2(\tau)\right)f(\tau)=0$ on $(0,t]$ for each $t\in(0,T]$, which leads to $a_1-a_2\equiv 0$ on $(0,T]$ due to $f>0$.

Lemma 1.20 *Suppose Assumption 1.2 be valid. If there exists a fixed point $a(t)\in\Omega$ of operator K, then the series $\{\bar{a}_n : n\in\mathbb{N}\}$ is contained by Ω and converges to $a(t)$.*

Proof $\bar{a}_0$ is the upper bound of Ω gives $a\leqslant\bar{a}_0$, which together with the monotonicity of K yields $Ka\leqslant K\bar{a}_0$, i.e., $a\leqslant\bar{a}_1$. Applying the monotonicity again to this inequality, it holds $Ka\leqslant K\bar{a}_1$, i.e., $a\leqslant\bar{a}_2$. Continuing this procedure, we deduce $a\leqslant\bar{a}_n,\ n\in\mathbb{N}$, which implies $a(t)$ is a lower bound of $\{\bar{a}_n : n\in\mathbb{N}\}$. On the other hand, lemma 4.1 gives $\bar{a}_1\leqslant\bar{a}_0$. Apply the monotonicity of K to it, there holds $K\bar{a}_1\leqslant K\bar{a}_0$, i.e., $\bar{a}_2\leqslant\bar{a}_1$. Repeat this argument again and again, we obtain $\bar{a}_{n+1}\leqslant\bar{a}_n,\ n\in\mathbb{N}$, which yields $\{\bar{a}_n : n\in\mathbb{N}\}$ is decreasing.

With the facts that $\{\bar{a}_n : n\in\mathbb{N}\}$ is decreasing, $\bar{a}_0$ is the upper bound of Ω and $a(t)\in\Omega$ is a lower bound of $\{\bar{a}_n : n\in\mathbb{N}\}$, the following results can be deduced: $\{\bar{a}_n : n\in\mathbb{N}\}$ is included by Ω and convergent in it. Denote the limit of $\{\bar{a}_n : n\in\mathbb{N}\}$ by $\bar{a}$. Clearly $\bar{a}$ is one of the fixed points of K and $\bar{a}(t)\geqslant a(t)$ due to $a(t)$ is a lower bound of $\{\bar{a}_n : n\in\mathbb{N}\}$. Then Lemma 1.19 gives $\bar{a}(t)=a(t)$, which completes this proof. The uniqueness of fixed points of K follows from Lemma 1.20 immediately.

Theorem 1.12 *Let Assumption 1.2 be fulfilled, then there is at most one fixed point of K in* Ω.

Proof Given two fixed points $a_1, a_2 \in \Omega$, Lemma 1.20 yields that $\bar{a}_n \to a_1$ and $\bar{a}_n \to a_2$, which leads to $a_1 = a_2$ and completes this proof.

We state the main theorem for the inverse problem which follows from Lemma 1.20 and Theorem 1.12 immediately.

Theorem 1.13 *Let Assumption 1.2 be valid. If there exists a fixed point of* K *in* Ω, *then it is unique and coincides with the limit of* $\{\bar{a}_n : n \in \mathbb{N}\}$.

1.4 Conclusion

In this chapter, two inverse problems of determining the reaction coefficient and one problem of finding the diffusion coefficient are investigated. In the first two problems, the direct problems are initial-boundary value problems with Dirichlet boundary conditions for the first case and periodic conditions for the second case. The overdetermination condition for both cases corresponds to the assignment of the heat flux value at the left end of the rod. Theorems are obtained that characterize the local existence and global uniqueness of inverse problems, as well as estimates of the conditional stability of the solution. In the third inverse problem, the problem of determining the time-dependent diffusion coefficient is studied. For the direct problem, the existence, uniqueness, and some regularity properties of a weak solution of the fractional diffusion equation with a given diffusion coefficient are proven. For the inverse problem, in order to restore the unknown coefficient, an operator is introduced and its monotonicity is shown. Using this property, the uniqueness of the determination of this coefficient is established. Regarding other problems for the fractional diffusion equation, in the work [52], known theoretical results and computational methods available in the literature are briefly described (see also references therein).

References

1. Ginoa, M., Cerbelli, S., Roman, H.E.: Fractional diffusion equation and relaxation in complex viscoelastic materials. Physica A **191**, 449–453 (2009)
2. Metzler, R., Klafter, J.: The random walk's guide to anomalous diffusion: a fractional dynamics approach. Phys. Rep. **339**(1), 1–77 (2000)
3. Nakagawa, J., Sakamoto, K., Yamamoto, M.: Overview to mathematical analysis for fractional diffusion equations-new mathematical aspects motivated by industrial collaboration. Journal of Math-industry. **2**, 99–108 (2010)
4. Wang, S., Zhao, M., Li, X.: Radial anomalous diffusion in an annulus. Physica A **390**, 3397–3403 (2011)
5. Kilbas, A.A., Srivastava, H.M., Trujillo, J.J.: Theory and application of fractional differetial equations. Elsevier, North-Holland Mathematical Studies, Amsterdam (2006)

6. Luchko, Y.: Some uniqueness and existence results for the initial-boundary-value problems for the generalized time-fractional diffusion equation. Comput. Math. Appl. **59**, 1766–1772 (2010)
7. Sakamoto, K., Yamamoto, M.: Initial value/boundary value problems for frac-tional diffusion-wave equations and applications to some inverse problems. J. Math. Anal. Appl. **382**, 426–447 (2011)
8. Gorenflo, R., Luchko, Y.F., Zabrejko, P.P.: On solvability of linear fractional differential equations in Banach spaces. Fractional Calculus and Applied Analysis **2**, 163–176 (1999)
9. Hanyga A., 2018. Multidimensional solutions of time-fractional diffusion-wave equations,*Proceedings of the Royal Society of London. Series A: Mathematical, Physical and Engineering Sciences*. 458: 933-957
10. Kochubei, A.N.: Diffusion of fractional order. Differential Equations. **26**(4), 485–492 (1990)
11. Kochubei, A.N.: The Cauchy problem for fractional-order evolution equations. Differential Equations. **25**(8), 1359–1368 (1986)
12. Pskhu, A.V.: Solution of Boundary Value Problems for the Fractional Diffusion Equation by the Green Function Method. Differential Equations. **39**(10), 1509–1513 (2003)
13. Metzler, R., Klafter, J.: Boundary value problems for fractional diffusion equations. Physic. **278**, 107–125 (2000)
14. Durdiev, D.K., Nuriddinov, Zh.Z.: Determination of a multidimensional kernel in some parabolic integro-differential equation. Journal of Siberian Federal University - Mathematics and Physics. **14**(1), 117–127 (2021)
15. Durdiev, D.K., Zhumaev, ZhZh.: Problem of determining a multidimensional thermal memory in a heat conductivity equation. Methods of Functional Analysis and Topology. **25**(3), 219–226 (2019)
16. Durdiev, D.K.: On the uniqueness of kernel determination in the integro-differential equation of parabolic type. J. Samara State Tech. Univ., Ser. Phys. Math. Sci. **19**(4), 658–666 (2015)
17. Durdiev, D.K., Turdiev, KhKh.: The problem of finding the kernels in the system of integro-differential Maxwell's equations. Sib. Zh. Ind. Math. **24**(2), 38–61 (2021)
18. Durdiev, D.K., Turdiev, KhKh.: An inverse problem for a first order hyperbolic system with memory. Differ. Equ. **56**(12), 1666–1675 (2020)
19. Durdiev, D.K., Jumaev, J.J.: Memory kernel reconstruction problems in the integro-differential equation of rigid heat conductor. Math. Meth. Appl. Sci. **45**(14), 8374–8388 (2022)
20. Durdiev, D.K., Zhumaev, ZhZh.: Problem of determining the thermal memory of a conducting medium. Differ. Equ. **56**(6), 785–796 (2020)
21. Miller, L., Yamamoto, M.: Coefficient inverse problem for a fractional diffusion equation. Inverse Probl. **29**(7), 075013 (2013)
22. Jiang, S., Wei, T.: Recovering a time-dependent potential function in a time fractional diffusion equation by using a nonlinear condition. Inverse Probl. Sci. Eng. **29**(2), 174–195 (2021)
23. Xiong, T.X., Zhou, Q., Hon, C.Y.: An inverse problem for fractional diffusion equation in 2-dimensional case: stability analysis and regularization. J. Math. Anal. Appl. **393**, 185–199 (2012)
24. Xiong, X., Guo, H., Liu, X.: An inverse problem for a fractional diffusion equation. J. Comput. Appl. Math. **236**, 4474–4484 (2012)
25. Durdiev, D.K., Rahmonov, A.A., Bozorov, Z.R.: A two-dimensional diffusion coefficient determination problem for the time-fractional equation. Math. Meth. Appl. Sci. **44**, 10753–10761 (2021)
26. Subhonova, Z.A., Rahmonov, A.A.: Problem of determining the time dependent coefficient in the fractional diffusion-wave equation. Lobachevskii J. Math. **42**(15), 3747–3760 (2022)
27. Durdiev, D.K.: Inverse coefficient problem for the time-fractional diffusion equation. Eurasian J. Math. Comput. Appl. **9**(1), 44–54 (2022)
28. Durdiev, D.K., Durdiev, D.D.: The Fourier spectral method for determining a heat capacity coefficient in a parabolic equation. Turk. J. Math. **46**(8), 3223–3233 (2022)
29. Durdiev, D.K., Durdiev, D.D.: An inverse problem of finding a time-dependent coefficient in a fractional diffusion equation. Turk. J. Math. **47**(5), 1437–1452 (2023)

30. Durdiev, D.K., Turdiev, H.H.: Inverse coefficient problem for a time-fractional wave equation with initial-boundary and integral type overdetermination conditions. Math. Methods Appl. Sci. **47**(6), 5329–5340 (2024)
31. Durdiev, D.K., Turdiev, H.H.: Determining of a space-dependent coefficient of fractional diffusion equation with the generalized Riemann-Liouville time derivative. Lobachevskii J. Math. **45**(2), 80–94 (2024)
32. Durdiev, U.D.: Problem of determining the reaction coefficient in a fractional diffusion equation. Differ. Equ. **57**(9), 1195–1204 (2021)
33. Aleroev, T.S., Kirane, M., Malik, S.A.: Determination of a source term for a time fractional diffusion equation with an integral type over-determining condition. Electron. J. Differ. Equ. **270**, 1–16 (2013)
34. Gel'fand, I.M., Shilov, G.E.: Generalized functions, vol. 1. Properties and operations. Academic Press Inc., New York (1964)
35. Henry, D.: Geometric Theory of Semi linear Parabolic Equations. Berlin, Germany (1981)
36. Budak, B.M., Samarskii, A.A., Tikhonov, A.N.: Collection of Problems on Mathematical Physics. Nauka, Moscow (in Russian) (1979)
37. Odibat, Z.M., Shawagfeh, N.T.: Generalized Taylor's formula. Appl. Math. Comput. **186**, 286–293 (2007)
38. Evans, L.: Partial Differential Equations, Grad. Stud. Math. 19. American Mathematical Society, Providence (2002)
39. Kolmogorov, A.N., Fomin, S.V.: Theory of Functions and Functional Analysis, Volume 1: Metric and Normed Spaces. Graylock Press, Rochester, N.Y (1957)
40. Montroll, E.W., Weiss, G.H.: Random walks on lattices: II. J. Math. Phys. **6**, 167–81 (1965)
41. Nigmatullin, R.: The realization of the generalized transfer equation in a medium with fractal geometry. Phys. Status Solidi b. **133**, 425–30 (1986)
42. Berkowitz, B., Cortis, A., Dentz, M., Scher, H.: Modeling non-fickian transport in geological formations as a continuous time random walk. Rev. Geophys. **44**, 2006 (2006)
43. Hatano, Y., Hatano, N.: Dispersive transport of ions in column experiments: an explanation of long-tailed profiles. Water Resour. Res. **34**, 1027–33 (1998)
44. Mainardi, F.: Fractional Calculus and Waves in Linear Viscoelasticity: An Introduction to Mathematical Models. Imperial College Press, London (2010)
45. Wharmby, A.W., Bagley, R.L.: Generalization of a theoretical basis for the application of fractional calculus to viscoelasticity. J. Rheol. (1978-present) **57**, 1429–40 (2013)
46. Wharmby, A.W., Bagley, R.L.: Modifying Maxwells equations for dielectric materials based on techniques from viscoelasticity and concepts from fractional calculus. Int. J. Eng. Sci. **79**, 59–80 (2014)
47. Jones, B.F.: The determination of a coefficient in a parabolic differential equation. I. Existence and uniqueness. J. Math. Mech. **11**, 907–18 (1962)
48. Jones, B.F.: Various methods for finding unknown coefficients in parabolic differential equations. Commun. Pure Appl. Math. **16**, 33–44 (1963)
49. Lopushanskyi, A.O., Lopushanska, H.P.: One inverse problem for the diffusion-wave equation in bounded domain. Ukrainian Math. J. **66**, 743–57 (2014)
50. Zhang, Z.: An undetermined coefficient problem for a fractional diffusion equation. Inverse Prob. **32**(1), 015011 (2016)
51. Zhang, Z.: An undetermined time-dependent coefficient in a fractional diffusion equation. Inverse Prob. Imaging. **11**(5), 875–900 (2017)
52. Jin, B., Rundell, W.: A tutorial on inverse problems for anomalous diffusion processes. Inverse Prob. **31**(3), 035003 (2015)

Chapter 2
Inverse Coefficient Problems with Nonlocal Conditions

This chapter deals with the inverse problems of determining the coefficient at the lower term of a fractional diffusion equation with nonlocal boundary and nonlocal overdetermination (integral type) conditions. In Sect. 2.1, an inverse problem of determining the coefficient at the lower term of a multidimensional fractional diffusion equation in the case the direct problem is the initial-boundary problem for this equation with nonlocal initial and homogeneous Dirichlet conditions is considered. To determine the unknown coefficient, an overdetermination condition of the integral form is specified for the solution of the direct problem. First the unique existence of the weak solution to direct problem and the asymptotic behavior as the time t goes to ∞ are established and the proofs are based on the eigenfunction expansions. Using Green's function for an ordinary fractional differential equation associated with direct problem and the Fourier method, the inverse problem is reduced to an equivalent problem. Further by using the fixed point argument in suitable Sobolev spaces, the global theorems of existence and uniqueness for the solution of the inverse problem are obtained. In Sect. 2.2, time-fractional diffusion equation is used with initial, periodic boundary conditions for the determination of two coefficients: the first of them is the reaction coefficient, and the second one is a time-dependent factor in source function. Two conditions of spatial integral type are specified as overdetermination conditions. The existence and uniqueness of the classical solution of this problem is reduced to fixed point principles by applying the Fourier method.

2.1 Global Solvability of an Inverse Coefficient Problem

The study of the Cauchy problem, initial-boundary value problems (direct problems), and inverse problems for fractional-order equations is a rapidly developing field of modern mathematical physics. The reader can get acquainted with the studies of direct and inverse problems for a fractional diffusion equation with a derivative in

D. K. Durdiev, *Inverse Problems for Fractional Diffusion Equations*,
Industrial and Applied Mathematics, https://doi.org/10.1007/978-981-96-5338-6_2

the sense of the Caputo in [1] (see also references therein). This book provides an overview of the theoretical results on direct problems and inverse source problems for diffusion equations with the Caputo time-fractional derivatives of order $\alpha \in (0, 1)$. This survey covers the initial-boundary problems and the following types of inverse problems: determination of time-dependent source functions; determination of space-dependent source functions; determination of time-dependent functions appearing in boundary conditions. The results on the unique solvability of the problems are presented. Studies of the uniqueness of the inverse problem of determining the spatial diffusivity coefficient are also given here.

In the first subsection, we formulate the direct and inverse problems which will be studied in this section. Here we will also provide the necessary definitions and assertions from functional analysis and the theory of fractional calculus. At the end of this subsection, we will formulate the main results. The second subsection is devoted to the construction of Green's function of the boundary problem for the linear ordinary fractional equation with nonlocal boundary conditions, which, instead with eigenfunctions of the uniform elliptic operator, will be necessary for constructing a solution to the direct problem. In the third subsection, we study the solvability of the direct problem and build a priori estimates, which will be used in the investigation of an inverse problem. Here, an equivalent form for the inverse problem is formulated. Using this equivalent form and contraction mapping principle, the result on the local in time existence and uniqueness of solutions is established in the fourth subsection. By the method of continuation of the solution in the fifth subsection, the main results are proved.

The results of this section were obtained by the author and his student in the work [2].

2.1.1 Generalized Statement of the Problem and Preliminaries

Let Ω be a bounded domain in $\mathbb{R}^n$ with sufficiently smooth boundary $\partial\Omega$. Consider the fractional diffusion equation in the domain $Q_0^T := \{(x, t) : x \in \Omega \subset \mathbb{R}^n,\ 0 < t \leq T\}$:

$$\left(\partial_t^\alpha u\right)(x, t) = Au(x, t) + q(t)u(x, t) + f(x, t), \quad (x, t) \in Q_0^T, \tag{2.1}$$

with the Caputo time-fractional derivative ∂_t^α of order $0 < \alpha < 1$, the operator $-A$ is a symmetric uniformly elliptic operator defined on $\mathcal{D}(-A) = H^2(\Omega) \cap H_0^1(\Omega)$ given by

$$Au(x, t) = \sum_{i=1}^{n} \frac{\partial}{\partial x_i}\left(\sum_{j=1}^{n} a_{ij}(x)\frac{\partial}{\partial x_j}u(x, t)\right), \quad (x, t) \in Q_0^T,$$

in which the coefficients satisfy

$$a_{ij} = a_{ji}, \quad 1 \le i, j \le n, \quad a_{ij} \in C^1(\overline{\Omega}),$$

$$\nu_1 \sum_{i=1}^{n} |\xi_i|^2 \le \sum_{i,j=1}^{n} a_{ij}(x)\xi_i\xi_j \le \nu_2 \sum_{i=1}^{n} |\xi_i|^2, \quad x \in \bar{\Omega},$$

$$\xi \in \mathbb{R}^n, \quad \nu_1, \nu_2 > 0.$$

We supplement the Eq. (2.1) with the nonlocal initial condition

$$u(x, 0) + \beta u(x, T) = \varphi(x), \quad x \in \Omega, \tag{2.2}$$

the boundary condition

$$u(x, t) = 0, \quad x \in \partial\Omega, \quad t \in (0, T), \tag{2.3}$$

and integral overdetermination condition of the first kind

$$\int_{\Omega} \omega(x)u(x, t)dx = h(t), \quad 0 \le t \le T, \tag{2.4}$$

where $\beta \ge 0$ and $f(x, t), \varphi(x), \omega(x), h(t)$ are known functions. Besides, $h(t)$ is the measurement data representing the average temperature on a small part of Ω, because the weight function $\omega(x)$ is usually chosen to satisfy $\operatorname{supp}(\omega) \subset \Omega$ in applied sciences [3]. Now we pass to a rigorous statement of the main problem of our paper. If function $q(t)$ is known, then the initial-boundary value problem (2.1)–(2.3) is called the direct problem. The initial-boundary value problem (2.1)–(2.3) together with extra condition (2.4) is called an inverse problem.

We investigate the following inverse problem:

Inverse problem. Find $(u, q) \in C\left([0, T];\ H^2(\Omega)\right) \times C[0, T]$ to satisfy (2.1)–(2.3) and the additional measurement (2.4).

Here we present the necessary definitions from functional analysis and fractional calculus theory.

For integers m, we denote $H^m(\Omega) = W^{m,2}(\Omega)$ (see [4]) and $H_0^m(\Omega)$ is the closure of $C_0^\infty(\Omega)$ in the norm of space $H^m(\Omega)$. For a given Banach space V on Ω, we use the notation $C^m([0, T];\ V)$ to denote the following space:

$$C^m([0, T]; V)$$

$$:= \left\{u : \ \|\partial_t^j u(t)\|_V \text{ is continuous in } t \text{ on } [0, T] \text{ for all } 0 \le j \le m\right\}.$$

We endow $C^m([0,T];V)$ with the following norm making it to be a Banach space:

$$\|u\|_{C^m([0,T];V)} = \sum_{j=0}^{m} \left(\max_{0\le t\le T} \|\partial_t^j u(t)\|_V \right).$$

In addition, we define Banach space X_0^T by

$$X_0^T := C([0,T]; H^2(\Omega)).$$

Furthermore, we set

$$Y_0^T = X_0^T \times C[0,T]$$

endowed with the norm

$$\|(u,q)\|_{Y_0^T} := \|u\|_{X_0^T} + \|q\|_{C[0,T]}.$$

It is well-known that the operator $-A$ has only real and simple eigenvalues λ_k, and with suitable numbering, we have $0 < \lambda_1 \le \lambda_2 \le \cdots$, $\lim_{k\to\infty} \lambda_k = \infty$. By e_k, we denote the eigenfunction corresponding to λ_k, which satisfies $\|e_k\|^2_{L^2(\Omega)} = (e_k, e_k) = 1$, where $(\cdot,\cdot)$ denotes the inner product in Hilbert space $L^2(\Omega)$ and λ_k, e_k satisfy $-Ae_k = \lambda_k e_k$, $e_k(x) = 0$, $x \in \partial\Omega$, $\{e_k\} \subset H^2(\Omega) \cap H_0^1(\Omega)$ is an orthonormal basis of $L^2(\Omega)$.

Now we define the fractional power operator $(-A)^\gamma$ for $\gamma \in \mathbb{R}$ (e.g., [5]). Then for $\gamma \in \mathbb{R}$, we define a Hilbert space $D((-A)^\gamma)$ by

$$D((-A)^\gamma) := \left\{ u \in L^2(\Omega) : \sum_{k=1}^{\infty} \lambda_k^{2\gamma} |(u, e_k)|^2 < \infty \right\},$$

$$(-A)^\gamma u = \sum_{k=1}^{\infty} \lambda_k^\gamma (u, e_k) e_k$$

with the norm

$$\|u\|_{D((-A)^\gamma)} = \left(\sum_{k=1}^{\infty} \lambda_k^{2\gamma} |(u, e_k)|^2 \right)^{1/2}.$$

We note that the norm $\|u\|_{D((-A)^\gamma)}$ is stronger than $\|u\|_{L^2(\Omega)}$ for $\gamma > 0$. Then, we have $D((-A)^\gamma) \subset H^{2\gamma}(\Omega)$ for $\gamma > 0$. In particular, $D((-A)^{\frac{1}{2}}) = H_0^1(\Omega)$. Since $D((-A)^\gamma) \subset L^2(\Omega)$, identifying the dual $(L^2(\Omega))'$ which itself, we have $D((-A)^\gamma) \subset L^2(\Omega) \subset (D((-A)^\gamma))'$. We set $D((-A)^{-\gamma}) = (D((-A)^\gamma))'$, which consists of bounded linear functionals on $D((-A)^\gamma)$. For $u \in D((-A)^\gamma)$ and $\varphi \in D((-A)^\gamma)$, the value obtained by operating u to φ is denoted by ${}_{-\gamma}\langle u, \varphi \rangle_\gamma$. $D((-A)^{-\gamma})$ is a Hilbert space with the norm:

$$\|\varphi\|_{D((-A)^{-\gamma})} = \left(\sum_{k=1}^{\infty} \lambda_k^{-2\gamma} \left|_{-\gamma} \langle u, e_n \rangle_\gamma \right|^2\right)^{\frac{1}{2}}.$$

We further note that

$$_{-\gamma} \langle u, \varphi \rangle_\gamma = (u, \varphi) \quad \text{if } u \in L^2(\Omega) \text{ and } \varphi \in D((-A)^\gamma)$$

(see, e.g., [6], Ch. V in [7]).

Moreover, we consider the Mittag-Leffler function (see (1.26)):

$$E_{\rho,\mu}(z) = \sum_{k=0}^{\infty} \frac{z^k}{\Gamma(\rho k + \mu)}, \quad z \in \mathbb{C}$$

with $\Re(\rho) > 0$ and $\mu \in \mathbb{C}$. It is known that $E_{\rho,\mu}(z)$ is an entire function in $z \in \mathbb{C}$. If the parameter $\mu = 1$, then we have the classical Mittag-Leffler function: $E_\rho(z) = E_{\rho,1}(z)$ (see (1.27)).

In what follows we need the asymptotic estimate of the Mittag-Leffler function with a sufficiently large negative argument. The well-known estimate has the form (see, e.g., [9, p.136])

$$|E_{\rho,\mu}(-t)| \le \frac{C}{1+t}, \quad t > 0. \tag{2.5}$$

This estimate essentially follows from the following asymptotic estimate (see, e.g., [9, p.134])

$$E_{\rho,\mu}(-t) = \frac{t^{-1}}{\Gamma(\mu-\rho)} + O(t^{-2}) \tag{2.6}$$

For the Mittag-Leffler function with two parameters $E_{\rho,\rho}(-t)$ one can obtain a better estimate than (2.5). Indeed, using the asymptotic estimate (see, e.g., [9, p.134])

$$E_{\rho,\rho}(-t) = -\frac{t^{-2}}{\Gamma(-\rho)} + O(t^{-3}), \tag{2.7}$$

and the fact that $E_{\rho,\rho}(t)$ is a real analytic, we can obtain the following inequality (see [10])

$$|E_{\rho,\rho}(-t)| \le \frac{C}{1+t^2}, \quad t > 0. \tag{2.8}$$

We will also use a coarser estimate with positive number λ and $0 < \varepsilon < 1$:

$$\left|t^{\rho-1} E_{\rho,\rho}(-\lambda t^\rho)\right| \le \frac{C t^{\rho-1}}{1 + (\lambda t^\rho)^2} \le C \lambda^{\varepsilon-1} t^{\varepsilon\rho-1}, \quad t > 0, \tag{2.9}$$

which is easy to verify (see [10]).

Proposition 2.1 *Let* $0 < \rho < 1$. *Then*

$$E_\rho(x) > 0, \quad \frac{d}{dx} E_\rho(x) > 0, \quad x \in \mathbb{R}.$$

Moreover, the Mittag-Leffler function of negative argument $E_\rho(-x)$ *is monotonically decreasing function for all* $0 < \rho < 1$ *and*

$$0 < E_\rho(-x) < 1.$$

Proposition 2.2 *(see [6]) For* $\lambda > 0$, $\alpha > 0$ *and positive integer* $m \in \mathbb{N}$, *we have*

$$\frac{d^m}{dt^m} E_{\alpha,1}(-\lambda t^\alpha) = -\lambda t^{\alpha-m} E_{\alpha,\alpha-m+1}(-\lambda t^\alpha), \; t > 0$$

and

$$\frac{d}{dt}\left(t E_{\alpha,2}(-\lambda t^\alpha)\right) = E_{\alpha,1}(-\lambda t^\alpha),$$

$$\partial_t^\alpha (E_{\alpha,1}(-\lambda t^\alpha)) = -\lambda E_{\alpha,1}(-\lambda t^\alpha), \; t \geq 0.$$

It is known (see [8, p. 96]) the following assertion:

Lemma 2.1 *If* $y(t)$ *belong to* $C^n[0, T]$ *or* $W^{n,1}(0, T)$, *then for* $c_j \in \mathbb{R}$ *we have*

$$\left(I_{0+}^\alpha \partial_t^\alpha y\right)(t) = y(t) - c_0 - c_1 t - c_2 t^2 - \ldots - c_{n-1} t^{n-1}, \quad n - 1 < \alpha \leq n.$$

We now give a similar definition of weak solution to (2.1)–(2.3), which is introduced by [6].

Definition 2.1 We call $u(x, t)$ a weak solution to (2.1)–(2.3) if (2.1) holds in $L^2(\Omega)$ and $u(\cdot, t) \in H_0^1(\Omega)$ for almost all $t \in (0, T)$ and $u \in C([0, T]; D((-A)^{-\gamma}))$,

$$\lim_{t\to 0} \|u(\cdot, t) + \beta u(\cdot, T - t) - \varphi\|_{D((-A)^{-\gamma})} = 0$$

with some $\gamma > 0$.

We make the following assumptions:

(K1) $\varphi \in H^2(\Omega) \cap H_0^1(\Omega)$, $f \in C([0, T]; D(-A)^\varepsilon)$ where $0 < \varepsilon < 1$;

(K2) $h(0) + \beta h(T) = (\omega, \varphi)$;

(K3) $\partial_t^\alpha h \in C[0, T]$ satisfies the following inequality:

$$|h(t)| \geq \frac{1}{h_0} > 0, \quad \text{for all} \quad t \in [0, T],$$

where h_0 is a positive constant;

(K4) $\gamma > \frac{n}{4} - \varepsilon - 1$ for $n > 3$;

(K5) $\omega(x) \in L^2(\Omega)$.

Remark 2.1 (K2) is the consistency condition for our problem (2.1)–(2.4), which guarantees that the inverse problem (2.1)–(2.4) is equivalent to (2.54) and (2.55) (see Lemma 2.8).

Our main results in this section are the following existence and uniqueness of our inverse problem.

Theorem 2.1 *Under hypotheses (K1)–(K5), there exists a solution* $(u, q) \in Y_0^T$ *of the inverse problem* (2.1)–(2.4) *for any* T.

Theorem 2.2 *Let* $T > 0$. *Under hypotheses (K1)–(K5), if the inverse problem* (2.1)–(2.4) *has two solutions* $(u_i, q_i) \in Y_0^T$ $(i = 1, 2)$, *then* $(u_1, q_1) = (u_2, q_2)$ *for* $0 \le t \le T$.

2.1.2 Construction of Green's Function for the Ordinary Fractional Differential Equation

In this section, we will construct a formal solution to the ordinary fractional differential equation with nonlocal boundary conditions, which will be used in the proof of our main results.

To study problem (2.1)–(2.4), we consider the following fractional differential equation

$$\left(\partial_t^\alpha y\right)(t) = \gamma(t)y(t), \quad 0 < t < T, \tag{2.10}$$

with the boundary conditions

$$y(0) + \beta y(T) = 0, \tag{2.11}$$

where $\beta \ge 0$ are fixed numbers, $\gamma(t) \in C[0, T]$ is given function, and $y = y(t)$ is desired function.

Now, we give a new construction method of Green's function for the boundary value problem (2.10) and (2.11).

Lemma 2.2 *The fractional boundary value problem* (2.10) *is equivalent to the integral equation*

$$y(t) = \int_0^T G_0(t, s)\gamma(s)y(s)ds, \tag{2.12}$$

where Green's function G_0 *is defined by*

$$G_0(t, s) = \frac{1}{\Gamma(\alpha)} \times \begin{cases} -\frac{1}{1+\beta}\left[\beta(T-s)^{\alpha-1} - (1+\beta)(t-s)^{\alpha-1}\right], & 0 \le s < t, \\ -\frac{\beta}{1+\beta}(T-s)^{\alpha-1}, & t < s \le T. \end{cases} \tag{2.13}$$

Proof Let $y \in C^1[0, T]$ or $W^{1,1}(0, T)$. Then, using Lemma 2.1, we have

$$y(t) = c_0 + \frac{1}{\Gamma(\alpha)} \int_0^t (t-s)^{\alpha-1} \gamma(s) y(s) ds, \tag{2.14}$$

where c_0 is a real constant. Now, by employing the condition (2.11) we can obtain the coefficient c_0 as follow

$$c_0 = -\frac{\beta}{(1+\beta)\Gamma(\alpha)} \int_0^T (T-s)^{\alpha-1} \gamma(s) y(s) ds.$$

Therefore, the solution of (2.10) is

$$\begin{aligned} y(t) = &-\frac{\beta}{(1+\beta)\Gamma(\alpha)} \int_0^T (T-s)^{\alpha-1} \gamma(s) y(s) ds \\ &+ \frac{1}{\Gamma(\alpha)} \int_0^t (t-s)^{\alpha-1} \gamma(s) y(s) ds = \\ = &\int_0^t \left(\frac{(t-s)^{\alpha-1}}{\Gamma(\alpha)} - \frac{\beta}{(1+\beta)\Gamma(\alpha)} (T-s)^{\alpha-1} \right) \gamma(s) y(s) ds \\ &- \frac{\beta}{(1+\beta)\Gamma(\alpha)} \int_t^T (T-s)^{\alpha-1} \gamma(s) y(s) ds = \int_0^T G_0(t,s) \gamma(s) y(s) ds. \end{aligned}$$

□

Lemma 2.3 *Suppose that the function* $\gamma(t)$ *is continuous on the interval* $[0, T]$. *If* $\beta \geq 0$ *and*

$$\rho_0 := \frac{1+2\beta}{(1+\beta)\Gamma(1+\alpha)} T^\alpha \|\gamma\|_{C[0,T]} < 1, \tag{2.15}$$

then problem (2.10) *has only a trivial solution.*

Proof From Lemma 2.1, we can conclude that the problem (2.10) is equivalent to the integral equation (2.12).

Let us introduce the notation

$$P(y(t)) = \int_0^T G_0(t,s) \gamma(s) y(s) ds. \tag{2.16}$$

Then Eq. (2.12) can be rewritten as

$$y(t) = P(y(t)). \tag{2.17}$$

The operator P is continuous in the space $C[0, T]$. Indeed, for $y(t) \in C[0, T]$, we have

$$|P(y(t))| \leq \int_0^T |G(t, s)\gamma(s)y(s)|\, ds \leq \rho_0 \|y\|_{C[0,T]},$$

that is, operator P is bounded. Hence, it is continuous.

Now we prove that P is a contraction operator in the space $C[0, T]$. It is easy to see that the inequality

$$\|P(y_1(t)) - P(y_2(t))\|_{C[0,T]} \leq \rho_0 \|y_1(t) - y_2(t)\|_{C[0,T]} \tag{2.18}$$

holds for any functions $y_1(t), y_2(t) \in C[0, T]$.

In view of (2.15) and (2.18) it is clear that the operator P is contractive in $C[0, T]$. Therefore, the operator P has a unique fixed point $y(t)$ in the space $C[0, T]$ which is a solution of Eq. (2.17). Thus, the integral equation (2.12) has a unique solution in $C[0, T]$. Consequently, problem (2.10) also has a unique solution in the indicated space. Since $y(t) = 0$ is a solution to problem (2.10), it follows that this problem has a unique trivial solution. □

Lemma 2.4 *Assume $\beta \geq 0$. Then, the solution to the boundary value problem*

$$\left(\partial_t^\alpha y_1\right)(t) + \lambda y_1(t) = 0, \quad t > 0, \tag{2.19}$$

$$y_1(0) + \beta y_1(T) = a, \quad a \in \mathbb{R} \tag{2.20}$$

with $0 < \alpha < 1$ and $\lambda \in \mathbb{R}$ has the form

$$y_1(t) = \frac{a}{\rho(T)} E_{\alpha,1}(-\lambda t^\alpha),$$

where

$$\rho(T) = 1 + \beta E_{\alpha,1}(-\lambda T^\alpha).$$

Proof According to Theorem 5.15 (see [8, p. 323]), the general solution to Eq. (2.19) is given by

$$y_1(t) = d_1 E_\alpha(-\lambda t^\alpha), \tag{2.21}$$

where d_1 is an arbitrary real constant.

Let $t = 0$ and $t = T$ in (2.21), then there are hold

$$y_1(0) = d_1, \quad y_1(T) = d_1 E_{\alpha,1}(-\lambda T^\alpha). \tag{2.22}$$

Substituting (2.22) into the boundary condition (2.20), one has

$$d_1 = \frac{a}{\rho(T)}, \tag{2.23}$$

Substituting the constant d_1 into (2.21), we arrive at the solution of (2.19), (2.20). □

Lemma 2.5 *Assume $\beta \geq 0$. Then, the solution to the boundary value problem*

$$\left(\partial_t^\alpha y_2\right)(t) + \lambda y_2(t) = f(t), \quad t > 0,$$

$$y_2(0) + \beta y_2(T) = 0$$

with $0 < \alpha < 1$ and $\lambda \in \mathbb{R}$ has the form

$$y_2(t) = \int_0^T G(t,s) f(s) ds,$$

where

$$G(t,s) = -\frac{1}{\rho(T)} \begin{cases} \beta (T-s)^{\alpha-1} E_{\alpha,\alpha}(-\lambda(T-s)^\alpha) E_{\alpha,1}(-\lambda t^\alpha) \\ \quad -\rho(T)(t-s)^{\alpha-1} E_{\alpha,\alpha}(-\lambda(t-s)^\alpha), & s \in [0,t) \\ \beta (T-s)^{\alpha-1} E_{\alpha,\alpha}(-\lambda(T-s)^\alpha) E_{\alpha,1}(-\lambda t^\alpha), & s \in (t,T]. \end{cases}$$

The prove of this is analogical to the proofs of Lemmas 2.2 and 2.4.

2.1.3 Generalized Solution of the Direct Problem

In this subsection, we will study the direct problem (2.1)–(2.3). We first consider the following initial-boundary problem

$$\begin{cases} \partial_t^\alpha u(x,t) - Au(x,t) = F(x,t), & (x,t) \in Q_0^T, \\ u(x,0) + \beta u(x,T) = \varphi(x), & x \in \Omega, \\ u(x,t) = 0, \quad x \in \partial\Omega, \quad 0 < t < T, \end{cases} \tag{2.24}$$

Let $u_k(t) = (u(\cdot,t), e_k)$, $k \geq 1$. Assume that problem (2.24) has a unique solution u which is given by

$$u(x,t) = \sum_{k=1}^{\infty} u_k(t) e_k(x), \tag{2.25}$$

where u_k, $(k = 1, 2, ...)$ are solutions of the nonlocal problems:

$$\left(\partial_t^\alpha u_k\right)(t) + \lambda_k u_k(t) = F_k(t), \quad 0 < t < T, \tag{2.26}$$

$$u_k(0) + \beta u_k(T) = \varphi_k, \tag{2.27}$$

where $F_k(t) = (F(\cdot,t), e_k)$, $\varphi_k = (\varphi, e_k)$. Using Lemma 2.4 and Lemma 2.5, we have

$$u_k(t) = \frac{\varphi_k}{\rho_k(T)} E_{\alpha,1}(-\lambda_k t^\alpha) + \int_0^T G_k(t,s) F_k(s) ds,$$

where $\rho_k(T)$ is $\rho(T)$ with λ_k instead of λ.

Thus, we have

$$u(x,t) = \sum_{k=1}^{\infty} \frac{(\varphi, e_k)}{\rho_k(T)} E_{\alpha,1}(-\lambda_k t^\alpha) e_k(x) + \sum_{k=1}^{\infty} \int_0^T G_k(t,s)(F, e_k)(s) ds \cdot e_k(x). \tag{2.28}$$

Based on the method of [6], we will prove the well-posedness of problem (2.24). We first split (2.24) into the following two initial and boundary value problems:

$$\begin{cases} \partial_t^\alpha v(x,t) - Av(x,t) = 0, & (x,t) \in Q_0^T, \\ v(x,0) + \beta v(x,T) = \varphi(x), & x \in \Omega, \\ v(x,t) = 0, & x \in \partial\Omega,\ 0 < t < T, \end{cases} \tag{2.29}$$

and

$$\begin{cases} \partial_t^\alpha w(x,t) - Aw(x,t) = F(x,t), & (x,t) \in Q_0^T, \\ w(x,0) + \beta w(x,T) = 0, & x \in \Omega, \\ w(x,t) = 0, & x \in \partial\Omega,\ 0 < t < T. \end{cases} \tag{2.30}$$

Lemma 2.6 *(i) Let $\varphi \in L^2(\Omega)$. Then there exists a unique weak solution $v \in C([0,T]; L^2(\Omega)) \cap C((0,T]; H^2(\Omega) \cap H_0^1(\Omega))$ to (2.29) with $\partial_t^\alpha v \in C((0,T]; L^2(\Omega))$. Moreover there exists a constant $c_1 > 0$ satisfying*

$$\begin{cases} \|v\|_{C([0,T];L^2(\Omega))} \le \|\varphi\|_{L^2(\Omega)}, \\ \|v(\cdot,t)\|_{H^2(\Omega)} + \|\partial_t^\alpha v(\cdot,t)\|_{L^2(\Omega)} \le c_1 t^{-\alpha} \|\varphi\|_{L^2(\Omega)}, \end{cases} \tag{2.31}$$

and we have

$$v(x,t) = \sum_{k=1}^{\infty} \frac{(\varphi, e_k)}{\rho_k(T)} E_{\alpha,1}(-\lambda_k t^\alpha) e_k(x) \tag{2.32}$$

in $C([0,T]; L^2(\Omega)) \cap C((0,T]; H^2(\Omega) \cap H_0^1(\Omega))$.

(ii) We assume that $\varphi \in H_0^1(\Omega)$. Then the unique weak solution v further belongs to $L^2(0,T; H^2(\Omega) \cap H_0^1(\Omega))$, $\partial_t^\alpha v \in L^2(Q_0^T)$ and there exists a constant $c_2 > 0$ satisfying the following inequality:

$$\|v\|_{L^2(0,T;H^2(\Omega))} + \|\partial_t^\alpha v\|_{L^2(Q_0^T)} \le c_2 \|\varphi\|_{H^1(\Omega)} \tag{2.33}$$

and we have (2.32) in the corresponding space on the right-hand side of (2.33).

(iii) Let $\varphi \in H^2(0,1) \cap H_0^1(0,1)$. *Then the unique weak solution* v *belong to* $C([0,T]; H^2(\Omega) \cap H_0^1(\Omega))$, $\partial_t^\alpha v \in C([0,T]; L^2(\Omega)) \cap C((0,T]; H_0^1(\Omega))$ *and the following inequality holds:*

$$\|v\|_{C([0,T];H^2(\Omega))} + \|\partial_t^\alpha v\|_{C([0,T];L^2(\Omega))} \le c_3 \|\varphi\|_{H^2(\Omega)} \tag{2.34}$$

and we have (2.32) *in the corresponding space on the right-hand side of* (2.34).

Proof (i) We will show that $v(x,t)$ certainly gives the weak solution to (2.29). It is easy to see that from the definition of β and Proposition 2.1, we have

$$\rho_k(T) = 1 + \beta E_{\alpha,1}(-\lambda_k T^\alpha) \ge 1, \quad \forall k \in \mathbb{N}.$$

Taking into account this relation and Proposition 2.1, we obtain

$$\|v(\cdot,t)\|^2_{L^2(\Omega)} = \sum_{k=1}^{\infty} \left| \frac{(\varphi, e_k)}{\rho_k(T)} E_{\alpha,1}(-\lambda_k t^\alpha) \right|^2 \le \|\varphi\|^2_{L^2(\Omega)} \tag{2.35}$$

and this is the first inequality of (2.31).

Moreover, by (2.5), we have

$$\|Av(\cdot,t)\|^2_{L^2(\Omega)} = \sum_{k=1}^{\infty} \left| \lambda_k \frac{(\varphi, e_k)}{\rho_k(T)} E_{\alpha,1}(-\lambda_k t^\alpha) \right|^2 \le C^2 t^{-2\alpha} \|\varphi\|^2_{L^2(\Omega)}, \tag{2.36}$$

for $t > 0$. In (2.35), since $\sum\limits_{k=1}^{\infty} \frac{(\varphi,e_k)}{\rho_k(T)} E_{\alpha,1}(-\lambda_k t^\alpha) e_k$ is convergent in $L^2(\Omega)$ uniformly in $t \in [0,T]$, we see that $u \in C([0,T]; L^2(\Omega))$. Moreover in (2.36), since $\sum\limits_{k=1}^{\infty} \lambda_k \frac{(\varphi,e_k)}{\rho_k(T)} E_{\alpha,1}(-\lambda_k t^\alpha) e_k$ is convergent in $L^2(\Omega)$ uniformly in $t \in [\varepsilon, T]$ with any given $\varepsilon > 0$, we see that $Av \in C((0,T]; L^2(\Omega))$, that is $v \in C((0,T]; H^2(\Omega) \cap H_0^1(\Omega))$. Therefore we obtain that $v \in C([0,T]; L^2(\Omega)) \cap C((0,T]; H^2(\Omega) \cap H_0^1(\Omega))$. By (2.29) we see that $\partial_t^\alpha v \in C((0,T]; L^2(\Omega))$.

We have to prove

$$\lim_{t \to 0} \|v(\cdot,t) + \beta v(\cdot, T-t) - \varphi\|_{L^2(\Omega)} = 0. \tag{2.37}$$

In fact,

$$\|v(\cdot,t) + \beta v(\cdot,T-t) - \varphi\|^2_{L_2(\Omega)} = \sum_{k=1}^{\infty} \left| \frac{(\varphi, e_k)}{\rho_k(T)} \right.$$

$$\times \left. \underbrace{\Big(E_{\alpha,1}(-\lambda_k t^\alpha) + \beta E_{\alpha,1}(-\lambda_k (T-t)^\alpha) - \rho_k(T) \Big)}_{S_{1k}(t)} \right|^2$$

and $\lim_{t\to 0} S_{1k}(t) = 0$ for each $k \in \mathbb{N}$ and

$$\sum_{k=1}^{\infty}(\varphi, e_k)^2 S_{1k}^2(t) \le 3\sum_{k=1}^{\infty}(\varphi, e_k)^2\Big[\Big(\frac{C}{1+\lambda_k t^\alpha}\Big)^2$$

$$+\Big(\frac{C\beta}{1+\lambda_k(T-t)^\alpha}\Big)^2 + \Big(1+\frac{C\beta}{1+\lambda_k T^\alpha}\Big)^2\Big] < \infty$$

for $0 \le t \le T$ and $\beta \ge 0$. Then, the Lebesgue theorem yields (2.37).

Next, we prove the uniqueness of the weak solution to (2.29) within the class given in Definition 2.1. Under the condition $\varphi = 0$, we have to prove that problem (2.29) has only a trivial solution. Since $e_k(x)$ is the eigenfunctions to the following eigenvalue problem:

$$(Ae_k)(x) = -\lambda_k e_k(x), \quad x \in \Omega, \quad e_k(x) = 0, \quad x \in \partial\Omega$$

in terms of the regularity of v, taking the duality pairing ${}_{-\gamma}\langle\cdot,\cdot\rangle_\gamma$ of (2.29) with e_k and setting $v_k(t) = {}_{-\gamma}\langle v(\cdot,t), e_k\rangle_\gamma$, we obtain

$$\partial_t^\alpha v_k(t) = -\lambda_k v_k(t), \quad \text{almost all} \quad t \in (0,T).$$

Since $v(\cdot,t) \in L^2(\Omega)$ for almost all $t \in (0,T)$ and $v_k(t) \equiv {}_{-\gamma}\langle v(\cdot,t), e_k\rangle_\gamma = (v(\cdot,t), e_k)$ where ${}_{-\gamma}\langle v(\cdot,t), e_k\rangle_\gamma$ denotes the duality pairing between $D((-A)^{-\gamma})$ and $D((-A)^\gamma)$, it follows from $\lim_{t\to 0}\|v(\cdot,t) + \beta v(\cdot, T-t)\|_{D((-A)^{-\gamma})} = 0$ that $v_k(0) + \beta v_k(T) = 0$. Due to the existence and uniqueness of the boundary value problem for the ordinary fractional differential equation (see, Lemma 2.3), we obtain that $v_k(t) = 0$, $k = 1, 2, \ldots$. Since $\{e_k\}_{k\in\mathbb{N}}$ is a complete orthonormal system in $L^2(\Omega)$, we have $v = 0$ in Q_0^T.

Moreover, by (2.5), we have

$$\|v(\cdot,t)\|^2_{H^2(\Omega)} \le c_1'\|Av(\cdot,t)\|^2_{L^2(\Omega)} \le$$
$$\le c_1'\sum_{k=1}^{\infty}\lambda_k^2\left|\frac{(\varphi,e_k)}{\rho_k(T)}E_{\alpha,1}(-\lambda_k t^\alpha)\right|^2 \le c_1' C^2 t^{-2\alpha}\|\varphi\|^2_{L^2(\Omega)}. \tag{2.38}$$

From (2.35) and (2.38), we get the estimates (2.31).

(ii) Let $\varphi \in H_0^1(\Omega)$. By (2.5), we have

$$\|v(\cdot,t)\|^2_{H^2(\Omega)} \le c_1'\|Av(\cdot,t)\|^2_{L^2(\Omega)} \le c_1'\sum_{k=1}^{\infty}\lambda_k^2\left|\frac{(\varphi,e_k)}{\rho_k(T)}E_{\alpha,1}(-\lambda_k t^\alpha)\right|^2$$

$$\le c_1' \sum_{k=1}^{\infty} \left| \lambda_k^{\frac{1}{2}} (\varphi, e_k)(\lambda_k t^\alpha)^{\frac{1}{2}} E_{\alpha,1}(-\lambda_k t^\alpha) \right|^2 t^{-\alpha}$$

$$\le c_1' C^2 \sum_{k=1}^{\infty} \left| ((-A)^{\frac{1}{2}}\varphi, e_k) \frac{(\lambda_k t^\alpha)^{\frac{1}{2}}}{1+\lambda_k t^\alpha} \right|^2 t^{-\alpha} \le \frac{c_1' C^2}{4} \|\varphi\|^2_{H^1(\Omega)} t^{-\alpha}. \tag{2.39}$$

By $0 < \alpha < 1$, we see $\|v\|_{L^2(0,T;H^2(\Omega))} \le c_2' \|\varphi\|_{H^1(\Omega)}$. Therefore we have $u \in L^2(0, T; H^2(\Omega) \cap H_0^1(\Omega))$.

Since $\partial_t^\alpha (E_{\alpha,1}(-\lambda_k t^\alpha)) = -\lambda_k E_{\alpha,1}(-\lambda_k t^\alpha)$ (e.g., [8, p. 98]), we have

$$\int_0^T \|\partial_t^\alpha v(\cdot,t)\|^2_{L^2(\Omega)} dt = \int_0^T \sum_{k=1}^{\infty} \frac{|(\varphi, e_k)|^2}{\rho_k^2(T)} \lambda_k^2 |E_{\alpha,1}(-\lambda_k t^\alpha)|^2 dt$$

$$\le C^2 \int_0^T \sum_{k=1}^{\infty} (\varphi, e_k)^2 \lambda_k \left(\frac{(\lambda_k t^\alpha)^{\frac{1}{2}}}{1+\lambda_k t^\alpha} \right)^2 t^{-\alpha} dt \le c_2'' \|\varphi\|^2_{H^1(\Omega)},$$

where we have used the following

$$\max_{y \ge 0} \frac{y^\theta}{1+y} = \frac{\left(\frac{\theta}{1-\theta}\right)^\theta}{1+\frac{\theta}{1-\theta}}, \quad 0 < \theta < 1. \tag{2.40}$$

By (2.29) we have $\partial_t^\alpha v = Av$, which yields $\partial_t^\alpha v \in L^2(Q_0^T)$. Thus the proof of Lemma 2.6(ii) is complete.

(iii) Let $\varphi \in H^2(\Omega) \cap H_0^1(\Omega)$. Then by Proposition 2.1, we have

$$\|v(\cdot,t)\|^2_{H^2(\Omega)} \le c_3' \|Av(\cdot,t)\|^2_{L^2(\Omega)} \le c_3' \sum_{k=1}^{\infty} \lambda_k^2 (\varphi, e_k)^2 E_{\alpha,1}(-\lambda_k t^\alpha)^2$$

$$\le c_3' \|\varphi\|^2_{H^2(\Omega)}, \quad t \ge 0.$$

By (2.29) obtain

$$\|\partial_t^\alpha v(\cdot,t)\|^2_{L^2(\Omega)} \le c_3'' \|\varphi\|^2_{H^2(\Omega)}, \quad t > 0.$$

Combining the last two estimates, we get (2.14). This completes the proof of Lemma 2.6. □

Using Lemma 2.6, we obtain the following result.

Corollary 2.1 *Let* $\varphi \in L^2(\Omega)$ *and* $F = 0$. *Then for the unique weak solution* $v \in C([0,\infty); L^2(\Omega)) \cap C((0,\infty); H^2(\Omega) \cap H_0^1(\Omega))$ *to* (2.29), *there exists a positive constant* c_4 *satisfying*

$$\|v(\cdot,t)\|_{L^2(\Omega)} \le \frac{c_4}{1+\lambda_1 t^\alpha}\|\varphi\|_{L^2(\Omega)}, \quad t \ge 0. \tag{2.41}$$

Moreover, there exists a positive constant c_5 such that

$$v \in C^\infty\left((0,\infty); L^2(\Omega)\right), \quad \|\partial_t^m v(\cdot,t)\|_{L^2(\Omega)} \le \frac{c_5}{t^m}\|\varphi\|_{L^2(\Omega)}, \quad t > 0, \ m \in \mathbb{N}. \tag{2.42}$$

Proof By (2.5), we have

$$\|v(\cdot,t)\|^2_{L^2(\Omega)} = \sum_{k=1}^\infty \frac{(\varphi,e_k)^2}{\rho_k^2(T)} E_{\alpha,1}(-\lambda_k t^\alpha)^2 \le \sum_{k=1}^\infty (\varphi,e_k)^2 \left(\frac{C}{1+\lambda_k t^\alpha}\right)^2$$

$$\le \left(\frac{C}{1+\lambda_1 t^\alpha}\right)^2 \|\varphi\|^2_{L^2(\Omega)}, \quad t \ge 0.$$

Further, by Proposition 2.2, we have

$$\partial_t^m v(\cdot,t) = -\sum_{k=1}^\infty \frac{(\varphi,e_k)}{\rho_k(T)} \lambda_k t^{\alpha-m} E_{\alpha,\alpha-m+1}(-\lambda_k t^\alpha) e_k$$

for $m \in \mathbf{N}$ and $t > 0$, so that

$$\|\partial_t^m v(\cdot,t)\|^2_{L^2(\Omega)} \le C^2 \sum_{k=1}^\infty (\varphi,e_k)^2 \left(\frac{\lambda_k t^\alpha}{1+\lambda_k t^\alpha}\right)^2 t^{-2m} \le \frac{C^2}{t^{2m}}\|\varphi\|^2_{L^2(\Omega)}.$$

□

Now, we study the problem of (2.30) and we have

Lemma 2.7 *Let $\varphi = 0$ and $F \in C([0,T]; D((-A)^\varepsilon))$, where $0 < \varepsilon < 1$. Then there exists a unique weak solution $w \in C([0,T]; H^2(\Omega) \cap H_0^1(\Omega))$ to (2.30) such that $\partial_t^\alpha w \in C([0,T]; L^2(\Omega))$. In particular, for any γ satisfying (K4), we have $w \in C([0,T]; D((-A)^{-\gamma}))$,*

$$\lim_{t\to 0} \|w(\cdot,t) + \beta w(\cdot,T-t)\|_{D((-A)^{-\gamma})} = 0 \tag{2.43}$$

and if $n = 1, 2, 3$, then

$$\lim_{t\to 0} \|w(\cdot,t) + \beta w(\cdot,T-t)\|_{L^2(\Omega)} = 0. \tag{2.44}$$

Moreover, there exists a constant $c_6 > 0$ such that

$$\|w\|_{C([0,T];H^2(\Omega))} + \|\partial_t^\alpha w\|_{C([0,T];L^2(\Omega))} \le c_6 \|F\|_{C([0,T];D((-A)^\varepsilon))} \tag{2.45}$$

for all $t \in [0, T]$ *and we have*

$$w(x,t) = \sum_{k=1}^{\infty} e_k(x) \int_0^T G_k(t,s)(F(\cdot,s), e_k)ds \tag{2.46}$$

in the corresponding space on the right-hand side of (2.45).

Proof Here we will show only regularity and estimate (2.45). We rewrite the standard form of (2.46) as follows:

$$w(x,t) = \sum_{k=1}^{\infty} e_k(x) \int_0^t (F(\cdot,s), e_k)(t-s)^{\alpha-1} E_{\alpha,\alpha}(-\lambda_k (t-s)^{\alpha}) ds$$

$$- \sum_{k=1}^{\infty} \frac{\beta E_{\alpha,1}(-\lambda_k t^{\alpha})}{\rho_k(T)} e_k(x) \int_0^T (T-s)^{\alpha-1} E_{\alpha,\alpha}(-\lambda_k (T-s)^{\alpha})$$

$$\times (F(\cdot,s), e_k) ds = I_1 + I_2.$$

By (2.5) and (2.9) for any $0 < \varepsilon < 1$, we have

$$\|I_1(\cdot,t)\|^2_{H^2(\Omega)} \le \sum_{k=1}^{\infty} \lambda_k^2 \left| \int_0^t (t-s)^{\alpha-1} E_{\alpha,\alpha}(-\lambda_k (t-s)^{\alpha}) (F(\cdot,s), e_k)\, ds \right|^2$$

$$\le \sum_{k=1}^{\infty} \max_{0 \le s \le t} |((-A)^{\varepsilon} F(\cdot,s), e_k))|^2 \left| \int_0^t s^{\varepsilon\alpha - 1} ds \right|^2$$

$$\le c_{10} \|F\|^2_{C([0,t];D((-A)^{\varepsilon}))} t^{2\varepsilon\alpha}, \quad t \in [0,T]. \tag{2.47}$$

Similarly, we have

$$\|I_2(\cdot,t)\|^2_{H^2(\Omega)} \le \sum_{k=1}^{\infty} \lambda_k^2 \left(\frac{\beta C}{1+\lambda_k t^{\alpha}} \right)^2 \left| \int_0^T \left((T-s)^{\alpha-1} \right.\right.$$

$$\left.\left. \times E_{\alpha,\alpha}(-\lambda_k (T-s)^{\alpha}) (F(\cdot,s), e_k)\, ds \right|^2$$

$$\le \left(\frac{\beta C}{1+\lambda_1 t^{\alpha}} \right)^2 \sum_{k=1}^{\infty} \max_{0 \le s \le T} |((-A)^{\varepsilon}[F](\cdot,s), e_k))|^2 \left| \int_0^T s^{\alpha\varepsilon - 1} ds \right|^2$$

$$\le c_{11} T^{2\varepsilon\alpha} \|F\|^2_{C([0,T];D((-A)^{\varepsilon}))}. \tag{2.48}$$

On the other hand, using Proposition 2.2 and the formula (3.1.34) in [8], we have

$$\partial_t^\alpha w(x,t) = \sum_{k=1}^{\infty} (F(\cdot,t), e_k)\, e_k(x) - \sum_{k=1}^{\infty} \lambda_k \int_0^t (F(\cdot,s), e_k)(t-s)^{\alpha-1}$$

$$\times E_{\alpha,\alpha}(-\lambda_k(t-s)^\alpha)ds \cdot e_k(x) +$$

$$+\beta \sum_{k=1}^{\infty} \frac{\lambda_k E_{\alpha,1}(-\lambda_k t^\alpha)}{\rho_k(T)} \int_0^T (T-s)^{\alpha-1} E_{\alpha,\alpha}(-\lambda_k(T-s)^\alpha)(F(\cdot,s), e_k)ds$$

$$\times e_k(x) = I_3 + I_4 + I_5.$$

Taking account this $D((-A)^\varepsilon) \subset L^2(\Omega)$, we have

$$\|I_3(\cdot,t)\|^2_{L^2(\Omega)} = \sum_{k=1}^{\infty} (F(\cdot,t), e_k)^2 = \|F\|^2_{L^2(\Omega)} \le c_9 \|F\|^2_{D((-A)^\varepsilon)}, \quad t \in [0,T]. \tag{2.49}$$

Furthermore, by (2.9) we have

$$\|I_4(\cdot,t)\|^2_{L^2(0,1)} \le \sum_{k=1}^{\infty} \left| \lambda_k \int_0^t (F(\cdot,s), e_k)(t-s)^{\alpha-1} E_{\alpha,\alpha}(-\lambda_k(t-s)^\alpha)ds \right|^2 \le$$

$$\le c_{10}' \sum_{k=1}^{\infty} \max_{0\le s\le t} |((-A)^\varepsilon F(\cdot,s), e_k)|^2 \left| \int_0^t s^{\varepsilon\alpha-1} ds \right|^2$$

$$\le c_{10} \|F\|^2_{C([0,t];D((-A)^\varepsilon))} t^{2\varepsilon\alpha}, \quad t \in [0,T]. \tag{2.50}$$

Argue similarly to the previous one, we get

$$\|I_5(\cdot,t)\|^2_{L^2(0,1)} \le c_{11} T^{2\varepsilon\alpha} \|F\|^2_{C([0,T];D((-A)^\varepsilon))}. \tag{2.51}$$

Summing up (2.47)–(2.51) yields (2.45).

Finally we have to prove (2.43). In fact, by Propositions 2.1 and 2.2, we have

$$\int_0^t \left| s^{\alpha-1} E_{\alpha,\alpha}(-\lambda_k s^\alpha) \right| ds = \int_0^t s^{\alpha-1} E_{\alpha,\alpha}(-\lambda_k s^\alpha) ds$$

$$= -\frac{1}{\lambda_k} \int_0^t \frac{d}{ds} E_{\alpha,1}(-\lambda_k s^\alpha) ds = \frac{1}{\lambda_k} \left(1 - E_{\alpha,1}(-\lambda_k t^\alpha)\right), \quad t > 0. \tag{2.52}$$

Using the (2.52), we have

$$\|w(\cdot,t)+\beta w(\cdot,T-t)\|^2_{D((-A)^{-\gamma})} = \sum_{k=1}^{\infty}\lambda_k^{-2\gamma}\Big|\int_0^t (F(\cdot,s),e_k)(t-s)^{\alpha-1}$$

$$\times\, E_{\alpha,\alpha}(-\lambda_k(t-s)^\alpha)ds$$

$$-\frac{\beta E_{\alpha,1}(-\lambda_k t^\alpha)}{\rho_k(T)}\int_0^T (F(\cdot,s),e_k)(T-s)^{\alpha-1}E_{\alpha,\alpha}(-\lambda_k(T-s)^\alpha)ds$$

$$+\beta\int_0^{T-t}(F(\cdot,s),e_k)(T-t-s)^{\alpha-1}E_{\alpha,\alpha}(-\lambda_k(T-t-s)^\alpha)ds$$

$$-\frac{\beta^2 E_{\alpha,1}(-\lambda_k(T-t)^\alpha)}{\rho_k(T)}\int_0^T (F(\cdot,s),e_k)(T-s)^{\alpha-1}E_{\alpha,\alpha}(-\lambda_k(T-s)^\alpha)ds\Big|^2$$

$$\le \sum_{k=1}^{\infty}\lambda_k^{-2(\gamma+\varepsilon+1)}\max_{0\le t\le T}|(-A)^\varepsilon F(\cdot,t),e_k|^2$$

$$\times\Big[\underbrace{1-E_{\alpha,1}(-\lambda_k t^\alpha)-\frac{\beta}{\rho_k(T)}E_{\alpha,1}(-\lambda_k t^\alpha)(1-E_{\alpha,1}(-\lambda_k T^\alpha))}$$

$$\underbrace{+\beta(1-E_{\alpha,1}(-\lambda_k T^\alpha))-\frac{\beta^2}{\rho_k(T)}E_{\alpha,1}(-\lambda_k(T-t)^\alpha)\left(1-E_{\alpha,1}(-\lambda_k T^\alpha)\right)}_{S_{2k}(t)}\Big]^2$$

$$\le c_{12}\|F\|^2_{C([0,T];D((-A)^\varepsilon))}\sum_{k=1}^{\infty}\lambda_k^{-2(\gamma+\varepsilon+1)}S_{2k}^2(t).$$

Since $\lambda_k \ge c'_{12}k^{\frac{2}{n}}$, $k\in\mathbb{N}$ (see [11, p. 407]), we have

$$\frac{1}{\lambda_k^{2(\gamma+\varepsilon+1)}} \le \frac{c''_{12}}{k^{\frac{4(\gamma+\varepsilon+1)}{n}}}.$$

By (K4), we have $\frac{4(\gamma+\varepsilon+1)}{n} > 1$, and $\sum\limits_{k=1}^{\infty}\lambda^{-2(\gamma+\varepsilon+1)}S_{2k}^2(t) < \infty$. Since $\lim\limits_{t\to 0} S_{2k}(t) = 0$ for each $n\in\mathbb{N}$, the Lebesgue theorem implies $\lim\limits_{t\to 0}\|w(\cdot,t)+\beta w(\cdot,T-t)\|_{D((-A)^{-\gamma})} = 0$. The uniqueness of the weak solution is verified similarly to Lemma 2.6. This completes the proof of Lemma 2.7.

□

As a result combining Lemmas 2.6, 2.7 and Corollary 2.1, we obtain the following result:

Corollary 2.2 *Let* $\varphi \in H^2(\Omega) \cap H_0^1(\Omega)$ *and* $F \in C([0,T]; D((-A)^{\varepsilon}))$, *where* $0 < \varepsilon < 1$. *Then there exists a unique weak solution* $u \in X_0^T$ *to (2.24), such that*

$$\|u\|_{X_0^T} \le c_{13}\Big(\|\varphi\|_{H^2(\Omega)} + T^{\alpha\varepsilon}\|F\|_{C([0,T];D((-A)^{\varepsilon}))}\Big). \tag{2.53}$$

At the end of this section, we give a lemma to show an equivalent form of our inverse problem.

Lemma 2.8 *Let (K1)–(K5) be held. Then our inverse problem (2.1)–(2.4) is equivalent to following problem:*

$$\begin{cases} \left(\partial_t^{\alpha} u\right)(x,t) - Au(x,t) = q(t)u(x,t) + f(x,t), & (x,t) \in Q_0^T, \\ u(x,0) + \beta u(x,T) = \varphi(x) \quad x \in \Omega, \\ u(x,t) = 0, \quad x \in \partial\Omega, \ 0 < t < T, \end{cases} \tag{2.54}$$

and

$$h(t)q(t) = \partial_t^{\alpha} h(t) - (f(\cdot,t),\omega) - (Au(\cdot,t),\omega), \quad t \in [0,T]. \tag{2.55}$$

Proof Obviously, the solution $(u(x,t), q(t)) \in X_0^T \times C[0,T]$ of our inverse problem (2.1)–(2.4) is also a solution to the problem (2.54), (2.55) in $X_0^T \times C[0,T]$. Because the problem (2.54) is the same as (2.1)–(2.3). Therefore, we should show only (2.55). Multiplying both sides of Eq. (2.1) by a function $\omega(x)$ and integrating from Ω with respect to x gives

$$\begin{aligned} &\left(\partial_t^{\alpha}\int_{\Omega}\omega(x)u(x,t)dx\right)(t) - \int_{\Omega}\omega(x)Au(x,t)dx = \\ &= q(t)\int_{\Omega}\omega(x)u(x,t)dx + \int_{\Omega}\omega(x)f(x,t)dx, \end{aligned} \tag{2.56}$$

for all $0 < t < T$. Taking into account the condition (K3), and fractional differentiating (2.4) α th order, we have

$$\partial_t^{\alpha}\left(\int_{\Omega}\omega(x)u(x,t)dx\right)(t) = \left(\partial_t^{\alpha}h\right)(t), \quad 0 < t < T. \tag{2.57}$$

From (2.56), taking into account (2.4) and (2.57) we arrive at (2.55).

Now, suppose that $(u(x,t), q(t)) \in X_0^T \times C[0,T]$ is a solution to the problem (2.54), (2.55). In order to prove that (u,q) is also a solution of (2.1)–(2.4), it suffices to prove that (u,q) satisfies (2.4). By Eq. (2.54), we have:

$$\int_\Omega \omega(x)\partial_t^\alpha u(x,t)dx - \int_\Omega \omega(x)Au(x,t)dx$$
$$= q(t)\int_\Omega \omega(x)u(x,t)dx + \int_\Omega \omega(x)f(x,t)dx, \tag{2.58}$$

for $0 < t < T$. Together with (2.55) and (K2), we obtain that $y(t) = \int_\Omega \omega(x)u(x,t)dx - h(t)$ satisfies

$$\begin{cases} \partial_t^\alpha y(t) = q(t)y(t), & 0 < t < T, \\ y(0) + \beta y(T) = 0. \end{cases} \tag{2.59}$$

Therefore, we have

$$y(t) = \int_0^T G_0(t,s)q(s)y(s)ds$$

for all $t \in [0, T]$.

Lemma 2.3 enables us to conclude that the problem (2.59) has only a trivial solution satisfying (2.15). Then, $\int_\Omega \omega(x)u(x,t)dx - h(t) = 0,\ 0 \le t \le T$, i.e., the condition (2.4) is satisfied. This completes the proof of this lemma. □

2.1.4 Local Existence and Uniqueness of the Solution to an Inverse Problem

We are now in a position to prove the existence of a solution to our inverse problem, i.e., Theorem 2.1, which proceeds by a fixed point argument. First, we define the function set

$$B_{r,T} = \Big\{(\bar u, \bar q) \in Y_0^T : \ \bar u(x,0) + \beta \bar u(x,T) = \varphi(x), \ x \in \Omega,$$

$$\bar u(x,t) = 0, \ x \in \partial\Omega, \ t \in (0,T), \ \|\bar u\|_{X_0^T} + \|\bar q\|_{C[0,T]} \le r\Big\}.$$

Here r is a large constant depending on the initial data φ, measurement data h and number β. Throughout, we use M to denote a constant which depends on Ω, α, β, the initial data φ, the known functions f, ω and measurement data h, but independent of r and T.

For given $(\bar u, \bar q) \in B_{r,T}$, we consider

$$\begin{cases} \partial_t^\alpha u(x,t) - Au = F(x,t), & (x,t) \in Q_0^T, \\ u(x,0) + \beta u(x,T) = \varphi(x), & x \in \Omega, \\ u(x,t) = 0, & x \in \Omega, \ t \in (0,T), \end{cases} \tag{2.60}$$

and

$$h(t)q(t) = \left(\partial_t^\alpha h\right)(t) - \int_\Omega \omega(x) f(x,t)dx + \sum_{k=1}^\infty \lambda_k(\omega, e_k)(u(\cdot,t), e_k). \quad (2.61)$$

where

$$F(x,t) = \bar{q}(t)\bar{u}(x,t) + f(x,t).$$

By Corollary 2.2, the unique solution $u \in X_0^T$ of problem (4.1), given by (3.5) satisfies (3.30). Furthermore

$$\begin{aligned} \|\bar{q}(t)\bar{u}(\cdot,t)\|^2_{C([0,T];D((-A)^\varepsilon))} &= \max_{0\le t\le T}\left|\sum_{k=1}^\infty \lambda_k^{2\varepsilon}(\bar{q}(t)\bar{u}(\cdot,t), e_k)^2\right| \\ &\le \|\bar{q}\|^2_{C[0,T]}\|\bar{u}\|^2_{C([0,T];D((-A)^\varepsilon))}. \end{aligned} \quad (2.62)$$

Using this result together with $f \in C([0,T]; D((-A)^\varepsilon))$, we have

$$\bar{q}(t)\bar{u}(x,t) + f(x,t) \in C([0,T]; D((-A)^\varepsilon)). \quad (2.63)$$

Therefore, Corollary 2.2 ensures that there exists a unique solution $u \in X_0^T$ to (2.60). Then (2.61) defines the function $q(t)$ in terms of u. Furthermore, by (2.61) we have

$$\begin{aligned} \|q\|_{C[0,T]} \le h_0\Big[\|\partial_t^\alpha h\|_{C[0,T]} &+ \max_{0\le t\le T}\left|\int_\Omega \omega(x) f(x,t)dx\right| \\ &+ \max_{0\le t\le T}\left|\sum_{k=1}^\infty \lambda_k(\omega, e_k)(u(\cdot,t), e_k)\right|\Big] \le \\ \le h_0\Big[\|\partial_t^\alpha h\|_{C[0,T]} &+ \left|\int_\Omega \omega(x)\|f(x,\cdot)\|_{C[0,T]}dx\right| \\ &+ \left(\sum_{k=1}^\infty |(\omega, e_k)|^2\right)^{1/2} \max_{0\le t\le T}\left(\sum_{k=1}^\infty \lambda_k^2|(u, e_k)|^2\right)^{1/2}\Big] \le \\ \le h_0\Big[\|\partial_t^\alpha h\|_{C[0,T]} &+ \left(\int_\Omega |\omega(x)|^2 dx\right)^{1/2}\left(\int_\Omega \|f(x,\cdot)\|^2_{C[0,T]}dx\right)^{1/2} \\ &+ \|\omega\|_{L^2(\Omega)}\|u\|_{X_0^T}\Big] \\ \le h_0\Big[\|\partial_t^\alpha h\|_{C[0,T]} &+ \|\omega\|_{L^2(\Omega)}\|f\|_{C([0,T];L^2(\Omega))} + \|\omega\|_{L^2(\Omega)}\|u\|_{X_0^T}\Big] + \end{aligned}$$

$$\leq h_0\Big[\|\partial_t^\alpha h\|_{C[0,T]} + \|\omega\|_{L^2(\Omega)}\|f\|_{C([0,T];D((-A)^\varepsilon))} \\ +\|\omega\|_{L^2(\Omega)}\|u\|_{X_0^T}\Big]. \tag{2.64}$$

Based on (2.53), (2.62) and (2.63) this implies that $q \in C[0, T]$. Thus, the mapping

$$S : B_{r,T} \to Y_0^T, \qquad (\bar{u}, \bar{q}) \mapsto (u, q) \tag{2.65}$$

given by (2.60) and (2.61), is well defined.

The next lemma shows that S is a contraction map on $B_{r,T}$.

Lemma 2.9 *Let (K1)-(K5) be held. Then there exists a sufficiently small τ and a suitable large r such that S is a contraction map on $B_{r,T}$ for all $T \in (0, \tau]$, where τ and r are two positive constants depending on α, β and the known functions φ, f and the measurement data h.*

Proof First, we prove $S(B_{r,T}) \subset B_{r,T}$ for sufficiently small T and suitable large r. To simplify the calculations, we restrict $T \in (0, 1]$. By (2.53), we obtain

$$\|u\|_{X_0^T} \leq M + T^{\alpha\varepsilon}M + T^{\alpha\varepsilon}r^2. \tag{2.66}$$

On the other hand, by (K3), (2.62), and (2.65), we have

$$\|q\|_{C[0,T]} \leq h_0\Big\{\|\partial_t^\alpha h\|_{C[0,T]} + \|\omega\|_{L^2(\Omega)}\|f\|_{C([0,T];D((-A)^\varepsilon))} \\ + \|\omega\|_{L^2(\Omega)}\|u\|_{X_0^T}\Big\} \leq M + M^2 + M\|u\|_{X_0^T}. \tag{2.67}$$

Hence, by (2.66) and (2.67), we have

$$\|(u, q)\|_{Y_0^T} \leq M\zeta_1(T)\left(r^2 + 1\right) + M, \tag{2.68}$$

where

$$\zeta_1(T) = T^{\alpha\varepsilon}$$

and therefore satisfies $\lim_{T\to+0} \zeta_1(T) = 0$. Now we take r, such that $r = 2M$ with the constant M in (2.68). Then there exists a sufficiently small $\tau_1 > 0$, such that

$$\|(\bar{u}, \bar{q})\|_{Y_0^T} \leq r, \tag{2.69}$$

for all $T \in (0, \tau_1]$, that is, S maps $B_{r,T}$ into itself for each fixed $T \in (0, \tau_1]$.

Next, we estimate the increment of operator S. To this end, we deduce the differences $(u - U, q - Q)$ from (2.60), (2.61) to yield

$$\begin{cases} \partial_t^\alpha (u-U)(x,t) - A(u-U)(x,t) = \bar{q}(\bar{u}-\bar{U}) + \bar{U}(\bar{q}-\bar{Q}), & (x,t) \in Q_0^T, \\ (u-U)(x,0) + \beta(u-U)(x,T) = 0, & x \in \Omega, \\ (u-U)(x,t)0, & x \in \partial\Omega,\ t \in (0,T), \end{cases} \tag{2.70}$$

and

$$h(t)\,(q-Q)\,(t) = \sum_{k=1}^{\infty} \lambda_k(\omega, e_k)\big(u(\cdot,t) - U(\cdot,t), e_k\big). \tag{2.71}$$

Using Corollary 2.2 and (2.65), we get

$$\|u-U\|_{X_0^T} \le c_{13} r T^{\alpha\varepsilon}(\|\bar{u}-\bar{U}\|_{X_0^T} + \|\bar{q}-\bar{Q}\|_{C[0,T]}), \tag{2.72}$$

Further, by (K3), (2.71) and (2.72), we have

$$\|q-Q\|_{C[0,T]} \le h_0 \left(\sum_{k=1}^{\infty} (\omega, e_k)^2\right)^{1/2}$$

$$\max_{0\le t\le T} \left(\sum_{k=1}^{\infty} \lambda_k^2 \big(u(\cdot,t) - U(\cdot,t), e_k\big)^2\right)^{1/2} \le$$

$$\le h_0 \|\omega\|_{L^2(\Omega)} \|u-U\|_{C([0,T];H^2(\Omega))} \le h_0 \|\omega\|_{L^2(\Omega)} \|u-U\|_{X_0^T} \le$$

$$\le c_{13} T^{\alpha\varepsilon} r h_0 \|\omega\|_{L^2(\Omega)} (\|\bar{u}-\bar{U}\|_{X_0^T} + \|\bar{q}-\bar{Q}\|_{C[0,T]}). \tag{2.73}$$

Therefore, by (2.72) and (2.73), we obtain

$$\|(u-U, q-Q)\|_{Y_0^T} \le M r T^{\alpha\varepsilon} \|(\bar{u}-\bar{U}, \bar{q}-\bar{Q})\|_{Y_0^T}. \tag{2.74}$$

Hence we can choose sufficiently small $\tau_2 > 0$ such that

$$M r T^{\alpha\varepsilon} \le \frac{1}{2} \tag{2.75}$$

for all $T \in (0, \tau_2]$ to obtain

$$\|S(\bar{u}, \bar{q}) - S(\bar{U}, \bar{Q})\|_{Y_0^T} \le \frac{1}{2} \|(\bar{u}-\bar{U}, \bar{q}-\bar{Q})\|_{Y_0^T}. \tag{2.76}$$

Estimates (2.69) and (2.76) show that S is a contraction map on $B_{r,T}$ for all $T \in (0, \tau]$, if we choose $\tau \le \min\{\tau_1, \tau_2\}$. □

2.1.5 Global Solvability of the Problem: Proof of the Main Results

In this section, we give proof of the global in time uniqueness of solutions to our inverse problem, i.e., Theorem 2.1.

Proof (*Theorem* 2.1) For A is a contraction map on $B_{r,T}$ for all $T \in (0, \tau]$, the Banach fixed point theorem concludes that there exists a unique solution $(u, q) \in X_0^T \times C[0, T]$ of the inverse problem (2.54) and (2.55).

Next we show that we could extend the solution (u, q) in $(0, \tau]$ to a larger interval $[\tau, 2\tau]$. To do this, we consider

$$\begin{cases} \partial_t^\alpha v(x,t) - Av(x,t) = p(t)v(x,t) + f(x,t), & (x,t) \in Q_\tau^T, \\ v(x,\tau) + \beta v(x,T) = u(x,\tau) + \beta u(x,T), & x \in \Omega \\ v(x,t) = 0, \quad x \in \partial\Omega, \quad t \in (\tau, T), \end{cases} \tag{2.77}$$

and

$$h(t)p(t) = \left(\partial_t^\alpha h\right)(t) - \int_\Omega \omega(x) f(x,t)dx+$$
$$+ \sum_{k=1}^\infty \lambda_k(\omega, e_k)(v(\cdot,t), e_k), \; t \in [\tau, T]. \tag{2.78}$$

Obviously, if we prove that there exists a solution $(v, p) \in Y_\tau^T$ with some $T \geq 2\tau$, then $(\tilde{u}, \tilde{q})$ defined by

$$(\tilde{u}, \tilde{q}) = \begin{cases} (u, q), & t \in [0, \tau], \\ (v, p), & t \in [\tau, 2\tau], \end{cases} \tag{2.79}$$

is a solution of the inverse problem (2.54) and (2.55) on the larger interval $[0, 2\tau]$.

We repeat a similar fixed pointed argument to prove the existence of (v, p). Define an operator

$$K : \tilde{B}_{\tilde{r},T} \to Y_\tau^T, \qquad (\bar{v}, \bar{p}) \to (v, p) \tag{2.80}$$

with $(\bar{v}, \bar{p}) \in \tilde{B}_{\tilde{r},T}$, where

$$\tilde{B}_{\tilde{r},T} = \Big\{(\bar{v}, \bar{p}) \in Y_\tau^T : \; \bar{v}(x,\tau) + \beta\bar{v}(x,T) = u(x,\tau) + \beta u(x,T), \; x \in \Omega$$
$$\bar{v}(x,t) = 0, \; x \in \partial\Omega, \; t \in (\tau, T), \; \|\bar{v}\|_{X_\tau^T} + \|\bar{p}\|_{C[\tau,T]} \leq \tilde{r}\Big\}.$$

Here v is the solution to the initial and boundary value problem

$$\begin{cases} \partial_t^\alpha v(x,t) - Av(x,t) = \bar{p}(t)\bar{v}(x,t) + f(x,t), & (x,t) \in Q_\tau^T, \\ v(x,\tau) + \beta v(x,T) = u(x,\tau) + \beta u(x,T), & x \in \Omega \\ v(x,t) = 0, \quad x \in \partial\Omega, \ t \in (\tau, T). \end{cases} \tag{2.81}$$

Furthermore, p is the solution of (2.78) in terms of v. Additionally, we have $F(x,t) = \bar{p}(t)\bar{v}(x,t) + f(x,t) \in C([0,T]; D((-A)^\varepsilon)), u(\cdot,\tau), u(\cdot,T) \in H^2(\Omega)$. In fact, the first property comes from (2.63), according to Corollary 2.2 the functions $u(\cdot,\tau), u(\cdot,T)$ the same as (2.63) at $t \in \{\tau, T\}$, therefore we can conclude that $u(\cdot,\tau) \in H^2(\Omega)$.

Therefore, applying (2.34) again, then we get

$$\|u_t(\cdot,\tau)\|_{H^1(0,1)} \le M\Big[(\tau^{-1} + T^{-1})\|\varphi\|_{H^2(0,1)} + \|\psi\|_{H^1(0,1)} + \tau^{\alpha-1}\|F\|_{C[0,T];D((-\mathcal{L})^{\frac{1}{\alpha}})} + \left(T^{\alpha-1} + T^2\right)\|F\|_{C[0,T];D((-\mathcal{L})^{\frac{1}{\alpha}})}\Big]. \tag{2.82}$$

Similarly, we have

$$\|u_t(\cdot,T)\|_{H^1(0,1)} \le M\Big[T^{-1}\|\varphi\|_{H^2(0,1)} + \|\psi\|_{H^1(0,1)} + \left(T^{\alpha-1} + T^2\right)\|F\|_{C[0,T];D((-\mathcal{L})^{\frac{1}{\alpha}})}\Big]. \tag{2.83}$$

Besides, by (2.53) we have

$$\|v\|_{X_\tau^T} \le c_{13}\Big[\|u(\cdot,\tau)\|_{H^2(\Omega)} + \beta\|u(\cdot,T)\|_{H^2(\Omega)} + (T-\tau)^{\alpha\varepsilon}(\tilde{r}^2 + M)\Big] \tag{2.84}$$

and from (2.78) via (K1), (K3), (K5) we get

$$\|p\|_{C[\tau,T]} \le h_0\Big\{\|\partial_t^\alpha h\|_{C[\tau,T]} + \|\omega\|_{L^2(\Omega)}\Big[\|f\|_{C([\tau,T];D((-A)^\varepsilon))} + c_{13}(\|u(\cdot,\tau)\|_{H^2(\Omega)} + \beta\|u(\cdot,T)\|_{H^2(\Omega)} + (T-\tau)^{\alpha\varepsilon}(\tilde{r}^2 + M))\Big]\Big\}. \tag{2.85}$$

We set $T - \tau \le 1$. Then using (2.84) and (2.85), we have

$$\|K(\tilde{v},\tilde{p})\|_{Y_\tau^T} \le h_0\big(\|\partial_t^\alpha h\|_{C[\tau,T]} + \|\omega\|_{L^2(\Omega)}\|f\|_{C([\tau,T];D((-A)^\varepsilon))}\big) + + c_{13}(h_0\|\omega\|_{L^2(\Omega)} + 1)\Big[\|u(\cdot,\tau)\|_{H^2(\Omega)} + \beta\|u(\cdot,T)\|_{H^2(\Omega)} + (M + \tilde{r}^2)(T-\tau)^{\alpha\varepsilon}\Big] \tag{2.86}$$

and by a similar calculation to (2.66), we have

$$\|K(\bar{v}_1, \bar{p}_1) - K(\bar{v}_2, \bar{p}_2)\|_{Y_\tau^T} \leq M(T-\tau)^{\alpha\varepsilon}\|(\bar{v}_1 - \bar{v}_2, \bar{p}_1 - \bar{p}_2)\|_{Y_\tau^T}, \tag{2.87}$$

where M is the same as the ones in Lemma 2.9. We choose $\tilde{r}$ such that $\tilde{r} \geq r$ and

$$h_0\big(\|\partial_{0t}^\alpha h\|_{C[\tau,T]} + \|\omega\|_{L^2(\Omega)}\|f\|_{C([\tau,T];D((-A)^\varepsilon))}\big) +$$

$$+ c_{13}(h_0\|\omega\|_{L^2(\Omega)} + 1)\big(\|u(\cdot,\tau)\|_{H^2(\Omega)} + \|u(\cdot,T)\|_{H^2(\Omega)}\big) \leq \frac{\tilde{r}}{2}.$$

It is easy to see that if we choose $\tilde{r}$ larger, then we could get larger $T-\tau$ to satisfy

$$M(T-\tau)^{\alpha\varepsilon}\Big(\|u(\cdot,\tau)\|_{H^2(\Omega)} + \|u(\cdot,T)\|_{H^2(\Omega)}\Big) \leq \frac{\tilde{r}}{2}. \tag{2.88}$$

Furthermore noticing that (2.88) and (2.75) we have the same structure, we can choose $T-\tau=\tau$ to satisfy (2.88), which yields $\|K(\tilde{u},\tilde{q})\|_{Y_\tau^T} \leq \tilde{r}$, i.e., $K(\tilde{B}_{\tilde{r},T}) \subset \tilde{B}_{\tilde{r},T}$. Additionally,

$$\|K(\tilde{u}_1,\tilde{q}_1) - K(\tilde{u}_2,\tilde{q}_2)\|_{Y_\tau^T} \leq \frac{1}{2}\|(\tilde{u}_1 - \tilde{u}_2, \tilde{q}_1 - \tilde{q}_2)\|_{Y_\tau^T} \tag{2.89}$$

for $T=2\tau$, because (2.87) is the same as (2.76), if we replace T in (2.76) by $T-\tau$. Hence we prove that K is a contraction operator on $\tilde{B}_{\tilde{r},T}$ for $T=2\tau$.

Repeating the extension process limited times, we could obtain a solution $(u,q) \in X_0^T \times C[0,T]$ of the inverse problem (2.54) and (2.55) for any T. Lemma 2.8 shows that the inverse problem (2.54) and (2.55) is equivalent to our inverse problem. Consequently, the inverse problem (2.1)–(2.4) also admits a unique solution (u,q) in the space $X_0^T \times C[0,T]$ for any T. □

Now, we prove Theorem 2.2.

Proof By Lemma 2.8, we know that our inverse problem is equivalent to the inverse problem (2.54) and (2.55). Hence in the following proof, we turn to prove the uniqueness of the solution to (2.54) and (2.55).

We set $(\hat{u},\hat{q}) = (u_1 - u_2, q_1 - q_2)$ and

$$\sigma = \inf\left\{t \in [0,T] : \|(\hat{u},\hat{q})\|_{Y_0^t} > 0\right\}. \tag{2.90}$$

It suffices to prove that $\sigma = T$.

If (2.90) is not true, then we have $\sigma < T$. Choose l such that $0 < l < T-\sigma$. Next, we consider

$$\begin{cases} \partial_t^\alpha \hat{u}(x,t) - A\hat{u}(x,t) = \hat{F}(x,t), & (x,t) \in Q_\sigma^{\sigma+l}, \\ \hat{u}(x,\sigma) + \beta\hat{u}(x,\sigma+l) = 0, & x \in \Omega, \\ \hat{u}(x,t) = 0, & x \in \partial\Omega, \ t \in (\sigma,\sigma+l), \end{cases} \tag{2.91}$$

and

$$h(t)\hat{q}(t) = \sum_{k=1}^{\infty} \lambda_k(\omega, e_k)(\hat{u}(\cdot,t), e_k), \quad t \in [\sigma, \sigma+l], \tag{2.92}$$

where

$$\hat{F}(x,t) = q_1(t)\hat{u}(x,t) + \hat{q}(t)u_2(x,t).$$

By (2.46), we can write the solution $\hat{u}$ as

$$\hat{u}(x,t) = \sum_{k=1}^{\infty} \int_0^T G_k(t,s)(\hat{F}(\cdot,s), e_k)ds, \quad t \in [\sigma,\sigma+l]. \tag{2.93}$$

Then similar to the proof of Corollary 2.2, we have

$$\|\hat{u}\|_{X_\sigma^{\sigma+\varepsilon}} \le c_{13} l^{\alpha\varepsilon} \|\hat{F}\|_{C([\sigma,\sigma+l];D((-A)^\varepsilon)} \tag{2.94}$$

and

$$\max_{\sigma \le t \le \sigma+l} \left| \sum_{k=1}^{\infty} \lambda_k(\omega, e_k)(\hat{u}(\cdot,t), e_k) \right| \le \|\omega\|_{L^2(\Omega)} \|\hat{u}\|_{C([\sigma,\sigma+l];H^2(\Omega))} \le$$

$$\le \|\omega\|_{H^2(\Omega)} \|\hat{u}\|_{X_\sigma^{\sigma+l}} \le M l^{\alpha\varepsilon} \|\hat{F}\|_{C([\sigma,\sigma+l];D((-A)^\varepsilon))} \tag{2.95}$$

with the same M in Lemma 2.7 and estimates (2.62). By (2.62), and noting that $(u_1, q_1) = (u_2, q_2)$ on $[0, \sigma]$, we have

$$\|\hat{F}\|_{C([\sigma,\sigma+l];D((-A)^\varepsilon))} \le M(\|\hat{u}\|_{X_\sigma^{\sigma+l}} + \|\hat{q}\|_{C[\sigma,\sigma+l]}), \tag{2.96}$$

which yields, together with (2.94), that

$$\|\hat{u}\|_{X_\sigma^{\sigma+l}} \le M l^{\alpha\varepsilon}(\|\hat{u}\|_{X_\sigma^{\sigma+l}} + \|\hat{q}\|_{C[\sigma,\sigma+l]}). \tag{2.97}$$

On the other hand, by (2.92), (2.95), and (2.96) we have the following estimate for $\hat{q}$

$$\|\hat{q}\|_{C[\sigma,\sigma+l]} \le M h_0 l^{\alpha\varepsilon}(\|\hat{u}\|_{X_\sigma^{\sigma+l}} + \|\hat{q}\|_{C[\sigma,\sigma+l]}). \tag{2.98}$$

Therefore, by (2.97) and (2.98) we have

$$\|(\hat{u},\hat{q})\|_{Y_\sigma^{\sigma+l}} \le M l^{\alpha\varepsilon} \|(\hat{u},\hat{q})\|_{Y_\sigma^{\sigma+l}} \tag{2.99}$$

implying

$$\|(\hat{u}, \hat{q})\|_{Y_\sigma^{\sigma+l_0}} = 0$$

for some sufficiently small positive constant l_0. This means that $(u_1 - q_1, u_2 - q_2)$ vanishes in some neighborhood of σ. But this is not compatible with the definition of σ. We conclude that $\sigma = T$. □

At the end of this section, we consider the following examples.

Example 2.1 Let $\beta = 1$, $n = 1$, $Q_0^T = (0, 1) \times (0, T = \pi)$, and $A := -\partial_x^2$. We consider the following inverse problem

$$\begin{cases} \partial_t^\alpha u - \partial_x^2 u = q(t)u + 2\sin(\pi x)\Big[t^{1-\alpha}E_{2,2-\alpha}(-t^2) \\ \qquad\qquad +(2+\sin t)\left(\pi^2 - (1+t)e^{-t}\right)\Big], \\ u(x,0) + u(x,\pi) = 8\sin(\pi x), \quad u(0,t) = u(1,t) = 0, \quad t \in (0,\pi), \\ \int_0^1 \sin(\pi x)u(x,t)dx = 2 + \sin t, \qquad t \in [0,\pi]. \end{cases} \tag{2.100}$$

Note that the all given data satisfy conditions (K1)-(K5). Furthermore, according to Theorem 2.1, using the given data, there exists a unique solution to the above inverse problem, and from Lemma 2.8, it has the form

$$u(x,t) = 2(2+\sin t)\sin(\pi x), \quad q(t) = (1+t)e^{-t}. \tag{2.101}$$

Now, let $T = 2\pi$ and using (2.100) and (2.101), we consider the following problem

$$\begin{cases} \partial_t^\alpha v - \partial_x^2 v = p(t)v + 2\sin(\pi x)\Big[t^{1-\alpha}E_{2,2-\alpha}(-t^2) \\ \qquad\qquad +(2+\sin t)\left(\pi^2 - (1+t)e^{-t}\right)\Big], \\ v(x,\pi) + v(x,2\pi) = 8\sin(\pi x), \quad v(0,t) = v(1,t) = 0, \quad t \in (\pi, 2\pi), \\ \int_0^1 \sin(\pi x)v(x,t)dx = 2 + \sin t, \qquad t \in [\pi, 2\pi]. \end{cases} \tag{2.102}$$

We can easily see that, as above, a pair of functions $v(x,t) = 2(2+\sin t)\sin(\pi x)$, $p(t) = (1+t)e^{-t}$ is a solution of the problem (2.102). Therefore, we can continue the same process for any $T = \pi n$, where $n \in \mathbb{N}$.

2.2 An Inverse Coefficient Problem with Nonlocal Boundary and Overdetermination Conditions

Periodic boundary conditions (PBCs) are a set of boundary conditions which are often chosen for approximating a large (infinite) system by using a small part called a unit cell. PBCs are often used in computer simulations and mathematical models. The topology of two-dimensional PBC is equal to that of a world map of some video games; the geometry of the unit cell satisfies perfect two-dimensional tiling, and when an object passes through one side of the unit cell, it re-appears on the opposite side with the same velocity (see [12–14]).

PBCs arise from many important applications in heat transfer, natural sciences [15, 17–19]. In these works, the existence, uniqueness, and continuous dependence on data of the solution were proved, and the numerical study of the diffusion problem with periodic boundary conditions was developed. It is important to note that in the paper [19] the time-dependent thermal coefficient is determined from nonlocal overdetermination condition data. Besides, in [20–22] the coefficients of the heat equations are determined in the case of nonlocal boundary conditions.

In the present section, time-fractional diffusion equation is used with initial, periodic boundary conditions for the determination of coefficients. The existence and uniqueness of the classical solution of the problem (2.103)–(2.106) is reduced to fixed point principles by applying the Fourier method.

We consider the initial-periodic boundary value problem for the fractional diffusion equation

$$\partial_t^\alpha u - u_{xx} + a(t)u = f(x,t)g(t), \quad (x,t) \in D_T, \tag{2.103}$$

$$u(x,0) = \varphi(x), \quad x \in [0,1], \tag{2.104}$$

$$u(0,t) = u(1,t), \;\; u_x(0,t) = u_x(1,t), \;\; t \in [0,T], \tag{2.105}$$

where ∂_t^α is the Caputo fractional derivative of order $0 < \alpha \le 1$ in the time variable (see (1.4)), $a(t), g(t), t > 0$ are the source control terms, $f(x,t)$ is known source term, $\varphi(x)$ is the initial temperature, T is arbitrary positive number and $D_T := \{(x,t) : 0 < x < 1, 0 < t \le T\})$.

The problem of determining a function $u(x,t)$, $(x,t) \in D_T$, that satisfies (2.103)–(2.105) with known functions $a(t)$, $g(t)$, $f(x,t)$ and $\varphi(x)$ will be called *the direct problem.*

In *the inverse problem,* it is required to determine the coefficients $a(t), g(t), t > 0$, in (1) using overdetermination conditions about the solution of the direct problem (2.103)–(2.105):

$$\int_0^1 \omega_i(x)u(x,t)dx = h_i(t), \; i = 1,2, \;\; x \in [0,1], \tag{2.106}$$

where $\omega_i(x), h_i(t), i = 1,2$ are given functions.

Let $C^{2,\alpha}(D_T)$ be the class of functions that are twice continuously differentiable with respect to x, for which a continuous derivative ∂_t^α exists in D_T.

Definition 2.2 The triple of functions $\{u(x,t), a(t), g(t)\}$ from the class $C^{2,\alpha}(D_T) \cap C^{1,0}(\overline{D}_T) \times C[0,T] \times C[0,T]$ is said to be a classical solution of problem (2.103)–(2.106), if the functions $u(x,t)$, $a(t)$ and $g(t)$ satisfy the following conditions:

(1) the function $u(x,t)$ and its derivatives $\partial_t^\alpha u(x,t)$, $u_{xx}(x,t)$ are continuous in the domain D_T;

(2) the functions $a(t)$, $g(t)$ are continuous on the interval $[0,T]$;

(3) equation (2.103) and conditions (2.104)–(2.106) are satisfied in the classical sense.

Throughout this section the functions φ, f, ω_i and h_i $(i = 1, 2)$ are assumed to satisfy the following conditions:

$(A1)$ $\varphi(x) \in C^2[0,1]$; $\varphi'''(x) \in L_2(0,1)$; $\varphi(0) = \varphi(1)$; $\varphi'(0) = \varphi'(1)$; $\varphi''(0) = \varphi''(1)$;

$(A2)$ $f(x,t) \in C^{2,0}(\overline{D}_T)$; $f'''_{xxx}(x,t) \in C([0,T], L_2(0,1))$; $f(0,t) = f(1,t)$; $f'_x(0,t) = f'_x(1,t)$; $f''_{xx}(0,t) = f''_{xx}(1,t)$;

$(A3)$ $h_i(t) \in C^1[0,T]$; $\omega_i(x) \in C^2[0,1]$; $\int_0^1 \omega_i(x)\varphi(x)dx = h_i(0)$, $\omega_i(0) = \omega_i(1)$, $\omega_i'(0) = \omega_i'(1)$, $i = 1, 2$.

In the next subsection, we study the direct problem.

2.2.1 Existence and Uniqueness of a Solution to the Initial-Periodic Boundary Value Problem

The use of the Fourier method for solving problem (2.103)–(2.105) leads to the spectral problem for the operator given by the differential expression and boundary conditions

$$X_n''(x) + \lambda^2 X_n(x) = 0, \ x \in (0,1),$$

$$X_n(0) = X_n(1), \quad X_n'(0) = X_n'(1), n = 0, 1, 2, ... \tag{2.107}$$

In [22], it is known that the system of eigenfunctions

$$1, \cos\lambda_1 x, \sin\lambda_1 x, \cos\lambda_2 x, \sin\lambda_2 x, ..., \cos\lambda_n x, \ \ \sin\lambda_n x, ... \tag{2.108}$$

where $\lambda_n = 2\pi n$ $(n = 0, 1, ...)$, is a basis for $L_2(0,1)$. This system is eigenfunctions of spectral problem (2.107).

Since the system (2.108) form a basis in $L_2(0,1)$, we shall seek the classical solution $u(x,t)$ to the problem (2.103)–(2.105) in the form

$$u(x,t) = \sum_{n=0}^{\infty} u_{1n}(t)\cos\lambda_n x + \sum_{n=1}^{\infty} u_{2n}(t)\sin\lambda_n x,\ \lambda_n = 2\pi n, \tag{2.109}$$

$$f(x,t) = \sum_{n=0}^{\infty} f_{1n}(t)\cos\lambda_n x + \sum_{n=1}^{\infty} f_{2n}(t)\sin\lambda_n x,$$

where

$$u_{10}(t) = \int_0^1 u(x,t)dx,\quad u_{1n}(t) = 2\int_0^1 u(x,t)\cos\lambda_n x dx,$$

$$u_{2n}(t) = 2\int_0^1 u(x,t)\sin\lambda_n x dx;\qquad f_{10}(t) = \int_0^1 f(x,t)dx,$$

$$f_{1n}(t) = 2\int_0^1 f(x,t)\cos\lambda_n x dx,\quad f_{2n}(t) = 2\int_0^1 f(x,t)\sin\lambda_n x dx.$$

Then, applying the formal scheme of the Fourier method for determining of unknown coefficients $u_{10}(t)$ and $u_{in}(t)$ $(i = 1, 2; n = 1, 2, ...)$ of function $u(x,t)$, from (2.103) and (2.104) we have

$$\partial_t^\alpha u_{10}(t) = -a(t)u_{10}(t) + g(t)f_{10}(t), \tag{2.110}$$

$$u_{10}(t)|_{t=0} = \varphi_{10}, \tag{2.111}$$

$$\partial_t^\alpha u_{in}(t) + \lambda_n^2 u_{in}(t) = -a(t)u_{in}(t) + g(t)f_{in}(t), \tag{2.112}$$

$$u_{in}(t)|_{t=0} = \varphi_{in},\ \ i = 1, 2,\ \ n = 1, 2, ..., \tag{2.113}$$

where

$$\varphi_{10} = \int_0^1 \varphi(x)dx,\quad \varphi_{1n} = 2\int_0^1 \varphi(x)\cos\lambda_n x dx,\quad \varphi_{2n} = 2\int_0^1 \varphi(x)\sin\lambda_n x dx.$$

According to Theorem 1.1, the solutions of problems (2.110), (2.111) and (2.112), (2.113) satisfy the following integral equations

$$u_{10}(t) = \varphi_{10} + \frac{1}{\Gamma(\alpha)}\int_0^t (t-\tau)^{\alpha-1}(f_{10}(\tau)g(\tau) - a(\tau)u_{10}(\tau))d\tau, \tag{2.114}$$

and

$$u_{in}(t) = \varphi_{in}E_\alpha(-\lambda^2 t^\alpha)$$

$$+\int_0^t (t-\tau)^{\alpha-1}E_{\alpha,\alpha}(-\lambda^2(t-\tau)^\alpha)(g(\tau)f_{in}(\tau) - a(\tau)u_{in}(\tau))d\tau. \tag{2.115}$$

Estimating the functions $u_{10}(t)$, $u_{in}(t)$, we obtain the integral inequalities:

$$|u_{10}(t)| \le |\varphi_{10}| + \frac{t^\alpha}{\Gamma(\alpha+1)}\|g\|\|f_{10}\| + \frac{\|a\|}{\Gamma(\alpha)}\int_0^t (t-\tau)^{\alpha-1}|u_{10}(\tau)|d\tau,$$

$$|u_{in}(t)| \le |\varphi_{in}| + \frac{t^\alpha}{\Gamma(\alpha+1)}\|g\|\|f_{in}\| +$$

$$+\frac{\|a\|}{\Gamma(\alpha)}\int_0^t (t-\tau)^{\alpha-1}|u_{in}(\tau)|d\tau, \quad (i=1,2, \quad n=1,2,...)$$

where $\|g\| = \max_{t\in[0,T]}|g(t)|$, $\|a\| = \max_{t\in[0,T]}|a(t)|$. Applying Gronwall's Lemma 1.4, from last inequalities we obtain the following estimates:

$$|u_{10}(t)| \le \left(|\varphi_{10}| + \frac{t^\alpha}{\Gamma(\alpha+1)}\|g\|\|f_{10}\|\right)E_\alpha(\|a\|t^\alpha), \tag{2.116}$$

$$|u_{in}(t)| \le \left(|\varphi_{in}| + \frac{t^\alpha}{\Gamma(\alpha+1)}\|g\|\|f_{in}\|\right)E_\alpha(\|a\|t^\alpha). \tag{2.117}$$

Using equalities (2.110), (2.112) and (2.116), (2.117) we obtain estimates for $\partial_t^\alpha u_{10}(t)$, $\partial_t^\alpha u_{in}(t)$:

$$|\partial_t^\alpha u_{10}(t)| \le \|a\|\left(|\varphi_{10}| + \frac{t^\alpha}{\Gamma(\alpha+1)}\|g\|\|f_{10}\|\right)E_\alpha(\|a\|t^\alpha) + \|g\|\|f_{10}\|,$$

$$|\partial_t^\alpha u_{in}(t)| \le (\lambda^2 + \|a\|)\left(|\varphi_{in}| + \frac{t^\alpha}{\Gamma(\alpha+1)}\|g\|\|f_{in}\|\right)E_\alpha(\|a\|t^\alpha) + \|g\|\|f_{in}\|.$$

Thus we have proved the following lemma:

Lemma 2.10 *For any $t \in [0,T]$ and $n \in \mathbb{N}$, the following estimates are valid:*

$$|u_{10}(t)| \le \left(|\varphi_{10}| + \frac{T^\alpha}{\Gamma(\alpha+1)}\|g\|\|f_{10}\|\right)E_\alpha(\|a\|T^\alpha),$$

$$|u_{in}(t)| \le \left(|\varphi_{in}| + \frac{T^\alpha}{\Gamma(\alpha+1)}\|g\|\|f_{in}\|\right)E_\alpha(\|a\|T^\alpha),$$

$$|\partial_t^\alpha u_{10}(t)| \le \|a\|\left(|\varphi_{10}| + \frac{T^\alpha}{\Gamma(\alpha+1)}\|g\|\|f_{10}\|\right)E_\alpha(\|a\|T^\alpha) + \|g\|\|f_{10}\|,$$

$$|\partial_t^\alpha u_{in}(t)| \le (\lambda^2 + \|a\|)\left(|\varphi_{in}| + \frac{T^\alpha}{\Gamma(\alpha+1)}\|g\|\|f_{in}\|\right)E_\alpha(\|a\|T^\alpha) + \|g\|\|f_{in}\|.$$

Formally, from (2.109) by term-by-term differentiation we compose the series

$$\partial_t^\alpha u(x,t) = \sum_{n=0}^{\infty} \partial_t^\alpha u_{1n}(t)\cos\lambda_n x + \sum_{n=1}^{\infty} \partial_t^\alpha u_{2n}(t)\sin\lambda_n x, \tag{2.118}$$

$$u_{xx}(x,t) = -\sum_{n=0}^{\infty} \lambda_n^2 u_{1n}(t)\cos\lambda_n x - \sum_{n=1}^{\infty} \lambda_n^2 u_{2n}(t)\sin\lambda_n x. \tag{2.119}$$

In view of Lemma 2.10, if following series converge then the series (2.109), (2.118), and (2.119) to be converge for any $(x,t) \in D_T$

$$C_4 \sum_{n=1}^{\infty} (\lambda_n^2 |\varphi_{in}| + \lambda_n^2 \|f_{in}\|),$$

where the constant C_4 depends only on $T, \alpha, \|a\|, \|g\|$.

We hold the following auxiliary lemma:

Lemma 2.11 *If the conditions (A1), (A2) are valid then, there are equalities*

$$\varphi_{in} = \frac{1}{\lambda_n^3}\varphi_{in}^{(3)}, \quad f_{in}(t) = \frac{1}{\lambda_n^3} f_{in}^{(3)}, \ (i=1,2), \tag{2.120}$$

where

$$\varphi_{1n}^{(3)} = 2\int_0^1 \varphi'''(x)\sin\lambda_n x dx, \quad \varphi_{2n}^{(3)} = 2\int_0^1 \varphi'''(x)\cos\lambda_n x dx,$$

$$f_{1n}^{(3)}(t) = 2\int_0^1 f_{xxx}'''(x,t)\sin\lambda_n x dx, \quad f_{2n}^{(3)}(t) = 2\int_0^1 f_{xxx}'''(x,t)\cos\lambda_n x dx$$

with the following estimates:

$$\sum_{n=1}^{\infty} |\varphi_{in}^{(3)}|^2 \le \|\varphi'''\|_{L_2(0,1)}^2, \quad \sum_{n=1}^{\infty} |f_{in}^{(3)}(t)|^2 \le \|f_{xxx}'''\|_{L_2(0,1)\times C[0,T]}^2, \ (i=1,2). \tag{2.121}$$

Proof Equalities (2.120), under the conditions of Lemma 2.11, are easily obtained if we integrate by parts three times in the integrals for the Fourier coefficients of the functions $\varphi(x)$, $f(x,t)$.

Relations (2.121) are Bessel inequalities for the functions $\varphi'''(x)$, $f_{xxx}(x,t)$. □

If the functions $\varphi(x)$, $f(x,t)$ satisfy the conditions of Lemma 2.11, then due to representations (2.120) and (2.121) series (2.109), (2.118) and (2.119) converge uniformly in the rectangle D_T, therefore, function $u(x,t)$ satisfies relations (2.103)–(2.105).

Using the above results, we obtain the following assertion.

Lemma 2.12 *Let* $\{g(t),\ a(t)\} \in C[0,T]$ *and the conditions* $(A1)$, $(A2)$ *be satisfied. Then, there exists a unique solution to direct problem (1)–(3)* $u(x,t) \in C^{2,\alpha}(D_T) \cap C^{1,0}(\overline{D}_T)$.

Remark 2.2 To prove Lemma (2.12), we employed the rougher estimate $\left|E_{\alpha,\beta}\left(-\lambda_n t^{\alpha}\right)\right| \le C$, which follows from Proposition (1.1) for all $t \in [0,T]$. This shows that, under the conditions of Lemma (2.12), the solution $u(x,t)$ belongs to the class $C^{2,\alpha}(\overline{D}_T)$.

In what follows, we need an a priori estimate for the norm of the difference between the solution of the original integral equations (2.114), (2.115) and the solution of this equation with perturbed functions $\widetilde{a}$, $\widetilde{g}$, $\widetilde{\varphi}_{in}$, $\widetilde{f}_{in}$. Let $\widetilde{u}_{in}(t)$, $(i := 0, 1, 2)$ be solutions of the integral equation (2.114), (2.115) corresponding to the functions $\widetilde{a}$, $\widetilde{g}$, $\widetilde{\varphi}_{in}$, $\widetilde{f}_{in}$; i.e.,

$$\widetilde{u}_{10}(t) = \widetilde{\varphi}_{10} + \frac{1}{\Gamma(\alpha)}\int_0^t (t-\tau)^{\alpha-1}\left(\widetilde{f}_{10}(\tau)\widetilde{g}(\tau) - \widetilde{a}(\tau)\widetilde{u}_{10}(\tau)\right)d\tau, \tag{2.122}$$

$$\widetilde{u}_{in}(t) = \widetilde{\varphi}_{in}E_{\alpha}(-\lambda^2 t^{\alpha})$$

$$+\int_0^t (t-\tau)^{\alpha-1}E_{\alpha,\alpha}(-\lambda^2(t-\tau)^{\alpha})\left(\widetilde{a}(\tau)\widetilde{u}_{in}(\tau) + \widetilde{g}(\tau)\widetilde{f}_{in}(\tau)\right)d\tau. \tag{2.123}$$

In view of the equations (2.114), (2.122), (2.115) (2.123) we compose the difference $u_{in} - \widetilde{u}_{in}$. Introducing the notations $u_{in} - \widetilde{u}_{in} = \overline{u}_{in}$, $a - \widetilde{a} = \overline{a}$, $g - \widetilde{g} = \overline{g}$, $f_{in} - \widetilde{f}_{in} = \overline{f}_{in}$, we obtain the integral equations

$$\overline{u}_{10}(t) = \overline{\varphi}_{10} + \frac{1}{\Gamma(\alpha)}\int_0^t (t-\tau)^{\alpha-1}(f_{10}(\tau)\overline{g}(\tau) + \widetilde{g}(\tau)\overline{f}_{10}(\tau))d\tau -$$

$$-\frac{1}{\Gamma(\alpha)}\int_0^t (t-\tau)^{\alpha-1}(\overline{a}(\tau)u_{10}(\tau) + \widetilde{a}(\tau)\overline{u}_{10}(\tau))d\tau,$$

$$\overline{u}_{in}(t) = \overline{\varphi}_n E_{\alpha,1}(-\lambda_n^2 t^{\alpha}) + \int_0^t (t-\tau)^{\alpha-1}E_{\alpha,\alpha}\left(-\lambda_n^2(t-\tau)^{\alpha}\right)\overline{g}(\tau)f_{in}(\tau)d\tau +$$

$$+\int_0^t (t-\tau)^{\alpha-1}E_{\alpha,\alpha}\left(-\lambda_n^2(t-\tau)^{\alpha}\right)\widetilde{g}(\tau)\overline{f}_{in}(\tau)d\tau -$$

$$-\int_0^t (t-\tau)^{\alpha-1}E_{\alpha,\alpha}\left(-\lambda_n^2(t-\tau)^{\alpha}\right)\overline{a}(\tau)u_{in}(\tau)d\tau -$$

$$-\int_0^t (t-\tau)^{\alpha-1}E_{\alpha,\alpha}\left(-\lambda_n^2(t-\tau)^{\alpha}\right)\widetilde{a}(\tau)\overline{u}_{in}(\tau)d\tau, \tag{2.124}$$

from which are derived the following linear integral inequalities for $|\overline{u}_{10}(t)|$, $|\overline{u}_{in}(t)|$:

$$|\overline{u}_{10}(t)| \le |\overline{\varphi}_{10}| + \frac{t^\alpha \|\overline{f}_{10}\| \|\widetilde{g}\|}{\Gamma(\alpha+1)} + \frac{t^\alpha \| f_{10}\| \|\overline{g}\|}{\Gamma(\alpha+1)} +$$

$$+ \frac{\|\overline{a}\| t^\alpha}{\Gamma(\alpha+1)} \Big(|\varphi_{10}| + \frac{t^\alpha}{\Gamma(\alpha+1)} \|g\| \|f_{10}\| \Big) E_\alpha(\|a\| t^\alpha) + \frac{\|\widetilde{a}\|}{\Gamma(\alpha)} \int_0^t (t-\tau)^{\alpha-1} |\overline{u}_{10}(\tau)| d\tau.$$

$$|\overline{u}_{in}(t)| \le |\overline{\varphi}_{in}| + \frac{t^\alpha \|\overline{f}_{in}\| \|\widetilde{g}\|}{\Gamma(\alpha+1)} + \frac{t^\alpha \| f_{in}\| \|\overline{g}\|}{\Gamma(\alpha+1)} +$$

$$+ \frac{\|\overline{a}\| t^\alpha}{\Gamma(\alpha+1)} \Big(|\varphi_{in}| + \frac{t^\alpha}{\Gamma(\alpha+1)} \|g\| \|f_{in}\| \Big) E_\alpha(\|a\| t^\alpha) + \frac{\|\widetilde{a}\|}{\Gamma(\alpha)} \int_0^t (t-\tau)^{\alpha-1} |\overline{u}_{in}(\tau)| d\tau.$$

Using Lemma 2.10 from last inequality, we arrive at the estimate:

$$|\overline{u}_{10}(t)| \le \Big\{ \overline{\varphi}_{10}| + \frac{t^\alpha \|\overline{f}_{10}\| \|\widetilde{g}\|}{\Gamma(\alpha+1)} + \frac{t^\alpha \| f_{10}\| \|\overline{g}\|}{\Gamma(\alpha+1)} +$$

$$+ \frac{\|\overline{a}\| t^\alpha}{\Gamma(\alpha+1)} \Big(|\varphi_{10}| + \frac{t^\alpha}{\Gamma(\alpha+1)} \|g\| \| f_{10}\| \Big) E_\alpha(\|a\| t^\alpha) \Big\} E_\alpha(\|\widetilde{a}\| t^\alpha). \tag{2.125}$$

$$|\overline{u}_{in}(t)| \le \Big\{ \overline{\varphi}_{in}| + \frac{t^\alpha \|\overline{f}_{in}\| \|\widetilde{g}\|}{\Gamma(\alpha+1)} + \frac{t^\alpha \| f_{in}\| \|\overline{g}\|}{\Gamma(\alpha+1)} +$$

$$+ \frac{\|\overline{a}\| t^\alpha}{\Gamma(\alpha+1)} \Big(|\varphi_{in}| + \frac{t^\alpha}{\Gamma(\alpha+1)} \|g\| \| f_{in}\| \Big) E_\alpha(\|a\| t^\alpha) \Big\} E_\alpha(\|\widetilde{a}\| t^\alpha). \tag{2.126}$$

In next subsection we study the inverse problem as the problem of determining of functions $a(t)$, $g(t)$ from relations (2.103)–(2.106), using Banach's contraction mapping principle.

2.2.2 Inverse Problem of Determining Two Unknown Coefficients

Let us multiply (2.103) by $\omega_i(x)$, $(i = 1, 2)$ and integrate over x from 0 to 1:

$$\int_0^1 \omega_i(x) \partial_t^\alpha u(x,t) dx - \int_0^1 \omega_i(x) u_{xx} dx + a(t) \int_0^1 \omega_i(x) u(x,t) dx =$$

$$= g(t) \int_0^1 \omega_i(x) f(x,t) dx, \;\; i = 1, 2, \;\; (x,t) \in D_T.$$

Note that, by Remark (2.2), all the integrals in this equality are well defined. After integrating by parts, in view of conditions (2.104)–(2.106) and $(A3)$, we obtain the equality

$$\partial_t^\alpha h_i(t) - \int_0^1 \omega_i''(x)u(x,t)dx + a(t)h_i(t) = g(t)\int_0^1 \omega_i(x)f(x,t)dx, \ i = 1, 2. \tag{2.127}$$

Solving the system (2.127) with respect to the unknown functions $a(t)$ and $g(t)$, we obtain the following integral equations:

$$g(t) = \frac{1}{\Delta(t)}\sum_{\substack{k,j=1\\ k\neq j}}^{2}(-1)^j h_j(t)$$
$$\times\left[\partial_t^\alpha h_k(t) - \sum_{n=0}^{\infty} u_{1n}(t;a;g)\lambda_n^2\omega_{k1n}^{(2)} - \sum_{n=1}^{\infty} u_{2n}(t;a;g)\lambda_n^2\omega_{k2n}^{(2)}\right], \tag{2.128}$$

$$a(t) = \frac{1}{\Delta(t)}\sum_{\substack{k,j=1\\ k\neq j}}^{2}(-1)^j F_j(t)$$
$$\times\left[\partial_t^\alpha h_k(t) - \sum_{n=0}^{\infty} u_{1n}(t;a;g)\lambda_n^2\omega_{k1n}^{(2)} - \sum_{n=1}^{\infty} u_{2n}(t;a;g)\lambda_n^2\omega_{k2n}^{(2)}\right], \tag{2.129}$$

where

$$\Delta(t) = h_1(t)F_2(t) - h_2(t)F_1(t);\ \ F_k(t) = \sum_{n=0}^{\infty} f_{1n}(t)\omega_{k1n} + \sum_{n=1}^{\infty} f_{2n}(t)\omega_{k2n};$$

$$\omega_{k1n} = 2\int_0^1 \omega_k(x)\cos(\lambda_n x)dx;\ \ \omega_{k1n}^{(2)} = 2\int_0^1 \omega_k''(x)\cos(\lambda_n x)dx.$$

$$\omega_{k2n} = 2\int_0^1 \omega_k(x)\sin(\lambda_n x)dx;\ \ \omega_{k2n}^{(2)} = 2\int_0^1 \omega_k''(x)\sin(\lambda_n x)dx, \ \ k = 1, 2.$$

The main result of this section is presented as follows:

Theorem 2.3 *Let* $(A1) - (A3)$ *be satisfied and* $|\Delta(t)| > m_0 > 0$, $t \in [0, T]$, m_0 *is a known number. Then, there exists* $T^* \in (0, T)$, *such that there exists a unique solution* $a(t)$, $g(t) \in C[0, T^*]$ *of the inverse problem* (2.103)–(2.106).

Proof We consider the operator equation

$$\psi = \Lambda[\psi], \tag{2.130}$$

where $\psi = (\psi_1, \psi_2) := (a(t); g(t))$ is unknown vector function. The components (Λ_1, Λ_2) of Λ are defined by the right-hand sides of equations (2.128), (2.129):

$$\Lambda_1[\psi](t) = \psi_{01}(t)$$

$$+\frac{1}{\Delta(t)}\sum_{\substack{k,j=1\\k\neq j}}^{2}(-1)^j F_j(t)\Big[\sum_{n=0}^{\infty}u_{1n}(t;a;g)\omega_{k1n}^{(2)}+\sum_{n=1}^{\infty}u_{2n}(t;a;g)\omega_{k2n}^{(2)}\Big], \quad (2.131)$$

$$\Lambda_2[\psi](t) = \psi_{02}(t)$$

$$+\frac{1}{\Delta(t)}\sum_{\substack{k,j=1\\k\neq j}}^{2}(-1)^j h_j(t)\Big[\sum_{n=0}^{\infty}u_{1n}(t;a;g)\omega_{k1n}^{(2)}+\sum_{n=1}^{\infty}u_{2n}(t;a;g)\omega_{k2n}^{(2)}\Big]. \quad (2.132)$$

Let $\psi_0 := (\psi_{01}, g_{02})$, where

$$\psi_{01}(t) = \frac{1}{\Delta(t)}\sum_{\substack{k,j=1\\k\neq j}}^{2}(-1)^j F_j(t)\partial_{0+,t}^{\alpha}h_k(t),$$

$$\psi_{02}(t) = \frac{1}{\Delta(t)}\sum_{\substack{k,j=1\\k\neq j}}^{2}(-1)^j h_j(t)\partial_{0+,t}^{\alpha}h_k(t).$$

Consider the functional space of vector functions $\psi \in C(D_T)$ with the norm given by the relation

$$\|\psi\| = \max\left\{\max_{t\in[0,T]}|\psi_1(t)|,\ \max_{t\in[0,T]}|\psi_2(t)|\right\}.$$

Fix a number $\rho > 0$ and consider the ball

$$\Phi^T(g_0,\rho) := \left\{\psi : \|\psi - \psi_0\|_{C[0,T]} \leq \rho\right\}.$$

Let us prove for an enough small $T > 0$ the operator Λ maps the ball $\Phi^T(\psi_0, \rho)$ into itself. For this purpose, using the estimates (2.116), (2.117) for u_{1n}, u_{2n} respectively, we find following estimates

$$\|\Lambda_1[\psi](t) - \psi_{01}(t)\|$$

$$\leq \frac{2F_0\omega_0}{m_0}\left|\sum_{n=0}^{\infty}u_{1n}(T;\psi_1;\psi_2)+\sum_{n=1}^{\infty}u_{2n}(T;\psi_1;\psi_2)\right| \leq \frac{2\omega_0 F_0}{m_0}\Big[\Big(|\varphi_{10}|$$

$$+\frac{T^\alpha}{\Gamma(\alpha+1)}\|\psi_1\|\|f_{10}\|\Big)+\sum_{i=1}^{2}\Big(|\varphi_{in}|+\frac{T^\alpha}{\Gamma(\alpha+1)}\|\psi_1\|\|f_{in}\|\Big)\Big]E_\alpha(\|\psi_2\|T^\alpha),$$

$$\|\Lambda_2[\psi](t)-\psi_{02}(t)\|$$

$$\leq\frac{2\omega_0h_0}{m_0}\left|\sum_{n=0}^{\infty}u_{1n}(T;\psi_1;\psi_2)+\sum_{n=1}^{\infty}u_{2n}(T;\psi_1;\psi_2)\right|\leq\frac{2\omega_0h_0}{m_0}\Big[\Big(|\varphi_{10}|$$

$$+\frac{T^\alpha}{\Gamma(\alpha+1)}\|\psi_1\|\|f_{10}\|\Big)+\sum_{i=1}^{2}\sum_{n=1}^{\infty}\Big(|\varphi_{in}|+\frac{T^\alpha}{\Gamma(\alpha+1)}\|\psi_1\|\|f_{in}\|\Big)\Big]E_\alpha(\|\psi_2\|T^\alpha),$$

where $F_0=\max\left\{\|F_1\|_{C[0,T]},\ \|F_2\|_{C[0,T]}\right\}$, $h_0=\max\left\{\|h_1\|_{C[0,T]},\ \|h_2\|_{C[0,T]}\right\}$, $\|f_{in}\|=\max_{t\in[0,T]}|f_{in}(t)|$, $i=0,1,2,\ldots$, $\omega_0=\max\left\{\|\omega_1\|_{C^2[0,1]},\ \|\omega_2\|_{C^2[0,1]}\right\}$.

According to Lemmas 2.10 and 2.11, the above series are convergent. Note that the functions occurring on the right-hand side in these inequalities are monotone increasing with T, and the fact that the function $\psi(t)$ belongs to the ball $\Phi^T(\psi_0,\rho)$ implies the inequality

$$\|\psi\|\leq\rho+\|\psi_0\|. \tag{2.133}$$

Therefore, we only strengthen the inequality if we replace $\|\psi\|$ in these inequalities with the relation $\rho+\|\psi_0\|$. Performing these replacements, we obtain the estimate

$$\|\Lambda_1[\psi](t)-\psi_{01}(t)\|\leq\frac{2F_0\omega_0}{m_0}\Big[(|\varphi_{10}|+\frac{T^\alpha}{\Gamma(\alpha+1)}(\rho+\|\psi_0\|)\|f_{10}\|)+$$

$$+\sum_{i=1}^{2}\sum_{n=1}^{\infty}\Big(|\varphi_{in}|+\frac{T^\alpha}{\Gamma(\alpha+1)}(\rho+\|\psi_0\|)\|f_{in}\|\Big)\Big]E_\alpha((\rho+\|\psi_0\|)T^\alpha),$$

$$\|\Lambda_2[\psi](t)-\psi_{02}(t)\|\leq\frac{2\omega_0h_0}{m_0}\Big[(|\varphi_{10}|+\frac{T^\alpha}{\Gamma(\alpha+1)}(\rho+\|\psi_0\|)\|f_{10}\|)+$$

$$+\sum_{i=1}^{2}\sum_{n=1}^{\infty}\Big(|\varphi_{in}|+\frac{T^\alpha}{\Gamma(\alpha+1)}(\rho+\|\psi_0\|)\|f_{in}\|\Big)\Big]E_\alpha((\rho+\|\psi_0\|)T^\alpha).$$

These relations together with (2.130), (2.131) and (2.132) imply the estimates

$$\|\Lambda[\psi](t)-\psi_0(t)\|=\max\{\|\Lambda_1[\psi](t)-\psi_{01}(t)\|,\ \|\Lambda_2[\psi](t)-\psi_{02}(t)\|\}\leq$$

$$\leq\max\left\{\frac{2F_0\omega_0}{m_0},\ \frac{2\omega_0h_0}{m_0}\right\}\Big[(|\varphi_{10}|+\frac{T^\alpha}{\Gamma(\alpha+1)}(\rho+\|\psi_0\|)\|f_{10}\|)+$$

$$+\sum_{i=1}^{2}\sum_{n=1}^{\infty}\left(|\varphi_{in}|+\frac{T^{\alpha}}{\Gamma(\alpha+1)}(\rho+\|\psi_0\|)\|f_{in}\|\right)\Big]E_{\alpha}((\rho+\|\psi_0\|)T^{\alpha}).$$

Suppose $\rho>\max\left\{\frac{2F_0\omega_0}{m_0},\ \frac{2\omega_0 h_0}{m_0}\right\}\left(|\varphi_{10}|+\sum_{n=1}^{\infty}|\varphi_{1n}|+\sum_{n=1}^{\infty}|\varphi_{2n}|\right)$.

Let T_1 be a positive root of the equation

$$z_1(T):=\max\left\{\frac{2F_0\omega_0}{m_0},\ \frac{2\omega_0 h_0}{m_0}\right\}\Big[(|\varphi_{10}|+\frac{T^{\alpha}}{\Gamma(\alpha+1)}(\rho+\|\psi_0\|)\|f_{10}\|)+$$

$$+\sum_{i=1}^{2}\sum_{n=1}^{\infty}\left(|\varphi_{in}|+\frac{T^{\alpha}}{\Gamma(\alpha+1)}(\rho+\|\psi_0\|)\|f_{in}\|\right)\Big]E_{\alpha}((\rho+\|\psi_0\|)T^{\alpha})=\rho.$$

Since $z_1(0)<\rho$ and $z_1(T)$ is the monotonically increasing function of $T>0$, then the equation $z_1(T)=\rho$ has a unique positive root.

Then, for $T\in[0,T_1]$, we have $\Lambda[\psi](t)\in\Phi^T(\psi_0,\rho)$.

Now, we consider two functions $\psi(t)$ and $\widetilde{\psi}(t)$ belonging $\Phi^T(\psi_0,\rho)$ and estimate the distance between their images $\Lambda[\psi](t)$ and $\Lambda[\widetilde{\psi}](t)$ in the space $C[0,T]$. The function $\widetilde{u}_n(t)$ corresponding to $\widetilde{\psi}(t)$ satisfies the integral equation (2.122), (2.123) with the functions $\varphi_n=\widetilde{\varphi}_n$ and $f_n=\widetilde{f}_n$. Composing the difference $\Lambda[\psi](t)-\Lambda[\widetilde{\psi}](t)$ involving the equations (2.114), (2.115), (2.122), (2.123) and then estimating its norm, we get

$$\|\Lambda_1[\psi](t)-\Lambda_1[\widetilde{\psi}](t)\|\le\frac{2F_0\omega_0}{m_0}\left(\sum_{n=0}^{\infty}\|u_{1n}(T;\psi_1;\psi_2)-\widetilde{u}_{1n}(T;\widetilde{\psi}_1;\widetilde{\psi}_2)\|\right.$$

$$\left.+\sum_{n=1}^{\infty}\|u_{2n}(T;\psi_1;\psi_2)-\widetilde{u}_{2n}(T;\widetilde{\psi}_1;\widetilde{\psi}_2)\|\right)\le$$

$$\le\frac{2F_0\omega_0}{m_0}\Big\{|\overline{\varphi}_{10}|+\frac{T^{\alpha}\|\overline{f}_{10}\|\|\widetilde{\psi}_2\|}{\Gamma(\alpha+1)}+\frac{T^{\alpha}\|f_{10}\|\|\overline{\psi}_2\|}{\Gamma(\alpha+1)}$$

$$+\frac{\|\overline{\psi}_2\|t^{\alpha}}{\Gamma(\alpha+1)}\left(|\varphi_{10}|+\frac{T^{\alpha}}{\Gamma(\alpha+1)}\|\psi_2\|\|f_{10}\|\right)E_{\alpha}(\|\psi_1\|T^{\alpha})$$

$$+\sum_{i=1}^{2}\sum_{n=1}^{\infty}\Big\{|\overline{\varphi}_{in}|+\frac{T^{\alpha}\|\overline{f}_{in}\|\|\widetilde{\psi}_2\|}{\Gamma(\alpha+1)}+\frac{T^{\alpha}\|f_{in}\|\|\overline{\psi}_2\|}{\Gamma(\alpha+1)}$$

$$+\frac{\|\overline{\psi}_1\|T^{\alpha}}{\Gamma(\alpha+1)}\left(|\varphi_{in}|+\frac{T^{\alpha}}{\Gamma(\alpha+1)}\|\psi_2\|\|f_{in}\|\right)E_{\alpha}(\|\psi_1\|T^{\alpha})\Big\}\Big\}E_{\alpha}(\|\widetilde{\psi}_1\|T^{\alpha});$$

$$\|\Lambda_2[\psi](t)-\Lambda_2[\widetilde{\psi}](t)\| \le \frac{2h_0\omega_0}{m_0}\Bigg(\sum_{n=0}^{\infty}\|u_{1n}(T;\psi_1;\psi_2)-\widetilde{u}_{1n}(T;\widetilde{\psi}_1;\widetilde{\psi}_2)\|$$

$$+\sum_{n=1}^{\infty}\Big\|u_{2n}(T;\psi_1;\psi_2)-\widetilde{u}_{2n}(T;\widetilde{\psi}_1;\widetilde{\psi}_2)\Big\|\Bigg)$$

$$\le \frac{2h_0\omega_0}{m_0}\Big\{|\overline{\varphi}_{10}|+\frac{T^\alpha\|\overline{f}_{10}\|\|\widetilde{\psi}_2\|}{\Gamma(\alpha+1)}+\frac{T^\alpha\|f_{10}\|\|\overline{\psi}_2\|}{\Gamma(\alpha+1)}$$

$$+\frac{\|\overline{\psi}_2\|T^\alpha}{\Gamma(\alpha+1)}\Big(|\varphi_{10}|+\frac{t^\alpha}{\Gamma(\alpha+1)}\|\psi_2\|\|f_{10}\|\Big)E_\alpha(\|\psi_1\|T^\alpha)$$

$$+\sum_{i=1}^{2}\sum_{n=1}^{\infty}\Big\{|\overline{\varphi}_{in}|+\frac{T^\alpha\|\overline{f}_{in}\|\|\widetilde{\psi}_2\|}{\Gamma(\alpha+1)}+\frac{T^\alpha\|f_{in}\|\|\overline{\psi}_2\|}{\Gamma(\alpha+1)}$$

$$+\frac{\|\overline{\psi}_1\|T^\alpha}{\Gamma(\alpha+1)}\Big(|\varphi_{in}|+\frac{T^\alpha}{\Gamma(\alpha+1)}\|\psi_2\|\|f_{in}\|\Big)E_\alpha(\|\psi_1\|T^\alpha)\Big\}\Big\}E_\alpha(\|\widetilde{\psi}_1\|T^\alpha). \tag{2.134}$$

Using inequality (2.116), (2.117) and the estimate (2.125) with $\|\overline{\psi}\|=\|\psi-\widetilde{\psi}\|=\max\{\|\psi_1-\widetilde{\psi}_1\|,\|\psi_2-\widetilde{\psi}_2\|\}$, $\varphi_n=\widetilde{\varphi}_n$ and $f_n=\widetilde{f}_n$, we continue the previous inequality in following form:

$$\|\Lambda_1[\psi](t)-\Lambda_1[\widetilde{\psi}](t)\| \le \frac{2F_0\omega_0}{m_0}\Big[\frac{T^\alpha\|f_{10}\|}{\Gamma(\alpha+1)}+$$

$$+\frac{T^\alpha}{\Gamma(\alpha+1)}\Big(|\varphi_{10}|+\frac{T^\alpha}{\Gamma(\alpha+1)}\|\psi_2\|\|f_{10}\|\Big)E_\alpha(\|\psi_1\|T^\alpha)+\sum_{i=1}^{2}\sum_{n=1}^{\infty}\Big\{\frac{T^\alpha\|f_{in}\|}{\Gamma(\alpha+1)}+$$

$$+\frac{T^\alpha}{\Gamma(\alpha+1)}\Big(|\varphi_{in}|+\frac{T^\alpha}{\Gamma(\alpha+1)}\|\psi_2\|\|f_{in}\|\Big)E_\alpha(\|\psi_1\|T^\alpha)\Big\}\Big]E_\alpha(\|\widetilde{\psi}_1\|T^\alpha)\|\psi-\widetilde{\psi}\|, \tag{2.135}$$

$$\|\Lambda_2[\psi](t)-\Lambda_2[\widetilde{\psi}](t)\| \le \frac{2h_0\omega_0}{m_0}\Big[\frac{T^\alpha\|f_{10}\|}{\Gamma(\alpha+1)}+$$

$$+\frac{T^\alpha}{\Gamma(\alpha+1)}\Big(|\varphi_{10}|+\frac{T^\alpha}{\Gamma(\alpha+1)}\|\psi_2\|\|f_{10}\|\Big)E_\alpha(\|\psi_1\|T^\alpha)+\sum_{i=1}^{2}\sum_{n=1}^{\infty}\Big\{\frac{T^\alpha\|f_{in}\|}{\Gamma(\alpha+1)}+$$

$$+\frac{T^\alpha}{\Gamma(\alpha+1)}\Big(|\varphi_{in}|+\frac{T^\alpha}{\Gamma(\alpha+1)}\|\psi_2\|\|f_{in}\|\Big)E_\alpha(\|\psi_1\|T^\alpha)\Big\}\Big]E_\alpha(\|\widetilde{\psi}_1\|T^\alpha)\|\psi-\widetilde{\psi}\|. \tag{2.136}$$

Because of $\psi(t)$ and $\widetilde{\psi}(t)$ belong to the ball $\Phi^T(g_0, \rho)$, then for these functions takes place the inequality (2.132). Note that the functions on the right-hand side in inequality (2.130) at the factor $\|\overline{\psi}\|$ is monotone increasing with $\|\psi\|$, $\|\widetilde{\psi}\|$, and T. Consequently, replacing $\|\psi\|$ and $\|\widetilde{\psi}\|$ in inequality (2.130) with $\rho + \|\psi_0\|$ will only strengthen the inequality. In this way, we obtain

$$\|\Lambda_1[\psi](t) - \Lambda_1[\widetilde{\psi}](t)\| \leq \frac{2F_0\omega_0}{m_0}\Big[\frac{T^\alpha\|f_{10}\|}{\Gamma(\alpha+1)} +$$

$$+\frac{T^\alpha}{\Gamma(\alpha+1)}\Big(|\varphi_{10}| + \frac{T^\alpha}{\Gamma(\alpha+1)}(\rho+\|\psi_0\|)\|f_{10}\|\Big)E_\alpha((\rho+\|\psi_0\|)T^\alpha)$$

$$+\sum_{i=1}^{2}\sum_{n=1}^{\infty}\Big\{\frac{T^\alpha\|f_{in}\|}{\Gamma(\alpha+1)} + \frac{T^\alpha}{\Gamma(\alpha+1)}\Big(|\varphi_{in}| + \frac{T^\alpha}{\Gamma(\alpha+1)}(\rho+\|\psi_0\|)\|f_{in}\|\Big)$$

$$\times E_\alpha((\rho+\|\psi_0\|)T^\alpha)\Big\}\Big]E_\alpha((\|\rho+\|\psi_0\|)T^\alpha)\|\overline{\psi}\|,$$

$$\|\Lambda_2[\psi](t) - \Lambda_2[\widetilde{\psi}](t)\| \leq \frac{2h_0\omega_0}{m_0}\Big[\frac{T^\alpha\|f_{10}\|}{\Gamma(\alpha+1)} +$$

$$+\frac{T^\alpha}{\Gamma(\alpha+1)}\Big(|\varphi_{10}| + \frac{T^\alpha}{\Gamma(\alpha+1)}(\rho+\|\psi_0\|)\|f_{10}\|\Big)E_\alpha((\rho+\|\psi_0\|)T^\alpha)$$

$$+\sum_{i=1}^{2}\sum_{n=1}^{\infty}\Big\{\frac{T^\alpha\|f_{in}\|}{\Gamma(\alpha+1)} + \frac{T^\alpha}{\Gamma(\alpha+1)}\Big(|\varphi_{in}| + \frac{T^\alpha}{\Gamma(\alpha+1)}(\rho+\|\psi_0\|)\|f_{in}\|\Big)$$

$$\times E_\alpha((\rho+\|\psi_0\|)T^\alpha)\Big\}\Big]E_\alpha((\|\rho+\|\psi_0\|)T^\alpha)\|\overline{\psi}\|.$$

Hence,

$$\|\Lambda[\psi](t) - \Lambda[\widetilde{\psi}](t)\| \leq \max\left\{\frac{2\omega_0F_0}{m_0}, \frac{2\omega_0h_0}{m_0}\right\}\Big[\frac{T^\alpha\|f_{10}\|}{\Gamma(\alpha+1)} +$$

$$+\frac{T^\alpha}{\Gamma(\alpha+1)}\Big(|\varphi_{10}| + \frac{T^\alpha}{\Gamma(\alpha+1)}(\rho+\|\psi 0\|)\|f_{10}\|\Big)E_\alpha((\rho+\|\psi_0\|)T^\alpha)$$

$$+\sum_{i=1}^{2}\sum_{n=1}^{\infty}\Big\{\frac{T^\alpha\|f_{in}\|}{\Gamma(\alpha+1)} + \frac{T^\alpha}{\Gamma(\alpha+1)}\Big(|\varphi_{in}| + \frac{T^\alpha}{\Gamma(\alpha+1)}(\rho+\|\psi_0\|)\|f_{in}\|\Big)$$

$$\times E_\alpha((\rho+\|\psi_0\|)T^\alpha)\Big\}\Big]E_\alpha((\|\rho+\|\psi_0\|)T^\alpha)\|\overline{\psi}\|.$$

Let T_2 be a positive root of the equation

$$z_2(T) := \max\left\{\frac{2\omega_0 F_0}{m_0},\ \frac{2\omega_0 h_0}{m_0}\right\}\Bigg[\frac{T^\alpha \|f_{10}\|}{\Gamma(\alpha+1)} +$$

$$+\frac{T^\alpha}{\Gamma(\alpha+1)}\Big(|\varphi_{10}| + \frac{T^\alpha}{\Gamma(\alpha+1)}(\rho + \|\psi_0\|)\|f_{10}\|\Big)E_\alpha((\rho + \|\psi_0\|)T^\alpha)$$

$$+\sum_{i=1}^{2}\sum_{n=1}^{\infty}\Bigg\{\frac{T^\alpha \|f_{in}\|}{\Gamma(\alpha+1)} + \frac{T^\alpha}{\Gamma(\alpha+1)}\Big(|\varphi_{in}| + \frac{T^\alpha}{\Gamma(\alpha+1)}(\rho + \|\psi_0\|)\|f_{in}\|\Big)$$

$$\times E_\alpha((\rho + \|\psi_0\|)T^\alpha)\Bigg\}\Bigg]E_\alpha((\|\rho + \|\psi_0\|)T^\alpha) = 1.$$

The above equation $z_2(T) = 1$ has a unique positive root, because $z_2(T)$ is the monotonically increasing function of $T > 0$, besides the point $(0, 0)$ belongs to the graph of this function.

Then, for $T \in (0,\ T_2)$, the operator Λ contracts the distance between the elements $\psi(t),\ \widetilde{\psi}(t) \in \Phi^T(\psi_0, \rho)$. Consequently, if we choose $T^* < \min(T_1,\ T_2)$, then the operator Λ is a contraction in the ball $\Phi^T(\psi_0, \rho)$. However, in accordance with the Banach theorem (see [24, pp. 87–97]), the operator Λ has a unique fixed point in the ball $\Phi^T(\psi_0, \rho)$; i.e., there exists a unique solution of Eq. (2.130). □

2.3 Conclusion

In this chapter, we studied an inverse problem concerning the determination of the coefficient in the lower order term of a fractional diffusion equation with nonlocal boundary and overdetermination conditions of integral type.

In Sect. 2.1, we examined a multidimensional fractional diffusion equation where the direct problem is formulated as an initial-boundary value problem with nonlocal initial conditions and homogeneous Dirichlet boundary conditions. To determine the unknown coefficient, we imposed an integral-type overdetermination condition on the solution of the direct problem. First, we established the unique existence of a weak solution for the direct problem, utilizing eigenfunction expansions in the proof. Next, by applying Green's function of an associated fractional differential equation and the Fourier method, we transformed the inverse problem into an equivalent form. Finally, using a fixed point argument in appropriate Sobolev spaces, we derived global existence and uniqueness theorems for the solution of the inverse problem.

In Sect. 2.2, we addressed a time-fractional diffusion equation with initial and periodic boundary conditions to determine two unknown coefficients: the reaction coefficient and a time-dependent factor in the source function. Two integral-type spatial conditions were introduced as overdetermination conditions. Using the Fourier method, we reduced the existence and uniqueness of the classical solution to a fixed point argument. Additionally, we identified sufficient conditions on the input data that ensure the existence and uniqueness of the solution within a sufficiently small range of variation of the unknown functions.

References

1. Kochubei, A., Luchko, Y.: Handbook of Fractional Calculus with Applications. Volume 2: Fractional Differential Equations, Walter de Gruyter GmbH, Berlin/Boston (2019)
2. Durdiev, D.K., Rahmonov, A.A.: Global solvability of inverse coefficient problem for one fractional diffusion equation with initial non-local and integral overdetermination conditions. Fractional Calculus and Applied Analysis (2025). https://doi.org/10.1007/s13540-024-00367-0
3. Wang, H., Wu, B.: On the well-posedness of determination of two coefficients in a fractional integrodifferential equation. Chin. Ann. Math. **35**(3), 447–468 (2014)
4. Adams, R.A.: Sobolev Spaces. Academic, New York (1975)
5. Pazy, A.: Semigroups of Linear Operators and Applications to Partial Differential Equations, vol. 44. Springer Science and Business Media, New York (2012)
6. Sakamoto, K., Yamamoto, M.: Initial value/boundary value problems for fractional diffusion-wave equations and applications to some inverse problems. J. Math. Anal. Appl. **382**, 426–447 (2011)
7. Brezis, H.: Functional Analysis, Sobolev Spaces and Partial Differential Equations. Springer (2010)
8. Kilbas, A.A., Srivastava, H.M., Trujillo, J.J.: Theory And Application of Fractional Differential Equations. North-Holland Mathematical Studies, Elsevier, Amsterdam (2006)
9. Djrbashian, M.M.: Integral transforms and representations of functions in the complex domain. (Russian). Nauka, Moscow (1996)
10. Ashurov, R., Fayziev, Y.: On the nonlocal problems in time for time-fractional subdiffusion equations. Fractal and Fractional **6**(1), 41 (2022)
11. Courant, R., Hilbert, D.: Methods of Mathematical Physics. Interscience, New York (1953)
12. Agarwal, R., Sharma, U.P., Agarwal, R.P.: Bicomplex Mittag-Leffler function and associated properties. J. Nonlinear Sci. Appl. **15**(1), 48–60 (2022)
13. Haddouchi, F., Guendouz, C., Benaicha, S.: Existence and multiplicity of positive solutions to a fourth-order multi-point boundary value problem. Mat. Vesn. **7**(1), 25–36 (2021)
14. Sharma, S.: Molecular Dynamics Simulation of Nanocomposites Using BIOVIA Materials Studio. Elsevier, Lammps and Gromacs (2019)
15. Baglan, I.: Determination of a coefficient in a quasilinear parabolic equation with periodic boundary condition. Inverse Prob. Sci. Eng. **23**(5), 1–17 (2014)
16. Nguyena, A.T., Nguyen, H.T., Dai, L.X., Nguyen, H.C.: On an inverse problem for a tempered fractional diffusion equation. Filomat **38**(19), 6809–6827 (2024)
17. Canca, F.: Inverse coefficient problem of the parabolic equation with periodic boundary and integral overdetermination conditions. Abstract Appl. Anal. V. 659804: 7 pages (2013)
18. Ivanchov, M.I.: Inverse problems for the heat-conduction equation with nonlocal boundary condition. Ukrain. Math. J. **45**(8), 1186–1192 (1993)
19. Kanca, F.: The inverse coefficient problem of the heat equation with periodic boundary and integral overdetermination conditions. J. In Equal. Appl. **2013**(108), 9 pages (2013)
20. Cannon, J.R., Lin, Y., Wang, S.: Determination of a control parameter in a parabolic partial differential equation. J. Austral. Math. Soc. Ser. B. **33**, 149–163 (1991)
21. Hazanee, A., Lesnic, D., Ismailov, M., Kerimov, N.: Inverse time-dependent source problems for the heat equation with nonlocal boundary conditions. Appl. Math. Comput. **346**, 800–815 (2019)

22. Huzyk, N.: Nonlocal inverse problem for a parabolic equation with degeneration. Ukr. Math. J. **65**(6), 847–863 (2013)
23. Budak, B.M., Samarskii, A.A., Tikhonov, A.N.: Collection of Problems on Mathematical Physics. Nauka, Moscow (in Russian) (1979)
24. Kolmogorov, A.N., Fomin, S.V.: Theory of Functions and Functional Analysis, Volume 1: Metric and Normed Spaces. Graylock Press, Rochester, N.Y (1957)

Chapter 3
Coefficient Determination Problems in Unbounded Domains

In this chapter, we study the inverse problem of determining the time-depending reaction-diffusion coefficient in the Cauchy problem for the time-fractional diffusion equation with the Caputo operator and the Hilfer operator (generalized Riemann-Liouville fractional derivative). An overdetermination condition is specified a single observation at the point $x = 0$ of the fractional diffusion process. Here also new formulas have been proposed for representing the solution of the first (Dirichlet), second (Neumann), and third (Robin) initial-boundary value problems for one-dimensional fractional diffusion equations with the time-fractional Caputo derivative in half-line.

In Sects. 3.1 and 3.2, the existence of a solution on the segment $[0, T^*]$, where T^* is a sufficiently small positive number, as well as the global uniqueness and the stability estimate of the solution to the inverse problems for fractional diffusion equations with Caputo and Hilfer derivatives, are proved. These results are established under the assumption that the direct problems are Cauchy problems. The theorems are constructive, i.e., in fact, the algorithm for finding T^* was given. In Sect. 3.3, the technique of the continuation method of the solution from the infinite axis was applied to derive exact solutions to the the first, second, and third initial-boundary problems for one-dimensional time-fractional diffusion equation with the Caputo fractional derivative of the order α $(0 < \alpha < 1)$. These formulas for solutions contain Green's functions corresponding to the Dirichlet, Neumann, and Robin boundary conditions. Green's functions of the problems are constructed in terms of the Fox $H-$ function, which is widely used in the theory of fractional calculus. It is shown that the obtained formulas coincide with the well-known formulas for solving the corresponding problems for $\alpha \to 1-$.

D. K. Durdiev, *Inverse Problems for Fractional Diffusion Equations*,
Industrial and Applied Mathematics, https://doi.org/10.1007/978-981-96-5338-6_3

3.1 Inverse Coefficient Problem for the Time-Fractional Diffusion Equation

The main idea of the method of Volterra operator equations used in this section as applied to inverse coefficient problems for fractional-order evolution equations is that for many fractional diffusion-wave equations integral representations of solutions are known in the form of integral equations of the Volterra type. Using these representations, as well as the overdetermination condition on the solution of the direct problem, one can obtain the Volterra operator equation with respect to the sought coefficients.

An expression for the fundamental solution of the Cauchy problem for time-fractional diffusion equation with the Laplace operator was found independently by Schneider and Wyss [1] and Kochubei [2]. In [3], using fundamental solutions, a solution to the Cauchy problem for the fractional diffusion-wave equation is constructed. In the work [4], an evolution equation with the regularized fractional derivative of an order $\alpha \in (0, 1)$ with respect to the time variable, and a uniformly elliptic operator with variable coefficients acting in the spatial variables is considered. Such equations describe diffusion on inhomogeneous fractals. A fundamental solution of the Cauchy problem is constructed and investigated.

In [5], the inverse problem of determining the time-depending reaction-diffusion coefficient in the Cauchy problem for the time-fractional diffusion equation by a single observation at the point $x = 0$ of the diffusion process. In the works [6–9], using research techniques of paper [5] the multidimensional inverse problems of identification of the coefficient at the lowest term of the fractional diffusion equation was investigated. Local existence theorems, global uniqueness, and conditional stability estimates for solutions to be obtained.

In this section, we investigate the local existence, global uniqueness, and the stability estimate for the inverse problem of determining time-dependent reaction coefficient in the time-fractional diffusive equation with the Caputo derivative by a single observation at the point $x = 0$ of the diffusion process. The results of this section were proven in [5].

3.1.1 Reduction of the Original Problem to an Equivalent Problem and Preliminaries

We consider the following one-dimensional anomalously diffusion equation:

$$\partial_t^\alpha u - u_{xx} + q(t)u(x,t) = f(x,t),\ \alpha \in (0,1),\ t > 0,\ x \in \mathbb{R}, \tag{3.1}$$

the solution of which satisfies the initial condition

$$u(x,0) = \varphi(x),\ x \in \mathbb{R}, \tag{3.2}$$

where $f(x,t)$, $\varphi(x)$ are given smooth functions.

Inverse problem is to find the function $q(t),\ t > 0$ in (3.1), if the solution of the Cauchy problem (3.1), (3.2) satisfies

$$u(0,t) = g(t),\ t > 0, \tag{3.3}$$

$g(t)$ is a given function.

Definition 3.1 A function $u(x,t)$ is called a classical solution to the Cauchy problem (3.1) and (3.2) if:

(1) $u(x,t)$ is twice continuously differentiable in x for each $t > 0$;
(2) for each $x \in \mathbb{R}$ $u(x,t)$ is continuous in t on $[0,T]$, and its fractional integral

$$\left(I_t^{1-\alpha}u\right)(x,t) = \frac{1}{\Gamma(1-\alpha)}\int_0^t \frac{u(x,\tau)d\tau}{(t-\tau)^{\alpha}} \tag{3.4}$$

is continuously differentiable in t for $t > 0$; $u(x,t)$ satisfies (3.1) and (3.2).

Let $u(x,t)$ be a classical solution to the Cauchy problem (3.1),(3.2) and $f,\ \varphi,\ g$ be enough smooth functions. We transform the inverse problem (3.1)–(3.3). Denote for this purpose the second derivative of $u(x,t)$ with respect to x, by $\vartheta(x,t)$, i.e., $\vartheta(x,t) := u_{xx}(x,t)$. Differentiating (3.1) and (3.2) twice in x, we get

$$\partial_t^{\alpha}\vartheta - \vartheta_{xx}(x,t) + q(t)\vartheta(x,t) = f_{xx}(x,t),\ t > 0,\ x \in \mathbb{R}, \tag{3.5}$$

$$\vartheta(x,0) = \varphi''(x),\ x \in \mathbb{R}. \tag{3.6}$$

To obtain an additional condition for the function $\vartheta(x,t)$, we note that the second term in (3.1) is $\vartheta(x,t)$. Setting $x = 0$ in (3.1) and using equality (3.3), we obtain

$$\vartheta(t,0) = q(t)g(t) + \partial_t^{\alpha}g(t) - f(0,t),\ t > 0. \tag{3.7}$$

When the matching condition $\varphi(0) = g(0)$ is fulfilled, it is easy to derive from (3.5)–(3.7) the equations (3.1)–(3.3) (see Sect. 1.1).

Given functions $q(t),\ f(x,t),\ \varphi(x)$ and a number $\alpha \in (0,1)$, the problem of determining the solution to the Cauchy problem (3.5) and (3.6) we call as *Direct problem.*

By $R_T := \{(x,t) : x \in \mathbb{R}, 0 < t < T\}$ we denote a strip with the thickness T, where $T > 0$ is any fixed number.

Let $C^{1-\alpha,m}(R_T)$ be the class of the m times continuously differentiable with respect to $x \in \mathbb{R}$ variable, continuous in t and its fractional integral (3.4) of the order $1-\alpha$ is continuously differentiable in t on $(0,T)$ functions.

By $C^m(\mathbb{R})$ we denote the class of the m times continuously differentiable with respect to $x \in \mathbb{R}$ variable functions, that are bounded in $\mathbb{R}$ together with all derivatives up to the order of m. When $m = 0$, $C^0(\mathbb{R}) =: C(\mathbb{R})$ and this is usual space of continuous and bounded functions.

We also introduce the class $H^l(\mathbb{R})$ of bounded Hölder continuous functions on $\mathbb{R}$ with exponent $l \in (0, 1)$. Define the space $H^{m+l}(\mathbb{R})$ (m is a nonnegative integer) as a subspace of the space of functions $y(x) \in C^m(\mathbb{R})$ such that $y^m(x) \in H^l(\mathbb{R})$.

The norm in $H^{m+l}(\mathbb{R})$ is defined as

$$|y|^{m+l} := \sum_{i=0}^{m} \sup_{x\in\mathbb{R}} \left|y^{(i)}(x)\right| + \sup_{(x_1, x_2)\in\mathbb{R},\, x_1\neq x_2} \frac{\left|y^{(m)}(x_1) - y^{(m)}(x_2)\right|}{|x_1 - x_2|^l}.$$

This, at $m = 0$, coincides with the norm in $H^l(\mathbb{R})$.

By $C\left([0, T], H^l(\mathbb{R})\right)$ we denote the class of continuous with respect to t variable on the segment $[0, T]$ with values in $H^l(\mathbb{R})$ functions. For a fixed t, the norm of the function $f(x, t)$ in the $H^l(\mathbb{R})$ will be denoted by $|f|^l(t)$. The norm of a function $f(x, t)$ in $C\left([0, T], H^l(\mathbb{R})\right)$ is defined by the equality

$$\|f\|^l := \max_{t\in[0, T]} |f|^l(t).$$

We present well-known definitions, assertions, and formulas that will be used for proof of the main results. They will mainly deal with fractional calculus. We will start by definition of the Fox $H-$function.

Definition 3.2 The Fox $H-$ function is a generalized hypergeometric function, defined by means of the Mellin-Barnes type contour integral [10]

$$H^{m,n}_{p,q}\left[z\left|\begin{matrix}(a_j, A_j)_1^p\\(b_j, B_j)_1^q\end{matrix}\right.\right] = \frac{1}{2\pi i}\int_\Omega \mathcal{H}^{m,n}_{p,q}(s) z^{-s} ds,$$

where

$$\mathcal{H}^{m,n}_{p,q}(s) = \frac{\prod\limits_{j=1}^{m}\Gamma(b_j + B_j s)\prod\limits_{i=1}^{n}\Gamma(1 - a_i - A_i s)}{\prod\limits_{j=m+1}^{q}\Gamma(1 - b_j - B_j s)\prod\limits_{i=n+1}^{p}\Gamma(a_i + A_i s)}$$

with complex variable $z \neq 0$ and a contour Ω in the complex domain; the orders (m, n, p, q) are nonnegative integers so that $0 \leq m \leq q$, $0 \leq n \leq p$, the parameters $A_i > 0$, $B_j > 0$ are positive and a_i, b_j, $i = 1, ..., p$; $j = 1, ..., q$ are arbitrary complex such that

$$A_i(b_j + l) \neq B_j(a_i - l' - 1), \quad l, l' = 0, 1, 2, ..., \; i = 1, ..., n, \; j = 1, ..., m. \tag{3.8}$$

Fox H-function has the following properties [10, pp. 11, 12]:

$$H^{m,n}_{p,q}\left[z\left|\begin{matrix}(a_j, A_j)_1^{p-1}, (b_1, B_1)\\(b_1, B_1), (b_j, B_j)_2^q\end{matrix}\right.\right] = H^{m-1,n}_{p-1,q-1}\left[z\left|\begin{matrix}(a_j, A_j)_1^{p-1}\\(b_j, B_j)_2^q\end{matrix}\right.\right], \tag{3.9}$$

$$H^{m,n}_{p,q}\left[z\left|\begin{matrix}(a_j,A_j)_1^p\\(b_j,B_j)_1^q\end{matrix}\right.\right]=H^{n,m}_{q,p}\left[\frac{1}{z}\left|\begin{matrix}(1-b_j,B_j)_1^q\\(1-a_j,A_j)_1^p\end{matrix}\right.\right]. \tag{3.10}$$

$$H^{m,n}_{p,q}\left[z\left|\begin{matrix}(a_j,A_j)_1^p\\(b_j,B_j)_1^q\end{matrix}\right.\right]=kH^{m,n}_{p,q}\left[z^k\left|\begin{matrix}(a_j,kA_j)_1^p\\(b_j,kB_j)_1^q\end{matrix}\right.\right] \tag{3.11}$$

for $k > 0$.

The details on the properties of the Fox $H-$function and types of contour Ω can be found in [10], where its behavior is described in terms of the following parameter:

$$\kappa := \sum_{j=1}^{q} B_j - \sum_{j=1}^{p} A_i.$$

Proposition 3.1 *Let $\kappa \geq 0$ and let the condition (3.8) be satisfied. Then the $H-$function has the asymptotic expansion at zero given by [10, p.19]*

$$H^{m,n}_{p,q}(z) = O(z^c), \quad |z| \to 0,$$

where

$$c := \min_{1\leq j\leq m}\left[\frac{\Re(b_j)}{B_j}\right],$$

$\Re(b_j)$ denotes the real part of the complex number b_j.

We also need the asymptotic behavior of $H-$ functions for $z \to \infty$. It will need only one result of this kind, for a specific class of H-functions, for a real argument, with only the leading term of the asymptotic expansion:

Proposition 3.2 *There holds the following result:*

$$H^{m,0}_{p,m}\left[z\left|\begin{matrix}(a_j,A_j)_1^p\\(b_j,B_j)_1^m\end{matrix}\right.\right] \sim Cz^{\frac{1-\varepsilon}{\rho}}\exp\left(-\chi^{\frac{1}{\rho}}\rho z^{\frac{1}{\rho}}\right),\ z\to\infty,$$

where $C = const$, $\varepsilon := \sum_{j=1}^{p} a_j - \sum_{j=1}^{m} b_j + (1/2)(m-p+1)$, $\chi := \prod_{j=1}^{p} A_j^{A_j}$ $\prod_{j=1}^{m} B_j^{-B_j}$, $\rho := \sum_{j=1}^{m} B_j - \sum_{j=1}^{p} A_j$.

Now we give formulas for differentiating the $H-$ functions of a special form. For the proofs and further details see [11, 12]. They have the following forms:

$$\frac{d}{dz}H^{2,0}_{1,2}\left[z\left|\begin{matrix}(a_1,A_1)\\(b_1,B_1),(b_2,B_2)\end{matrix}\right.\right]=-\frac{1}{z}H^{3,0}_{2,3}\left[z\left|\begin{matrix}(a_1,A_1),(0,1)\\(b_1,B_1),(b_2,B_2),(1,1),\end{matrix}\right.\right], \tag{3.12}$$

$$\frac{d}{dz}H_{2,3}^{3,0}\left[z\left|\begin{matrix}(a_1,A_1),(a_2,A_2)\\(b_1,B_1),(b_2,B_2),(b_3,B_3)\end{matrix}\right.\right]=-\frac{1}{z}H_{3,4}^{4,0}\left[z\left|\begin{matrix}(a_1,A_1),(a_2,A_2),(0,1)\\(b_1,B_1),(b_2,B_2),(b_3,B_3),(1,1)\end{matrix}\right.\right]. \quad (3.13)$$

In a similar way $\frac{d}{dz}H_{3,4}^{4,0}$ is expressed through $H_{4,5}^{5,0}$ and etc.

Here we also give the definition of the Yang $Y-$ function, which is an extended version of the Fox H function based on the Pincherle-Mellin-Barnes type integral (see [13, 14]).

Definition 3.3 The Yang $Y-$ function is called a generalized hypergeometric function, defined by means of the following contour integral [13]:

$$Y_{p,q}^{m,n}\left[z;\,y;\,x\left|\begin{matrix}(a_j,A_j)_1^p\\(b_j,B_j)_1^q\end{matrix}\right.\right]=\frac{1}{2\pi i}\int_\Omega \mathcal{Y}_{p,q}^{m,n}(s)z^{-s}e^{sy}s^x ds,$$

where

$$\mathcal{Y}_{p,q}^{m,n}(s)=\mathcal{H}_{p,q}^{m,n}(-s)=\frac{\prod\limits_{j=1}^{m}\Gamma(b_j-B_js)\prod\limits_{i=1}^{n}\Gamma(1-a_i+A_is)}{\prod\limits_{j=m+1}^{q}\Gamma(1-b_j+B_js)\prod\limits_{i=n+1}^{p}\Gamma(a_i-A_is)}$$

with complex variables $x,\ y,\ z$ and a contour Ω in the complex plan; the orders (m,n,p,q), as in the case of Definition 3.2, are nonnegative integers so that $0\le m\le q,\ 0\le n\le p$, the parameters $A_i>0,\ B_j>0$ are positive and $a_i,\ b_j,\ i=1,\ldots,p;\ j=1,\ldots,q$ are arbitrary complex satisfying conditions (3.8).

Taking $y=0$ and $x=0$, the relationship between the Y and H functions reads

$$Y_{p,q}^{m,n}\left[z;\,0;\,0\left|\begin{matrix}(a_j,A_j)_1^p\\(b_j,B_j)_1^q\end{matrix}\right.\right]=H_{p,q}^{m,n}\left[\frac{1}{z}\left|\begin{matrix}(a_j,A_j)_1^p\\(b_j,B_j)_1^q\end{matrix}\right.\right]. \quad (3.14)$$

Recall that the function $E_{\alpha,\beta}(z)$ (see (1.26)) can be rewritten in terms of the Fox H-function as [10, p. 24]:

$$E_{\alpha,\beta}(z)=H_{1,2}^{1,1}\left[-z\left|\begin{matrix}(0,1)\\(0,1),(1-\beta,\alpha)\end{matrix}\right.\right]. \quad (3.15)$$

Proposition 3.3 *For arbitrary $\alpha>0$, complex number β, $\mu>0$ and $a\in\mathbb{R}$, the following formula is valid [15]:*

$$\int\limits_{\mathbb{R}^n} e^{i\xi\cdot x}E_{\alpha,\beta}^{(m)}(-a|\xi|^\mu)d\xi$$

$$=(2\pi)^{n/2}|x|^{1-n/2}\int\limits_0^\infty |\xi|^{n/2}E_{\alpha,\beta}^{(m)}(-a|\xi|^\mu)J_{\frac{n}{2}-1}(|x||\xi|)d|\xi|.$$

Here $J_{\frac{n}{2}-1}(\cdot)$ is a Bessel function and $E^{(m)}_{\alpha,\beta}(z)$ denotes m-th derivatives of the Mittag-Leffler function. m-th derivatives of the Mittag-Leffler function can be expressed in terms of the Fox H-function as

$$E^{(m)}_{\alpha,\beta}(z) = H^{1,1}_{1,2}\left[-z \left|{}^{(-m,1)}_{(0,1),(1-(\alpha m+\beta),\alpha)}\right.\right].$$

Now, we present definitions of the Fourier [16, p. 12], Mellin [17, p. 20] and Hankel transforms (see, for example, [10, p. 56], [16, p. 12]).

Definition 3.4 The $n-$dimensional Fourier transform of a function $f(x,\cdot)$ for $x \in \mathbb{R}^n$ is defined by the formula

$$F[f(x,\cdot)](\xi) = \tilde{f}(\xi,\cdot) := \int\limits_{\mathbb{R}^n} f(x,\cdot)e^{-i\xi x}dx, \quad \xi \in \mathbb{R}^n,$$

where

$$x \in \mathbb{R}^n, \qquad \xi \in \mathbb{R}^n, \qquad \xi x = \sum_{j=1}^{n} \xi_i x_i, \qquad dx = dx_1 dx_2 \cdots dx_n,$$

while the corresponding inverse Fourier transform is given by

$$F^{-1}[f(\xi,\cdot)](x) = \frac{1}{2\pi}\tilde{f}(-x,\cdot) := \frac{1}{(2\pi)^n}\int\limits_{\mathbb{R}^n} f(\xi,\cdot)e^{i\xi x}d\xi, \quad x \in \mathbb{R}^n.$$

Definition 3.5 The Mellin transform of a function $h(t)$ for $t > 0$ is defined by the formula

$$(\mathcal{M}h)(s) = h^*(s) := \int\limits_0^\infty t^{s-1}h(t)dt, \quad s \in \mathbb{C}$$

and the inverse Mellin transform is given for $x > 0$ by the formula

$$\left(\mathcal{M}^{-1}g\right)(x) = \frac{1}{2\pi i}\int\limits_{\sigma-i\infty}^{\sigma+i\infty} x^{-s}g(s)dt, \quad \sigma = \Re(s).$$

Definition 3.6 The Hankel transform of an one-dimensional function $f(x)$ is defined as follows

$$g(\lambda) = \int\limits_0^\infty (\lambda x)^{\frac{1}{2}} J_\nu(\lambda x) f(x)dx, \qquad \lambda > 0.$$

The inverse Hankel transform is given by

$$f(x) = \int_0^\infty (x\lambda)^{\frac{1}{2}} J_\nu(x\lambda) g(\lambda) d\lambda, \qquad \Re(\nu) > 0,$$

where $J_\nu(z)$ is a Bessel function of the first kind and defined by

$$J_\nu(z) = \sum_{n=0}^{\infty} \frac{(-1)^n}{\Gamma(1+\nu+n)n!} \left(\frac{z}{2}\right)^{\nu+2n}.$$

The Hankel transform the H-function is defined in the following assertion [10, p. 57]:

Proposition 3.4 *Suppose that $\alpha > 0$ or $\alpha = \mu = 0$ and $\Re(\delta) < -1$. If ρ, ν, $b \in \mathbb{C}$, $\sigma > 0$ satisfy the conditions*

$$\Re(\rho) + \Re(\nu) + \sigma \min_{1 \le j \le m} \left[\frac{\Re(b_j)}{B_j}\right] > -1,$$

and

$$\Re(\rho) + \sigma \max_{1 \le j \le n} \left[\frac{1}{A_j} - \frac{\Re(a_j)}{A_j}\right] < \frac{3}{2},$$

for $\Re(\nu) > -\frac{1}{2}$, then for $a, b > 0$ there holds the formula

$$\int_0^\infty x^{\rho-1} J_\nu(ax) H_{p,q}^{m,n}\left[bx^\sigma \left|\begin{matrix}(a_p, A_p)\\(b_q, B_q)\end{matrix}\right.\right] dx$$

$$= \frac{2^{\rho-1}}{a^\rho} H_{p+2,q}^{m,n+1}\left[b\left(\frac{2}{a}\right)^\sigma \left|\begin{matrix}(1-\frac{\rho+\nu}{2}, \frac{\sigma}{2}), (a_p, A_p), (1-\frac{\rho-\nu}{2}, \frac{\sigma}{2})\\(b_q, B_q)\end{matrix}\right.\right]. \tag{3.16}$$

Definition 3.7 The Mellin convolution operator of two functions $h(t)$ and $g(t)$, given for $t > 0$, is defined for $x > 0$ by the integral [17, p. 22]

$$(h * g)(y) := \int_0^\infty h\left(\frac{y}{t}\right) g(t) \frac{dt}{t},$$

which has the commutative property $h * g = g * h$.

The convolution theorem for Mellin transform can be written:

$$(\mathcal{M}(h * g))(s) = (\mathcal{M}h)(s)(\mathcal{M}g)(s), \tag{3.17}$$

which holds for suitable functions h and g.

When deriving an estimate for the conditional stability of the solution of the inverse problem, we need the following statement as a generalized analog of Gronwall's inequality [18, p. 189]:

Proposition 3.5 *Suppose* $b \geq 0,\ \alpha > 0,\ \gamma > 0,\ \alpha + \gamma > 1$ *and* $a(t)$ *nonnegative function locally integrable on* $0 \leq t < T$ *and suppose* $t^{\gamma-1}u(t)$ *is nonnegative and locally integrable on* $0 \leq t < T$ *with*

$$u(t) \leq a(t) + b \int_0^t (t-s)^{\alpha-1} s^{\gamma-1} u(s) ds,$$

then there holds estimate

$$u(t) \leq a(t) W_{\alpha,\gamma}\left((b\Gamma(\alpha))^{\frac{1}{\alpha+\gamma-1}} t \right),$$

where $W_{\alpha,\gamma}(t) = \sum\limits_{m=0}^{\infty} c_m t^{m(\alpha+\gamma-1)},\ c_0 = 1,\ \frac{c_{m+1}}{c_m} = \frac{\Gamma(m(\alpha+\gamma-1)+\gamma)}{\Gamma(m(\alpha+\gamma-1)+\alpha+\gamma)}$ *for* $m \geq 0$. *As* $t \to +\infty$ $W_{\alpha,\gamma}(t) = O\left(t^{\frac{1}{2}(\frac{\alpha+\gamma-1}{\alpha}-\gamma)} \exp\left(\frac{\alpha}{\alpha+\gamma-1} t^{\frac{\alpha+\gamma-1}{\alpha}} \right) \right)$.

3.1.2 About Solvability of the Cauchy Problem

In [4], it was found the representation for the solution in terms of the fundamental solution to the following Cauchy problem

$$\partial_t^\alpha w - Bw(x,t) = F(x,t),\ t > 0,\ x \in \mathbb{R}^n,$$

$$w(x,0) = w_0(x),\ x \in \mathbb{R}^n,$$

where

$$B := \sum_{i,j=1}^n a_{ij}(x) \frac{\partial^2}{\partial x_i \partial x_j} + \sum_{j=1}^n b_j(x) \frac{\partial}{\partial x_j} + c(x)$$

is a uniformly second-order elliptic differential operator with bounded continuous real-valued coefficients. In the case $B \equiv \Delta$, where Δ is $n-$dimensional Laplacian, for any bounded continuous function $w_0(x)$ (locally Hölder continuous, if $n > 1$) and any bounded continuous with respect to the both variables $x,\ t$ and locally Hölder continuous in x function $F(x,t)$, it has the form

$$w(x,t) = \int_{\mathbb{R}^n} Z(t, x-\xi) w_0(\xi) d\xi + \int_0^t \int_{\mathbb{R}^n} Y(t-\tau, x-\xi) F(\xi,\tau) d\xi d\tau, \tag{3.18}$$

where

$$Z(t, x-\xi) = \pi^{-n/2}\,|x-\xi|^{-n}\,H_{1,2}^{2,0}\left[\frac{1}{4}t^{-\alpha}|x-\xi|^2\Big|_{(n/2,1/2),(1,1)}^{(1,\alpha)}\right],$$

$$Y(t-\tau, x-\xi)$$
$$= \pi^{-n/2}\,|x-\xi|^{-n}(t-\tau)^{\alpha-1}H_{1,2}^{2,0}\left[\frac{1}{4}(t-\tau)^{-\alpha}|x-\xi|^2\Big|_{(n/2,1),(1,1)}^{(\alpha,\alpha)}\right],$$

H is Fox's $H-$function (see Definition 3.2). Actually, $Y(t,x)$ is the Riemann-Liouville derivative of $Z(x,t)$ with respect to t of the order $1-\alpha$ (for $x \neq 0$ $Z(t,x) \to 0$ as $t \to 0$, so that the Riemann-Liouville derivative coincides in this case with the Caputo derivative, i.e., $Y(t,x) = \partial_t^{1-\alpha} Z(t,x)$) [4].

In (3.5), introducing the notation $f_{xx}(x,t) - q(t)\vartheta(x,t) =: F(x,t)$ and applying the formula (3.18) to direct problem (3.5), (3.6) for $n=1$, we obtain the integral equation for determining $\vartheta(x,t)$:

$$\vartheta(x,t) = \vartheta_0(x,t) - \int_0^t q(\tau)\int_{\mathbb{R}} Y(t-\tau, x-\xi)\vartheta(\tau,\xi)d\tau d\xi, \tag{3.19}$$

where

$$\vartheta_0(x,t) := \int_{\mathbb{R}} Z(t,x-\xi)\varphi''(\xi)d\xi + \int_0^t\int_{\mathbb{R}} Y(t-\tau,x-\xi)f_{\xi\xi}(\tau,\xi)d\tau d\xi. \tag{3.20}$$

It is true the following assertion:

Lemma 3.1 *If* $q(t) \in C[0,T]$, $f(x,t) \in C\left([0,T],\ H^{2+l}(\mathbb{R})\right)$, $l \in (0,1)$, $\varphi(x) \in C^2(\mathbb{R})$, *then there exists a unique solution* $\vartheta(x,t) \in C^{1-\alpha,2}(R_T)$ *of the integral equation (3.19).*

Proof For proof of Lemma 3.1 we use the method of successive approximations and consider the sequence of functions defined recursively by the formulas:

$$\vartheta_n(x,t) = -\int_0^t q(\tau)\int_{\mathbb{R}} Y(t-\tau,x-\xi)\vartheta_{n-1}(\tau,\xi)d\tau d\xi, n = 1,2,\cdots, \tag{3.21}$$

where $\vartheta_0(x,t)$ is determined by the equality (3.20).

Further we use the following estimates [4]:

$$\left|\frac{\partial^m}{\partial x^m}Z(x,t)\right| \le Ct^{-\frac{\alpha(1+m)}{2}}\exp\left\{-\mu_m t^{-\frac{\alpha}{2-\alpha}}x^{\frac{2}{2-\alpha}}\right\},\ m = 0,1,2,3, \tag{3.22}$$

$$\left|\frac{\partial^m}{\partial x^m}Y(x,t)\right| \le Ct^{-\frac{\alpha(1+m)}{2}-1+\alpha}\exp\left\{-\mu_m t^{-\frac{\alpha}{2-\alpha}}x^{\frac{2}{2-\alpha}}\right\},\ m = 0,1,2,3, \tag{3.23}$$

for $x^2 \geq t^\alpha$; where $\mu_0 := (2-\alpha)\alpha^{\alpha/(2-\alpha)} \cdot 2^{-2/(2-\alpha)}$, and as μ_m can be taken any positive number less than μ_0;

$$\left|\frac{\partial^m}{\partial x^m} Z(x,t)\right| \leq C t^{-\frac{\alpha(1+m)}{2}}, \; m = 0, 1, 2, 3, \tag{3.24}$$

$$\left|\frac{\partial^m}{\partial x^m} Y(x,t)\right| \leq C t^{-\frac{\alpha(m-1)}{2}-1}, \; m = 0, 1, 2, 3, \tag{3.25}$$

for $x^2 \leq t^\alpha$, $(x,t) \in R_T$. In (3.22)–(3.25) the letter C denotes various positive constants.

We also note that it follows from the construction of the function $Z(x,t)$:

$$\int_{\mathbb{R}} Z(x,t)dx = 1 \tag{3.26}$$

and it is true the equality [4]

$$\int_{\mathbb{R}} Y(x,t)dx = C_0 t^{\alpha-1}, \; t \in [0,T], \tag{3.27}$$

where C_0 depends only on α.

Set $\|q\| = \max_{t\in[0,T]} |q(t)|$, $\varphi_0 := \sup_{x\in\mathbb{R}} |\varphi''(x)|$ and $f_0 := \|f\|^{2+l}$. Using (3.21), (3.26) and (3.27) we estimate the modulus of $\vartheta_n(x,t)$ in the domain R_T as

$$|\vartheta_0(x,t)| \leq \varphi_0 + C_0 f_0 T^\alpha \frac{1}{\alpha} =: \varphi_{00},$$

$$|\vartheta_1(x,t)| \leq C_0\|q\|\varphi_{00} \int_0^t (t-\tau)^{\alpha-1} d\tau = C_0\|q\|\varphi_{00}\frac{t^\alpha}{\alpha} = C_0\|q\|\varphi_{00}\frac{\Gamma(\alpha)}{\Gamma(1+\alpha)} t^\alpha,$$

$$\begin{aligned}|\vartheta_2(x,t)| &\leq \varphi_{00}\left(C_0\|q\|\Gamma(\alpha)\right)^2 \frac{1}{\Gamma(1+\alpha)}\frac{1}{\Gamma(\alpha)} \int_0^t \frac{\tau^\alpha d\tau}{(t-\tau)^{1-\alpha}} \\ &= \varphi_{00}\left(C_0\|q\|\Gamma(\alpha)\right)^2 \frac{1}{\Gamma(1+\alpha)} I_t^\alpha t^\alpha,\end{aligned}$$

where $I_t^\alpha t^\alpha$ is the Riemann-Liouville fractional integral of the power function t^α. It can be verified by direct calculation that the formula

$$I_t^\alpha t^{n\alpha} = \frac{\Gamma(1+n\alpha)}{\Gamma(1+(n+1)\alpha)} t^{(n+1)\alpha}, \; n = 0, 1, 2, ...$$

is valid. In accordance with this formula we continue to estimate $\vartheta_2(x,t)$:

$$|\vartheta_2(x,t)| \leq \varphi_{00}\left(C_0\|q\|\Gamma(\alpha)\right)^2 \frac{1}{\Gamma(1+\alpha)} I_t^{\alpha} t^{\alpha} = \varphi_{00}\left(C_0\|q\|\Gamma(\alpha)\right)^2 \frac{1}{\Gamma(1+2\alpha)} t^{2\alpha}.$$

For arbitrary $n = 0, 1, 2, \ldots,$ we have

$$|\vartheta_n(x,t)| \leq \varphi_{00}\left(C_0\|q\|\Gamma(\alpha)\right)^n \frac{1}{\Gamma(1+n\alpha)} t^{n\alpha}.$$

It follows from the above estimates that the series

$$\vartheta(x,t) = \sum_{n=1}^{\infty} \vartheta_n(x,t)$$

converges uniformly in R_T, since it can be majorized in R_T by the convergent numerical series

$$\varphi_{00} \sum_{n=1}^{\infty} \frac{(C_0\|q\|\Gamma(\alpha)T^{\alpha})^n}{\Gamma(1+n\alpha)}.$$

This means the following estimate for the solution of the integral equation (3.19) takes place:

$$|\vartheta(x,t)| \leq \varphi_{00} \sum_{n=1}^{\infty} \frac{(C_0\|q\|\Gamma(\alpha)T^{\alpha})^n}{\Gamma(1+n\alpha)} = \varphi_{00} E_{\alpha}\left(C_0\|q\|\Gamma(\alpha)T^{\alpha}\right),\ (x,t) \in R_T, \tag{3.28}$$

where $E_{\alpha}(\cdot)$ is the Mittag-Leffler function of a nonnegative real argument (1.27).

Under the assumptions of Lemma and on the bases of estimates (3.22)–(3.25) (see also [4]) one has the inclusion $\vartheta_n(x,t) \in C^{1-\alpha,2}(R_T)$. According to the general theory of integral equations, this implies that the same property will be possessed by the function $\vartheta(x,t)$ in R_T.

Now, we derive a stability estimate of the solution to the direct problem (3.5), (3.5). Let $\tilde{\vartheta}(x,t)$ be a solution of the integral equation (3.19) corresponding to the perturbed functions $\tilde{q}$, $\tilde{f}_{xx}$ and $\tilde{\varphi}''$; i.e.,

$$\tilde{\vartheta}(x,t) = \tilde{\vartheta}_0(x,t) - \int_0^t \tilde{q}(\tau) \int_{\mathbb{R}} Y(t-\tau, x-\xi)\tilde{\vartheta}(\tau,\xi)d\tau d\xi, \tag{3.29}$$

where in accordance with (3.10)

$$\tilde{\vartheta}_0(x,t) := \int_{\mathbb{R}} Z(t, x-\xi)\tilde{\varphi}''(\xi)d\xi + \int_0^t \int_{\mathbb{R}} Y(t-\tau, x-\xi)\tilde{f}_{\xi\xi}(\tau,\xi)d\tau d\xi. \tag{3.30}$$

We compose the difference $\vartheta - \tilde{\vartheta}$ with the help of the equations (3.19), (3.29) and introduce the notations $\vartheta - \tilde{\vartheta} = \overline{\vartheta}$, $\vartheta_0 - \tilde{\vartheta}_0 = \overline{\vartheta}_0$, $q - \tilde{q} = \overline{q}$. Then, we obtain the integral equation

$$\overline{\vartheta}(x,t)=\overline{\vartheta}_0(x,t)-\int_0^t \overline{q}(\tau)\int_{\mathbb{R}} Y(t-\tau,x-\xi)\vartheta(\tau,\xi)d\tau d\xi$$

$$-\int_0^t \tilde{q}(\tau)\int_{\mathbb{R}} Y(t-\tau,x-\xi)\overline{\vartheta}(\tau,\xi)d\tau d\xi, \tag{3.31}$$

from which, is derived the following linear integral inequality for $\left|\overline{\vartheta}(x,t)\right|$:

$$\left|\overline{\vartheta}(x,t)\right| \leq \left|\overline{\vartheta}_0(x,t)\right| + \varphi_{00}C_0\frac{T^\alpha}{\alpha}E_\alpha\left(C_0\|q\|\Gamma(\alpha)T^\alpha\right)\|\overline{q}\|_{C[0,T]}$$

$$+\tilde{q}_0\int_0^t\int_{\mathbb{R}} Y(t-\tau,x-\xi)\left|\overline{\vartheta}(\tau,\xi)\right| d\tau d\xi, \tag{3.32}$$

where $\tilde{q}_0 := \|\tilde{q}(t)\|_{C[0,T]}$. It follows from the equalities (3.20) and (3.30) the estimate

$$\left|\overline{\vartheta}_0(x,t)\right| \leq \|\overline{\varphi}''\|_{C(\mathbb{R})} + C_0\frac{T^\alpha}{\alpha}\left\|\overline{f}_{xx}\right\|^l,$$

where $\overline{\varphi}'' = \varphi'' - \tilde{\varphi}''$, $\overline{f}_{xx} = f_{xx} - \tilde{f}_{xx}$.

Let $\lambda = \lambda(\alpha, T, \|q\|, \varphi_0, f_0) = \max\left\{1, C_0\frac{T^\alpha}{\alpha}, \varphi_{00}C_0\frac{T^\alpha}{\alpha}E_\alpha\left(C_0\|q\|\Gamma(\alpha)T^\alpha\right)\right\}$. Applying the successive approximation method to inequality (3.32) with the help of the scheme

$$\left|\overline{\vartheta}(x,t)\right|_0 \leq \lambda\left(\|\overline{\varphi}''\|_{C(\mathbb{R})} + \left\|\overline{f}_{xx}\right\|^l + \|\overline{q}\|_{C[0,T]}\right),$$

$$\left|\overline{\vartheta}(x,t)\right|_n \leq \tilde{q}_0\int_0^t\int_{\mathbb{R}} Y(t-\tau,x-\xi)\left|\overline{\vartheta}(\tau,\xi)\right|_{n-1} d\tau d\xi,\ n=1,2,...,$$

we arrive at the estimate

$$\left|\overline{\vartheta}(x,t)\right| \leq \lambda\varphi_{00}E_\alpha\left(C_0\|q\|\Gamma(\alpha)T^\alpha\right)\left(\|\overline{\varphi}''\|_{C(\mathbb{R})} + \left\|\overline{f}_{xx}\right\|^l + \|\overline{q}\|_{C[0,T]}\right), \tag{3.33}$$

which will be used in the next section.

Indeed the expression (3.33) is the stability estimate for the solution to the Cauchy problem (3.5) and (3.6). The uniqueness for this solution follows from (3.33). □

3.1.3 Investigation of the Inverse Problem by the Fixed Point Argument

Setting in (3.19) $x = 0$ and using additional condition (3.7), after simple converting, we get the following integral equation for determining $q(t)$:

$$q(t) = q_0(t) - \frac{1}{g(t)} \int_0^t q(\tau) \int_{\mathbb{R}} Y(t-\tau, \xi)\vartheta(\tau, \xi) d\tau d\xi, \tag{3.34}$$

where

$$q_0(t) := \frac{1}{g(t)} \Big[f(0,t) - \partial_t^{\alpha} g(t)$$

$$+ \int_{\mathbb{R}} Z(t,\xi)\varphi''(\xi) d\xi + \int_0^t \int_{\mathbb{R}} Y(t-\tau,\xi) f_{\xi\xi}(\tau,\xi) d\tau d\xi \Big].$$

The solution of integral equation (3.19) depends on q, i.e., $\vartheta = \vartheta(x, t; q)$. We introduce an operator G defining it by the right-hand side of (3.34)

$$G[q](t) = q_0(t) - \frac{1}{g(t)} \int_0^t q(\tau) \int_{\mathbb{R}} Y(t-\tau, \xi)\vartheta(\tau, \xi; q) d\tau d\xi. \tag{3.35}$$

Then the equation (3.34) is written in a more convenient form as

$$q(t) = G[q](t). \tag{3.36}$$

Let $q_{00} := \max_{t \in [0,T]} |q_0(t)|$. Fix a number $\rho > 0$ and consider the ball

$$\Phi^T(q_0, \rho) := \left\{ q(t) : q(t) \in C[0, T], \ \|q - q_0\|_{C[0,T]} \le \rho \right\}.$$

Theorem 3.1 *If $f(x,t) \in C\left([0, T], H^l(\mathbb{R})\right)$, $\varphi(x) \in C^2(\mathbb{R})$, $g(t) \in C^1[0, T]$, $|g(t)| \ge g_0 = const > 0$, g_0 is a given number, $\varphi(0) = g(0)$, then there exists a number $T^* \in (0, T)$, such that there exists a unique solution $q(t) \in C[0, T^*]$ of the inverse problem (3.1)–(3.3).*

Let us first prove that for an enough small $T > 0$ the operator G maps the ball $\Phi^T(q_0, \rho)$ into itself; i.e., the condition $q(t) \in \Phi^T(q_0, \rho)$ implies that $G[q](t) \in \Phi^T(q_0, \rho)$. Indeed, for any continuous function $q(t)$, the function $G[q(t)]$ calculated using formula (3.36) will be continuous. Moreover, estimating the norm of the differences, we find that

$$\|G[q](t) - q_0(t)\| \le \frac{1}{\alpha g_0} C_0 \|q\| T^{\alpha} \varphi_{00} E_{\alpha}\left(C_0 \|q\| \Gamma(\alpha) T^{\alpha}\right).$$

Here we have used the estimate (3.28). Note that the function occurring on the right-hand side in this inequality is monotone increasing with T, and the fact that the function $q(t)$ belongs to the ball $\Phi^T(q_0, \rho)$ implies the inequality

$$\|q\| \le \rho + \|q_0\|. \tag{3.37}$$

Therefore, we only strengthen the inequality if we replace $\|q\|$ in this inequality with the expression $\rho + \|q_0\|$. Performing these replacements, we obtain the estimate

$$\|G[q](t) - q_0(t)\| \leq \frac{1}{\alpha g_0} C_0 (\rho + \|q_0\|) T^\alpha \varphi_{00} E_\alpha (C_0 (\rho + \|q_0\|) \Gamma(\alpha) T^\alpha).$$

Let T_1 be a root of the equation

$$m_1(T) = \frac{1}{\alpha g_0} C_0 (\rho + \|q_0\|) T^\alpha \varphi_{00} E_\alpha (C_0 (\rho + \|q_0\|) \Gamma(\alpha) T^\alpha) = \rho.$$

Then, for $T \in [0, T_1]$ we have $G[q](t) \in \Phi^T (q_0, \rho)$.

Now consider two functions $q(t)$ and $\tilde{q}(t)$ belonging to the ball $\Phi^T (q_0, \rho)$ and estimate the distance between their images $G[q](t)$ and $G[\tilde{q}](t)$ in the space $C[0, T]$. The function $\tilde{\vartheta}(x, t)$ corresponding to $\tilde{q}(t)$ satisfies the integral equation (3.29) with the functions $\varphi'' = \tilde{\varphi}''$ and $f_{xx} = \tilde{f}_{xx}$. Composing the difference $G[q](t) - G[\tilde{q}](t)$ with the help of equations (3.19), (3.29) and then estimating its norm, we obtain

$$\|G[q](t) - G[\tilde{q}](t)\| \leq \frac{C_0 T^\alpha}{\alpha g_0} \left(\|\overline{q}\| \, \|\vartheta\| + \|\tilde{q}\| \left\|\overline{\vartheta}\right\| \right).$$

Using inequality (3.28) and the estimate (3.33) with $\varphi'' = \tilde{\varphi}''$, $f_{xx} = \tilde{f}_{xx}$, we continue the previous inequality in the following form:

$$\|G[q](t) - G[\tilde{q}](t)\| \leq \frac{C_0 T^\alpha}{\alpha g_0} \varphi_{00} E_\alpha (C_0 \|q\| \Gamma(\alpha) T^\alpha) (1 + \lambda \|\tilde{q}\|) \|\overline{q}\|. \quad (3.38)$$

The functions $q(t)$ and $\tilde{q}(t)$ belong to the ball $\Phi^T (q_0, \rho)$, and hence for each of these functions one has inequality (3.37). Note that the function on the right-hand side in inequality (3.38) at the factor $\|\overline{q}\|$ is monotone increasing with $\|q\|$, $\|\tilde{q}\|$, and T. Consequently, replacing $\|q\|$ and $\|\tilde{q}\|$ in inequality (3.38) (including in λ) with $\rho + \|q\|$ will only strengthen the inequality. Thus, we have

$$\|G[q](t) - G[\tilde{q}](t)\|$$

$$\leq \frac{C_0 T^\alpha}{\alpha g_0} \varphi_{00} E_\alpha (C_0 \Gamma(\alpha) T^\alpha (\rho + \|q\|)) (1 + \lambda (\rho + \|q\|)) \|\overline{q}\| \leq m_2(T) \|\overline{q}\|.$$

Let T_2 be a root of the equation

$$m_2(T) = \frac{C_0 T^\alpha}{\alpha g_0} \varphi_{00} E_\alpha (C_0 \Gamma(\alpha) T^\alpha (\rho + \|q\|)) (1 + \lambda (\rho + \|q\|)) = 1.$$

Then, for $T \in (0, T_2)$ we have that the distance between the functions $G[q](t)$ and $G[\tilde{q}](t)$ in the function space $C[0, T]$ is not greater than the distance between the functions $q(t)$ and $\tilde{q}(t)$ multiplied by $m_2(T) < 1$. Consequently, if we choose $\Phi^{T*} < \min(T_1, T_2)$, then the operator G is a contraction in the ball $\Phi^{T*} (q_0, \rho)$. However, in accordance with the Banach theorem, the operator G has a unique fixed point in the ball $\Phi^{T*}(q_0, \rho)$; i.e., there exists a unique solution of the Eq. (3.36). Theorem 3.1 is proven.

Let T be a positive fixed number. Consider the set $\Omega(\omega_0)$ ($\omega_0 > 0$ is some fixed number) of the given functions (f, φ, g) for which all conditions from Theorem 3.1 are fulfilled and $\max\left\{\|f\|^{2+l}, \|\varphi\|_{C^2(\mathbb{R})}, \|g\|_{C^1[0,T]}\right\} \le \omega_0$. By $Q(\omega_1)$ we denote the class of functions $q(t) \in C[0, T]$ satisfying the inequality $\|q\| \le \omega_1$ with some fixed positive number ω_1.

Theorem 3.2 *Let* $(f, \varphi, g) \in \Omega(\omega_0)$, $\left(\tilde{f}, \tilde{\varphi}, \tilde{g}\right) \in \Omega(\omega_0)$ *and* $(q, \tilde{q}) \in Q(\omega_1)$. *Then for the solution of the inverse problem (3.1)–(3.3) the following stability estimate holds:*

$$\|q - \tilde{q}\|_{C[0,T]} \le d\left(\|f - \tilde{f}\|^{2+l} + \|\varphi - \tilde{\varphi}\|_{C^2(\mathbb{R})} + \|g - \tilde{g}\|_{C^1[0,T]}\right), \tag{3.39}$$

where the constant d *depends only on* T, α, g_0, ω_0, ω_1.

To **prove** this theorem, using (3.34) we write down the equations for $\tilde{q}(t)$ and compose the difference $q(t) - \tilde{q}(t)$. Then, after evaluating this expression and using estimates (3.28), (3.33), we obtain

$$|q(t) - \tilde{q}(t)| \le d_0\left(\|f - \tilde{f}\|^{2+l} + \|\varphi - \tilde{\varphi}\|_{C^2(\mathbb{R})} + \|g - \tilde{g}\|_{C^1[0,T]}\right)$$
$$+ d_1 \int_0^t |q(\tau) - \tilde{q}(\tau)|\, d\tau, \quad t \in [0, T], \tag{3.40}$$

where d_0 and d_1 depend on the same constants as d. From (3.40) using Gronwall's inequality, we get the estimate

$$|q(t) - \tilde{q}(t)|$$
$$\le d_0 \exp(d_1 t)\left(\|f - \tilde{f}\|^{2+l} + \|\varphi - \tilde{\varphi}\|_{C^2(\mathbb{R})} + \|g - \tilde{g}\|_{C^1[0,T]}\right), \quad t \in [0, T].$$

This inequality implies the estimate (3.39), if we set $d = d_0 \exp(d_1 T)$.

From Theorem 3.2 readily follows the uniqueness theorem for any $T > 0$

Theorem 3.3 *Let the functions* q, f, φ, g *and* $\tilde{q}$, $\tilde{f}$, $\tilde{\varphi}$, $\tilde{g}$ *have the same meaning as in Theorem 3.2. Moreover, if* $f = \tilde{f}$, $\varphi = \tilde{\varphi}$, $g = \tilde{g}$ *for* $t \in [0, T]$, *then* $q(t) = \tilde{q}(t)$, $t \in [0, T]$.

3.2 Inverse Coefficient Problem for the Time-Fractional Diffusion Equation with Hilfer Operator

In this section, the inverse problem of determining the time-dependent reaction-diffusion coefficient in the Cauchy problem for the time-fractional diffusion equation with the Hilfer operator by a single observation at the point $x = 0$ of the diffusion process is studied.

In the literature, many authors focused on Riemann-Liouville and Caputo type derivatives in investigating fractional differential equations. In [19], Hilfer has introduced a generalized form of the Riemann-Liouville fractional derivative, known in literature as the Hilfer fractional derivative of order α and a type $\beta \in [0, 1]$, which interpolates the both Riemann-Liouville as well as Caputo sense.

We underline that the diffusion equation with Riemann-Liouville fractional derivative coincides with classical diffusion equation from both sides when $\alpha \to 1^-$ and when $\alpha \to 1^+$, while in the case of the derivative of Caputo, there is no coincidence with the classical case except for the left-side limit when $\alpha \to 1^-$, (see [20]). The Hilfer fractional derivative takes advantage of both properties. In addition, an example of the physical interest of this generalized derivative is explained in [21]. The results of this section are proven in the paper [22].

3.2.1 Formulation of the Cauchy Problem and Preliminary Constructions

In this subsection, we consider the Cauchy problem for the following generalized time-fractional diffusion equation

$$\partial_t^{\alpha,\beta} u - u_{xx} + q(t)u(x,t) = f(x,t), \; t > 0, \; x \in \mathbb{R}, \tag{3.41}$$

with the initial integral condition

$$I_t^{1-\gamma} u(x,0) = \varphi(x), \;\; \gamma = \alpha + \beta(1-\alpha), \;\; x \in \mathbb{R}, \tag{3.42}$$

where the Hilfer (generalized Riemann-Liouville) fractional differential operator $\partial_t^{\alpha,\beta}$ of the order $0 < \alpha < 1$ and type $0 \le \beta \le 1$ is defined by [19, p. 113]:

$$\partial_t^{\alpha,\beta} u(\cdot, t) := \left(I_t^{\beta(1-\alpha)} \frac{\partial}{\partial t} \left(I_t^{(1-\beta)(1-\alpha)} u \right) \right) (\cdot, t),$$

$f(x,t)$, $\varphi(x)$ are given smooth functions.

For the given functions $q(t)$, $f(x,t)$, $\varphi(x)$ and numbers $\alpha \in (0, 1)$, $\beta \in [0, 1]$ the problem of finding the solution to the Cauchy problem (3.41) and (3.42) will be called as the *direct problem.* A *classical solution* of this problem consists of determining the function $u(x,t)$ satisfying equalities (3.41), (3.42), such that

$$t^{1-\gamma}u \in C\left(\overline{D}_T\right),\ t^{1-\gamma}\left(\partial_t^{\alpha,\beta}u\right) \in C\left(\overline{D}_T\right),$$

where $D_T := \{(x,t) : x \in \mathbb{R}, 0 < t \leq T\}$, $T > 0$ is any fixed number.

We pose the inverse problem as follows: find the function $q(t),\ t > 0$ in (3.41), if the solution of the Cauchy problem (3.41), (3.42) satisfies the following additional condition:

$$u(0,t) = g(t),\ t > 0, \tag{3.43}$$

$g(t)$ is a given function.

Let $u(x,t)$ be a regular solution to the Cauchy problem (3.41),(3.42) and $f,\ \varphi,\ g$ be enough smooth functions. As in the previous section, we carry out the next transformation of the inverse problem (3.41)–(3.43). For this purpose, we denote the second partial derivative of $u(x,t)$ with respect to x, by $\vartheta(x,t)$, i.e., $\vartheta(x,t) := u_{xx}(x,t)$. Differentiating (3.41) and (3.42) twice in x, we have

$$\partial_t^{\alpha,\beta}\vartheta - \vartheta_{xx}(x,t) + q(t)\vartheta(x,t) = f_{xx}(x,t),\ t > 0,\ x \in \mathbb{R}, \tag{3.44}$$

$$I_t^{1-\alpha}\vartheta(x,0) = \varphi''(x),\ x \in \mathbb{R}. \tag{3.45}$$

To obtain an additional condition for the function $\vartheta(x,t)$, we note that the second term on the left side of (3.41) is $\vartheta(x,t)$. Setting $x = 0$ in (3.41) and using equality (3.43), we get

$$\vartheta(0,t) = q(t)g(t) + \partial_t^{\alpha,\beta}g(t) - f(0,t),\ t > 0. \tag{3.46}$$

Under fulfilling the matching condition $\varphi(0) = g(0)$, it is easy to derive the equations (3.41)–(3.43) from (3.44)–(3.46) (see, for example, Sect. 1.1).

Let $C_\gamma[a,b]$ be a weighted space of functions g defined on $[0,T]$ such that $t^{1-\gamma}g(t) \in C[0,T]$:

$$C_\gamma[0,T] = \left\{g(t) : \|g\|_{C_\gamma[0,T]} := \left\|t^{1-\gamma}g(t)\right\|_C < \infty\right\}.$$

The weighted space $C_\gamma^n[0,T]$ of functions g is defined by

$$C_\gamma^n[a,b] = \left\{g(t) : [0,T] \to \mathbb{R},\ g^{(n-1)}(t) \in C[a,b],\ g^{(n)}(t) \in C_\gamma[a,b]\right\}$$

with the norm

$$\|g\|_{C_\gamma^n[0,T]} = \sum_{k=0}^{n-1}\|g^{(k)}\|_{C[0,T]} + \left\|g^{(n)}\right\|_{C_\gamma[0,T]}.$$

In the following we will retain the definition of Sect. 3.1.1 for the class of functions $C^{1-\gamma,m}(D_T)$, $C^m(\mathbb{R})$, $H^l(\mathbb{R})$, $l \in (0,1)$, $H^{m+l}(\mathbb{R})$ (m is a nonnegative integer), $C([0,\ T],\ H^l(\mathbb{R}))$. The norm of a function $t^{1-\gamma}f(x,t)$ in $C([0,\ T],\ H^l(\mathbb{R}))$ is defined by the equality

$$\|f\|_\gamma^l := \max_{t\in[0,\,T]}\left[t^{1-\gamma}\,|f|^l\,(t)\right].$$

3.2.2 Direct Problem: Method of Successive Approximations

We consider the following Cauchy-type problem:

$$\partial_t^{\alpha,\beta}v - \Delta v = F(x,t),\ t>0,\ x\in\mathbb{R}^n \tag{3.47}$$

with the initial condition

$$I_t^{1-\alpha}u(x,0) = \Phi(x),\ \ \gamma = \alpha+\beta(1-\alpha),\ \ x\in\mathbb{R}^n, \tag{3.48}$$

where $\triangle$ is the n-dimensional Laplacian in x variable.

For finding the solution of (3.47) and (3.48) we apply the Fourier transform (see Definition 3.4) with respect to space variable x to this equations. As result, we have

$$\partial_t^{\alpha,\beta}\tilde{v}(\xi,t) + |\xi|^2\tilde{v}(\xi,t) = \tilde{F}(\xi,t),\ t>0,\ \xi\in\mathbb{R}^n \tag{3.49}$$

and the initial condition (3.48) takes the form

$$I_t^{1-\alpha}\tilde{v}(\xi,0) = \tilde{\Phi}(\xi),\ \ \xi\in\mathbb{R}^n, \tag{3.50}$$

where the symbol $\tilde{}$ over functions denotes the Fourier transform of the corresponding functions.

The solution to problem (3.49) and (3.50) has the following form (see [23]):

$$\tilde{v}(\xi,t) = t^{\gamma-1}E_{\alpha,\gamma}\left(-|\xi|^2t^\alpha\right)\tilde{\Phi}(\xi) +$$

$$+\int_0^t (t-\tau)^{\alpha-1}E_{\alpha,\alpha}\left(-|\xi|^2(t-\tau)^\alpha\right)\tilde{F}(\xi,\tau)d\tau,\ t>0,\ \xi\in\mathbb{R}^n. \tag{3.51}$$

Now we find the inverse Fourier transform (see Definition 3.4) of function $E_{\alpha,\alpha}\left(-|\xi|^2t^\alpha\right)$ in terms of the $H-$ function. Based on Definition 3.2, formula (3.15) and Proposition 3.3, we have the following relations:

$$F^{-1}\left[E_{\alpha,\gamma}\left(-|\xi|^2 t^\alpha\right)\right] = \frac{1}{(2\pi)^n}\int\limits_{\mathbb{R}^n} H^{1,1}_{1,2}\left[|\xi|^2 t^\alpha \left|{}^{(0,1)}_{(0,1),(1-\gamma,\alpha)}\right.\right] e^{i\xi\cdot x} d\xi =$$

$$= \frac{|x|^{1-\frac{n}{2}}}{(2\pi)^n}\int\limits_0^\infty y^{\frac{n}{2}} H^{1,1}_{1,2}\left[y^2 t^\alpha \left|{}^{(0,1)}_{(0,1),(1-\beta,\alpha)}\right.\right] J_{\frac{n}{2}-1}(|x|y)\,dy.$$

In view of Definition 3.6 of the Hankel integral transform, using Proposition 3.4 and formulas (3.9), (3.10), we continue the previous calculations as follows

$$F^{-1}\left[E_{\alpha,\gamma}\left(-|\xi|^2 t^\alpha\right)\right] = \frac{1}{(2\pi)^{\frac{n}{2}}|x|^n} H^{1,2}_{3,2}\left[\frac{4t^\alpha}{|x|^2}\left|{}^{(1-\frac{n}{2},1),(0,1),(0,1)}_{(0,1),(1-\gamma,\alpha)}\right.\right] =$$

$$= \frac{1}{(2\pi)^{\frac{n}{2}}|x|^n} H^{0,2}_{2,1}\left[\frac{4t^\alpha}{|x|^2}\left|{}^{(1-\frac{n}{2},1),(0,1)}_{(1-\alpha,\alpha)}\right.\right] = \frac{1}{(2\pi)^{\frac{n}{2}}|x|^n} H^{2,0}_{1,2}\left[\frac{|x|^2}{4t^\alpha}\left|{}^{(\gamma,\alpha)}_{(\frac{n}{2},1),(1,1)}\right.\right]. \tag{3.52}$$

Applying the inverse Fourier transform to (3.51), in view of (3.52) and the convolution property, we obtain

$$v(x,t) = \int\limits_{\mathbb{R}^n} G_{\alpha,\gamma}(x-\xi,t)\Phi(\xi)d\xi$$

$$+ \int_0^t \int\limits_{\mathbb{R}^n} G_{\alpha,\alpha}(x-\xi,t-\tau)F(\xi,\tau)d\xi d\tau,\ t>0,\ \xi\in\mathbb{R}^n, \tag{3.53}$$

where

$$G_{\alpha,\gamma}(x,t) = \frac{t^{\gamma-1}}{(\pi)^{\frac{n}{2}}|x|^n} H^{2,0}_{1,2}\left[\frac{|x|^2}{4t^\alpha}\left|{}^{(\gamma,\alpha)}_{(\frac{n}{2},1),(1,1)}\right.\right]. \tag{3.54}$$

Remark 3.1 As it is easy to see from (3.53), the function $G_{\alpha,\gamma}(x,t)$ is a generalized solution of the problem (3.47) and (3.48) with $F(x,t)=0$ and $\Phi(x)=\delta(x)$, where $\delta(\cdot)$ is the $n-$dimensional Dirac delta function, i.e., $G_{\alpha,\gamma}(x,t)$ is the fundamental solution of the n-dimensional fractional diffusion operator $\partial_t^{\alpha,\beta}-\Delta$ with the Hilfer derivative.

It should be noted that the fundamental solution of the operator $\partial_t^{\alpha,\beta}-\Delta$ can be also obtained in terms of the Yang $Y-$ function (see Definition 3.4). We denote this solution by $G^*_{\alpha,\gamma}(x,t)$. Thus, the equalities

$$\left(\partial_t^{\alpha,\beta}-\Delta\right)G^*_{\alpha,\gamma}(x,t)=0,\quad I_t^{1-\alpha}G*_{\alpha,\gamma}(x,0)=\delta(x)$$

are satisfied in a generalized sense.

According to (3.51) the Fourier transform of $G^*_{\alpha,\gamma}(x,t)$ reads

$$\tilde{G}^*_{\alpha,\gamma}(\xi,t) = t^{\gamma-1} E_{\alpha,\gamma}\left(-|\xi|^2 t^\alpha\right), \quad t > 0,\ \xi \in \mathbb{R}^n.$$

Taking Proposition 3.3 into account, from the last formula we have

$$G^*_{\alpha,\gamma}(x,t) = \frac{t^{\gamma-1}}{(2\pi)^n} \int\limits_{\mathbb{R}^n} e^{i\xi\cdot x} E_{\alpha,\gamma}\left(-|\xi|^2 t^\alpha\right) d\xi$$

$$= \frac{t^{\gamma-1}}{(2\pi)^{n/2}} |x|^{1-n/2} \int\limits_0^\infty r^{n/2} E_{\alpha,\beta}\left(-r^2 t^\alpha\right) J_{\frac{n}{2}-1}\left(|x| r\right) dr. \tag{3.55}$$

Now by using the technique of the Mellin transform (see Definition 3.5) we deduce a Mellin-Barnes representation of function $G^*_{\alpha,\gamma}(x,t)$. Equation (3.55) can be considered (without factor $t^{\gamma-1}$) as the Mellin convolution (3.17) of the functions

$$f_1(r) = E_{\alpha,\gamma}\left(-r^2 t^\alpha\right) \text{ and } f_2 = \frac{1}{(2\pi)^{n/2} |x|^n} r^{-1-\frac{n}{2}} J_{\frac{n}{2}-1}\left(\frac{1}{r}\right)$$

evaluated at the point $y = \frac{1}{|x|}$. By using the Mellin transform of the Mittag-Leffler function, the Bessel function and some elementary properties of the Mellin transform (see Definition 3.6), we get the following formulas:

$$f_1^*(s) = \frac{t^{-\alpha s/2}}{2} \frac{\Gamma\left(\frac{s}{2}\right)\Gamma\left(1-\frac{s}{2}\right)}{\Gamma\left(\gamma - \alpha\frac{s}{2}\right)}, \quad 0 < \Re(s) < 2,$$

$$f_2^*(s) = \frac{1}{(2\pi)^{n/2}|x|^n} \left(\frac{1}{2}\right)^{s-\frac{n}{2}} \frac{\Gamma\left(\frac{n}{2}-\frac{s}{2}\right)}{\Gamma\left(\frac{s}{2}\right)}, \quad \frac{n-1}{2} < \Re(s) < n.$$

Further, Definition 3.5, Definition 3.6 and the Mellin convolution formula (3.17) lead to the following Mellin-Barnes type integral representation of the fundamental solution $G^*_{\alpha,\gamma}(x,t)$ (compare with the formula (28) in [24]):

$$G^*_{\alpha,\gamma}(x,t) = \frac{t^{\gamma-1}}{2(\pi)^{n/2}|x|^n} \frac{1}{2\pi i} \int_{\sigma-i\infty}^{\sigma+i\infty} \frac{\Gamma\left(\frac{n}{2}-\frac{s}{2}\right)\Gamma\left(1-\frac{s}{2}\right)}{\Gamma\left(\gamma-\alpha\frac{s}{2}\right)} \left(\frac{2t^{\alpha/2}}{|x|}\right)^{-s} ds, \tag{3.56}$$

where $\frac{n-1}{2} < \sigma < \min(2, n)$. In view of Definition 3.5 for Yang $Y-$ function the formula (3.56) can be rewritten in the form

$$G^*_{\alpha,\gamma}(x,t) = \frac{t^{\gamma-1}}{2(\pi)^{n/2}|x|^n} Y^{2,0}_{1,2}\left[\frac{2t^{\frac{\alpha}{2}}}{|x|}; 0; 0 \,\middle|\, \begin{matrix} (\gamma, \frac{\alpha}{2}) \\ (\frac{n}{2}, \frac{1}{2}), (1, \frac{1}{2}) \end{matrix}\right].$$

To compare the functions $G_{\alpha,\gamma}(x,t)$ and $G^*_{\alpha,\gamma}(x,t)$, first we transform $G_{\alpha,\gamma}(x,t)$ determined by (3.54) according to the formula (3.15) for $k = 1/2$ to the form

$$G_{\alpha,\gamma}(x,t) = \frac{t^{\gamma-1}}{2(\pi)^{\frac{n}{2}}|x|^n} H^{2,0}_{1,2}\left[\frac{|x|}{2t^{\frac{\alpha}{2}}}\middle|^{(\gamma,\frac{\alpha}{2})}_{(\frac{n}{2},\frac{1}{2}),(1,\frac{1}{2})}\right].$$

Now if we take into account the relationship (3.14) between the Y and H functions, then it is easily to see $G_{\alpha,\gamma}(x,t) = G^*_{\alpha,\gamma}(x,t)$.

We will begin to study the properties of the solution of the integral equation (3.53) for $n = 1$, i.e., $x \in \mathbb{R}$. In view of Proposition 3.1, Proposition 3.2 and formulas of differentiation $H-$ functions (3.12) and (3.13), we get the following estimates:

if $x^2 \le t^\alpha$, $x \ne 0$, then

$$\left|\frac{\partial^k}{\partial x^k} G_{\alpha,\gamma}(x,t)\right| \le C t^{\gamma-1-\frac{(k-1)\alpha}{2}}, \quad k = 0, 1, 2. \tag{3.57}$$

if $x^2 \ge t^\alpha$, then

$$\left|\frac{\partial^k}{\partial x^k} G_{\alpha,\gamma}(x,t)\right| \le C t^{\gamma-1-\frac{(1+k)\alpha}{2}} \exp\left(-\sigma t^{-\frac{\alpha}{2-\alpha}} |x|^{\frac{2}{2-\alpha}}\right) \quad k = 0, 1, 2. \tag{3.58}$$

Here the letters C, σ denote various positive constants. For a comparison of these estimates for $\gamma = \alpha$, see [4].

We also note that it is true the equality [4]:

$$\int_{\mathbb{R}} G_{\alpha,\gamma}(x,t)dx = C_0 t^{\gamma-1}, \ t \in [0,T], \tag{3.59}$$

where C_0 depends only on α, β.

Putting in (3.53) $\Phi(x) = \varphi''(x)$ and $F(x,t) = f_{xx}(x,t) - q(t)\vartheta(x,t)$, we obtain the integral equation for the solution to the direct problem (3.44), (3.45):

$$v(x,t) = \vartheta_0(x,t) - \int_0^t q(\tau) \int_{\mathbb{R}} G_{\alpha,\alpha}(x-\xi, t-\tau)\vartheta(\tau,\xi)d\tau d\xi, \tag{3.60}$$

where

$$v_0(x,t) := \int_{\mathbb{R}} G_{\alpha,\gamma}(x-\xi,t)\varphi''(\xi)d\xi + \int_0^t \int_{\mathbb{R}} G_{\alpha,\alpha}(x-\xi,t-\tau) f_{\xi\xi}(\tau,\xi)d\tau d\xi. \tag{3.61}$$

It is true the following assertion:

Lemma 3.2 *If* $q(t) \in C[0,T]$, $t^{1-\gamma} f(x,t) \in C\left([0,T],\ H^{2+l}(\mathbb{R})\right)$, $l \in (0,1)$, $\varphi(x) \in C^2(\mathbb{R})$, *then there exists a unique solution of the integral equation (3.60) such that* $t^{1-\gamma}\vartheta(x,t) \in C^{1-\alpha,2}(D_T)$.

Proof We use the method of successive approximations and consider the sequence of functions defined recursively by the formulas:

$$\vartheta_n(x,t) = -\int_0^t q(\tau)\int_{\mathbb{R}} G_{\alpha,\alpha}(x-\xi, t-\tau)\vartheta_{n-1}(\tau,\xi)d\tau d\xi,\ n = 1, 2, \cdots, \tag{3.62}$$

where $\vartheta_0(x,t)$ is determined by the equality (3.61).

Set

$$\|q\| = \max_{t\in[0,T]} |q(t)|,\ \varphi_0 := \|\varphi\|_{C^2(\mathbb{R})} = \sum_{i=0}^{2}\sup_{x\in\mathbb{R}}\left|\frac{d^i\varphi(x)}{dx^i}\right|,\ f_0 := \|f\|_{\gamma}^{2+l}$$

and using (3.61), (3.62), (3.59), we estimate the modulus of $v_n(x,t)$ in the domain D_T as

$$|v_0(x,t)| \le C_0\left[\varphi_0 + f_0 T\gamma^{-1}\right]t^{\gamma-1} =: \varphi_{00}t^{\gamma-1},$$

$$|v_1(x,t)| \le \varphi_{00}C_0\|q\|\int_0^t \frac{\tau^{\gamma-1}}{(t-\tau)^{1-\alpha}}d\tau = \varphi_{00}C_0\|q\|\Gamma(\alpha)I_t^{\alpha}t^{\gamma-1}.$$

For $\mu \in \mathbb{R}$ by direct calculation we have the following formula

$$I_t^{\alpha}t^{\mu} = \frac{\Gamma(\mu+1)}{\Gamma(\alpha+\mu+1)}t^{\alpha+\mu}.$$

In accordance with this formula we continue to estimate $v_1(x,t)$:

$$|v_1(x,t)|\,\varphi_{00} \le C_0\|q\|\Gamma(\alpha)I_t^{\alpha}t^{\gamma-1} = \varphi_{00}C_0\|q\|\Gamma(\alpha)\frac{\Gamma(\gamma)}{\Gamma(\alpha+\gamma)}t^{\alpha+\gamma-1},$$

Similarly, we estimate $v_2(x,t)$ as

$$|\vartheta_2(x,t)| \le \varphi_{00}\left(C_0\|q\|\Gamma(\alpha)\right)^2\frac{\Gamma(\gamma)}{\Gamma(\alpha+\gamma)}I_t^{\alpha}t^{\alpha+\gamma-1}$$

$$= \varphi_{00}\left(C_0\|q\|\Gamma(\alpha)\right)^2\frac{\Gamma(\gamma)}{\Gamma(2\alpha+\gamma)}t^{2\alpha+\gamma-1}.$$

For arbitrary $n = 1, 2, \ldots$, we have

$$|v_n(x,t)| \le \varphi_{00}\left(C_0\|q\|\Gamma(\alpha)\right)^n\frac{\Gamma(\gamma)}{\Gamma(\alpha n+\gamma)}t^{\alpha n+\gamma-1}.$$

Multiplying the last inequality by $t^{1-\gamma}$, we find estimates

$$t^{1-\alpha}\left|v_n(x,t)\right| \le \varphi_{00}\Gamma\left(\gamma\right)\left(C_0\|q\|\Gamma(\alpha)\right)^n \frac{t^{\alpha n}}{\Gamma\left(\alpha n+\gamma\right)},$$

which are fulfilled for all $n = 0, 1, 2, \ldots$.

It follows from the above estimates that the series

$$t^{1-\gamma}v(x,t) = \sum_{n=0}^{\infty} t^{1-\gamma}v_n(x,t)$$

converges uniformly in $\overline{D}_T$, since it can be majorized in $\overline{D}_T$ by the convergent numerical series

$$\varphi_{00}\Gamma\left(\gamma\right)\sum_{n=0}^{\infty}\left(C_0\|q\|\Gamma(\alpha)\right)^n \frac{T^{\alpha n}}{\Gamma\left(\alpha n+\gamma\right)}.$$

This means the following estimate for the solution of the integral equation (3.60) takes place:

$$t^{1-\gamma}\left|v(x,t)\right| \le \varphi_{00}\Gamma\left(\alpha\right)\sum_{n=0}^{\infty}\left(C_0\|q\|\Gamma(\alpha)\right)^n \frac{T^{\alpha n}}{\Gamma\left(\alpha n+\gamma\right)}$$

$$= \varphi_{00}\Gamma(\gamma)E_{\alpha,\gamma}\left(C_0\|q\|\Gamma(\gamma)T^{\alpha}\right),\ (x,t)\in\overline{D}_T, \tag{3.63}$$

where $E_{\alpha,\gamma}(\cdot)$ is the Mittag-Leffler function of a nonnegative real argument. When $\beta = 1$ the fractional operator $\partial_t^{\alpha,\beta}$ coincides with the Caputo fractional differential operator ∂_t^{α}, and the estimate (3.63) and the corresponding estimate in (3.28) are the same.

Under the assumptions of Lemma 3.1 and on the bases of estimates (3.57), (3.58) (see also [4]) one has the inclusion $t^{1-\gamma}v_n(x,t) \in C^{1-\gamma,2}\left(\overline{D}_T\right)$. According to the general theory of integral equations, this implies that the same property will be possessed the function $t^{1-\gamma}v(x,t)$ in $\overline{D}_T$. As usual, this function is a solution of integral equation (3.60).

3.2.3 *Stability Estimate for a Solution to the Direct Problem*

Here we derive an estimate in the norm of $C_\gamma\left(\bar{D}_T\right)$ for the difference between the solution of the original integral equation (3.60) and the solution of this equation with perturbed functions $\tilde{q}$, $\tilde{f}_{xx}$ and $\tilde{\varphi}''$. Let $\tilde{\vartheta}(x,t)$ be a solution of the integral equation (3.60) corresponding to the functions $\tilde{q}$, $\tilde{f}_{xx}$ and $\tilde{\varphi}''$; i.e.,

$$\tilde{\vartheta}(x,t) = \tilde{\vartheta}_0(x,t) - \int_0^t \tilde{q}(\tau)\int_{\mathbb{R}} G_{\alpha,\alpha}(t-\tau,x-\xi)\tilde{\vartheta}(\tau,\xi)d\tau d\xi, \tag{3.64}$$

where in accordance with (3.61)

$$\tilde{\vartheta}_0(x,t) := \int_{\mathbb{R}} G_{\alpha,\gamma}(t, x-\xi)\tilde{\varphi}''(\xi)d\xi + \int_0^t \int_{\mathbb{R}} G_{\alpha,\alpha}(t-\tau, x-\xi)\tilde{f}_{\xi\xi}(\tau,\xi)d\tau d\xi. \tag{3.65}$$

Composing the difference $\vartheta - \tilde{\vartheta}$ with the help of the equations (3.60), (3.64) and introducing the notations $\vartheta - \tilde{\vartheta} = \overline{\vartheta}$, $\vartheta_0 - \tilde{\vartheta}_0 = \overline{\vartheta}_0$, $q - \tilde{q} = \overline{q}$, we obtain the integral equation

$$\overline{\vartheta}(x,t) = \overline{\vartheta}_0(x,t) - \int_0^t \overline{q}(\tau) \int_{\mathbb{R}} G_{\alpha,\alpha}(t-\tau, x-\xi)\vartheta(\tau,\xi)d\tau d\xi$$

$$- \int_0^t \tilde{q}(\tau) \int_{\mathbb{R}} G_{\alpha,\alpha}(t-\tau, x-\xi)\overline{\vartheta}(\tau,\xi)d\tau d\xi, \tag{3.66}$$

from which, using (3.59) at $\gamma = \alpha$ and (3.63) is obtained the following linear integral inequality for $\left|\overline{\vartheta}(x,t)\right|$:

$$\left|\overline{\vartheta}(x,t)\right| \le \left|\overline{\vartheta}_0(x,t)\right| + C_0\varphi_{00}\Gamma(\gamma)E_{\alpha,\gamma}\left(C_0\|q\|\Gamma(\gamma)T^\alpha\right) t^{\alpha+\gamma-1}B(\alpha,\gamma)\,\|\overline{q}\|_{C[0,T]}$$

$$+ \tilde{q}_0 \int_0^t \int_{\mathbb{R}} G_{\alpha,\alpha}(t-\tau, x-\xi)\left|\overline{\vartheta}(\tau,\xi)\right| d\tau d\xi,\ (x,t) \in D_T, \tag{3.67}$$

where $\tilde{q}_0 := \|\tilde{q}_0(t)\|_{C[0,T]}$, $B(\cdot,\cdot)$ is Euler's Beta function and to estimate the second term on the right side of (3.66) it was used the equality $\int_0^t \tau^{\gamma-1}(t-\tau)^{\alpha-1}d\tau = t^{\alpha+\gamma-1}B(\alpha,\gamma)$.

From the equalities (3.61) and (3.65) on based (3.59), (3.63) it follows the estimate for $(x,t) \in D_T$:

$$\left|\overline{\vartheta}_0(x,t)\right| \le C_0 t^{\gamma-1}\left[\|\overline{\varphi}''\|_{C(\mathbb{R})} + \frac{t^\alpha}{\alpha}\left\|\overline{f}_{xx}\right\|_\gamma^l\right],$$

where $\overline{\varphi}'' = \varphi'' - \tilde{\varphi}''$, $\overline{f}_{xx} = f_{xx} - \tilde{f}_{xx}$.

Let $\lambda = \lambda(\alpha, \beta, T, \|q\|, \varphi_0, f_0) = \max\left\{1, \frac{T^\alpha}{\alpha}, \varphi_{00}\Gamma(\gamma)E_{\alpha,\gamma}\left(C_0\|q\|\Gamma(\gamma)T^\alpha\right) T^\alpha B(\alpha,\gamma)\right\}$. Applying the successive approximation method to inequality (3.67) with the help of the scheme

$$\left|\overline{\vartheta}(x,t)\right|_0 \le \lambda C_0 t^{\gamma-1}\left(\|\overline{\varphi}''\|_{C(\mathbb{R})} + \left\|\overline{f}_{xx}\right\|_\gamma^l + \|\overline{q}\|_{C[0,T]}\right),$$

$$\left|\overline{\vartheta}(x,t)\right|_n \le \tilde{q}_0 \int_0^t \int_{\mathbb{R}} G_{\alpha,\alpha}(t-\tau, x-\xi)\left|\overline{\vartheta}(\tau,\xi)\right|_{n-1} d\tau d\xi,\ n = 1, 2, \ldots,$$

we obtain the estimate

$$\left|\overline{\vartheta}(x,t)\right| \leq t^{\gamma-1}\lambda C_0\varphi_{00}\Gamma(\gamma)E_{\alpha,\gamma}\left(C_0\|\tilde{q}\|\Gamma(\alpha)T^{\alpha}\right)$$

$$\times\left(\left\|\overline{\varphi}''\right\|_{C(\mathbb{R})}+\left\|\overline{f}_{xx}\right\|_{\gamma}^{l}+\|\overline{q}\|_{C[0,T]}\right), \tag{3.68}$$

which will be used in the next section.

Note that the expression (3.68) is the stability estimate for the solution $t^{1-\gamma}v(x,t)$ to the Cauchy problem (3.44) and (3.45) in the class $C_{\gamma}\left(\overline{D_T}\right)$. The uniqueness for this solution follows from (3.68). □

3.2.4 Recovering a Potential from a Simple Observation: Local Solvability

Setting in (3.60) $x=0$ and in view of overdetermination condition (3.46), after simple converting, we obtain the following integral equation of Volterra type with respect to $q(t)$:

$$q(t)=q_0(t)-\frac{1}{g(t)}\int_0^t q(\tau)\int_{\mathbb{R}}G_{\alpha,\alpha}(\xi,t-\tau)\vartheta(\tau,\xi)d\tau d\xi, \tag{3.69}$$

where

$$q_0(t):=\frac{1}{g(t)}\Bigg[f(0,t)-\partial_t^{\alpha,\beta}g(t)$$

$$+\int_{\mathbb{R}}G_{\alpha,\gamma}(\xi,t)\varphi''(\xi)d\xi+\int_0^t\int_{\mathbb{R}}G_{\alpha,\alpha}(\xi,t-\tau)f_{\xi\xi}(\tau,\xi)d\tau d\xi\Bigg].$$

The solution of integral equation (3.60) depends on q, i.e., $\vartheta=\vartheta(x,t;q)$. We introduce an operator Λ defining it by the right-hand side of (3.69):

$$\Lambda[q](t)=q_0(t)-\frac{1}{g(t)}\int_0^t q(\tau)\int_{\mathbb{R}}G_{\alpha,\alpha}(\xi,t-\tau)\vartheta(\tau,\xi;q)d\tau d\xi. \tag{3.70}$$

Then, the equation (3.69) is written in the following form:

$$q(t)=\Lambda[q](t). \tag{3.71}$$

The following assertion is the main result of this paper:

Theorem 3.4 *If* $t^{1-\gamma}f(x,t)\in C\left([0,T],\ H^{2+l}(\mathbb{R})\right)$, $\varphi(x)\in C^2(\mathbb{R})$, $g(t)\in C_{\gamma}^1[0,T]$, $t^{1-\gamma}|g(t)|\geq g_0=const>0$, g_0 *is a given number,* $\varphi(0)=g(0)$, *then there exists a number $T^* \in (0, T)$, such that there exists a unique solution $q(t) \in C[0, T^*]$ of the inverse problem (3.41)–(3.43).*

Under the conditions of Theorem 4.1, we have $q_0(t) \in C[0, T]$. Let $q_{00} := \max_{t\in[0,T]} |q_0(t)|$. Fix a number $\rho > 0$ and consider the ball

$$\Phi^T(q_0, \rho) := \left\{q(t) : q(t) \in C[0, T],\ \|q - q_0\|_{C[0,T]} \le \rho\right\}.$$

First we prove that for an enough small $T > 0$ the operator Λ maps the ball $\Phi^T(q_0, \rho)$ into itself; i.e., the condition $q(t) \in \Phi^T(q_0, \rho)$ implies that $\Lambda[q](t) \in \Phi^T(q_0, \rho)$. Indeed, for any continuous function $q(t)$, the function $\Lambda[q(t)]$ calculated using formula (3.69) will be continuous. Moreover, estimating the norm of the differences based on (3.69), we find

$$\|\Lambda[q](t) - q_0(t)\| \le \frac{C_0}{g_0}\|q\|\, T^\alpha B(\alpha, \gamma)\varphi_{00}\Gamma(\gamma)E_{\alpha,\gamma}\left(C_0\|q\|\Gamma(\gamma)T^\alpha\right).$$

Note that the function occurring on the right-hand side in this inequality is monotone increasing with T, and the fact that the function $q(t)$ belongs to the ball $\Phi^T(q_0, \rho)$ implies the inequality

$$\|q\| \le \rho + \|q_0\|. \tag{3.72}$$

Therefore, we only strengthen the inequality if we replace $\|q\|$ in it with the expression $\rho + \|q_0\|$. Performing these replacements, we obtain the estimate

$$\begin{aligned}&\|\Lambda[q](t) - q_0(t)\|\\ &\le \frac{C_0}{g_0}(\rho + \|q_0\|)\, T^\alpha B(\alpha, \gamma)\varphi_{00}\Gamma(\gamma)E_{\alpha,\gamma}\left(C_0(\rho + \|q_0\|)\,\Gamma(\gamma)T^\alpha\right).\end{aligned}$$

Let T_1 be a root of the equation

$$m_1(T) = \frac{C_0}{g_0}(\rho + \|q_0\|)\, T^\alpha B(\alpha, \gamma)\varphi_{00}\Gamma(\gamma)E_{\alpha,\gamma}\left(C_0(\rho + \|q_0\|)\,\Gamma(\gamma)T^\alpha\right) = \rho.$$

Then, for $T \in [0, T_1]$ we get $\Lambda[q](t) \in \Phi^T(q_0, \rho)$.

Now we consider two functions $q(t)$ and $\tilde{q}(t)$ belonging to the ball $\Phi^T(q_0, \rho)$ and estimate the distance between their images $\Lambda[q](t)$ and $\Lambda[\tilde{q}](t)$ in the space $C[0, T]$. The function $\tilde{\vartheta}(x, t)$ corresponding to $\tilde{q}(t)$ satisfies the integral equation (3.64) with the functions $\varphi'' = \tilde{\varphi}''$ and $f_{xx} = \tilde{f}_{xx}$. Composing the difference $\Lambda[q](t) - \Lambda[\tilde{q}](t)$ with the help of equations (3.60), (3.64) and then estimating it in modules, we find

$$|\Lambda[q](t) - \Lambda[\tilde{q}](t)| \le \frac{C_0 T^\alpha}{g_0} B(\alpha, \gamma)\left(\|\overline{q}\|\, t^{1-\gamma}|\vartheta| + \|\tilde{q}\|\, t^{1-\gamma}\left|\overline{\vartheta}\right|\right).$$

Using inequality (3.63) and the estimate (3.68) with $\varphi'' = \tilde{\varphi}''$, $f_{xx} = \tilde{f}_{xx}$, we continue the previous inequality in the following form:

$$\|\Lambda[q](t) - \Lambda[\tilde{q}](t)\|$$
$$\leq \frac{C_0 T^\alpha}{g_0} B(\alpha,\gamma)\varphi_{00}\Gamma(\gamma)E_{\alpha,\gamma}\left(C_0\|q\|\Gamma(\alpha)T^\alpha\right)\left(1+\lambda C_0\|\tilde{q}\|\right)\|\overline{q}\|. \tag{3.73}$$

The functions $q(t)$ and $\tilde{q}(t)$ belong to the ball $\Phi^T(q_0, \rho)$, and hence for each of these functions one has inequality (3.72). Note that the function on the right-hand side of (3.73) at the factor $\|\overline{q}\|$ is monotone increasing with $\|q\|$, $\|\tilde{q}\|$, and T. Consequently, replacing $\|q\|$ and $\|\tilde{q}\|$ in (3.73) (including in λ) in accordance with the right side of (3.72) will only strengthen the inequality. Thus, we have

$$\|\Lambda[q](t) - \Lambda[\tilde{q}](t)\|$$

$$\leq \frac{C_0 T^\alpha}{g_0} B(\alpha,\gamma)\varphi_{00}\Gamma(\gamma)E_{\alpha,\gamma}\left(C_0(\rho+\|q_0\|)\Gamma(\alpha)T^\alpha\right)\left(1+\lambda C_0(\rho+\|q_0\|)\right)\|\overline{q}\|$$
$$=: m_2(T)\|\overline{q}\|.$$

Let T_2 be a root of the equation

$$m_2(T) = 1.$$

Then, for $T \in (0, T_2)$ we have that the distance between the functions $\Lambda[q](t)$ and $\Lambda[\tilde{q}](t)$ in the functional space $C[0, T]$ is not greater than the distance between the functions $q(t)$ and $\tilde{q}(t)$ multiplied by $m_2(T) < 1$. Consequently, if we choose $T^* < \min(T_1, T_2)$, then the operator Λ is a contraction in the ball $\Phi^{T*}(q_0, \rho)$. However, in accordance with the Banach theorem, the operator Λ has a unique fixed point in the ball $\Phi^{T*}(q_0, \rho)$; i.e., there exists a unique solution of the Eq. (3.71). Theorem 3.1 is proven.

3.2.5 Global Stability and Uniqueness

Let T be a positive fixed number. Consider the set $\Omega(\omega_0)$ ($\omega_0 > 0$ is some fixed number) of the given functions (f, φ, g) for which all conditions of Theorem 3.1 are fulfilled and $\max\left\{\|f\|_\gamma^{2+l}, \|\varphi\|_{C^2(\mathbb{R})}, \|g\|_{C^1_\gamma[0,T]}\right\} \leq \omega_0$. By $Q(\omega_1)$ we denote the class of functions $q(t) \in C[0, T]$ satisfying the inequality $\|q\| \leq \omega_1$ with some fixed positive number ω_1.

Theorem 3.5 *Let* $(f, \varphi, g) \in \Omega(\omega_0)$, $\left(\tilde{f}, \tilde{\varphi}, \tilde{g}\right) \in \Omega(\omega_0)$ *and* $(q, \tilde{q}) \in Q(\omega_1)$. *If* $\alpha \in (\frac{1}{2}, 1)$, *then for the solution of the inverse problem (3.41)–(3.43) the following stability estimate holds:*

$$\|q - \tilde{q}\|_{C[0,T]} \le d\left(\|f - \tilde{f}\|_{\gamma}^{2+l} + \|\varphi - \tilde{\varphi}\|_{C^2(\mathbb{R})} + \|g - \tilde{g}\|_{C^1_\gamma[0,T]}\right), \tag{3.74}$$

where the constant d *depends only on* T, α, β, g_0, ω_0, ω_1.

To **prove** this theorem, using (3.69) we write down the equations for $\tilde{q}(t)$ and compose the difference $q(t) - \tilde{q}(t)$. Then, after evaluating this expression and using estimates (3.63), (3.68), we obtain

$$|q(t) - \tilde{q}(t)| \le d_0\left(\|f - \tilde{f}\|_{\gamma}^{2+l} + \|\varphi - \tilde{\varphi}\|_{C^2(\mathbb{R})} + \|g - \tilde{g}\|_{C^1_\gamma[0,T]}\right)$$

$$+ d_1 t^{1-\gamma} \int_0^t \tau^{\gamma-1}(t-\tau)^{\alpha-1}\, |q(\tau) - \tilde{q}(\tau)|\, d\tau, \ \ t \in [0, T], \tag{3.75}$$

where d_0 and d_1 depend on the same constants as d. From (3.75) using Proposition 3.5, we get the estimate

$$|q(t) - \tilde{q}(t)| \le d_0 W_{\alpha,\gamma}\left(\left(d_1 t^{1-\gamma}\Gamma(\alpha)\right)^{\frac{1}{\alpha+\gamma-1}} t\right)$$

$$\times\left(\|f - \tilde{f}\|_{\gamma}^{2+l} + \|\varphi - \tilde{\varphi}\|_{C^2(\mathbb{R})} + \|g - \tilde{g}\|_{C^1_\gamma[0,T]}\right), \ t \in [0, T].$$

This inequality implies the estimate (3.74), if we set $d = d_0 W_{\alpha,\gamma}\left(\left(d_1 T^{1-\gamma}\Gamma(\alpha)\right)^{\frac{1}{\alpha+\gamma-1}} T\right)$.

From Theorem 3.5 readily follows the uniqueness theorem for any $T > 0$:

Theorem 3.6 *Let the functions* q, f, φ, g *and* $\tilde{q}$, $\tilde{f}$, $\tilde{\varphi}$, $\tilde{g}$ *have the same meaning as in Theorem3.2. Moreover, if* $f = \tilde{f}$, $\varphi = \tilde{\varphi}$, $g = \tilde{g}$ *for* $t \in [0, T]$ *and* $\alpha \in (\frac{1}{2}, 1)$, *then* $q(t) = \tilde{q}(t)$, $t \in [0, T]$.

3.3 Initial-Boundary Value Problems in a Half-Line

The problem of the existence of a fundamental solution to various versions of time-fractional problems has already been addressed in the literature. In [25, 26], the author obtained the exact solution for a homogeneous fractional diffusion equation in half-space with the Dirichlet boundary condition in terms of the Wright function. In paper [27], the fractional diffusion equation in half-space was investigated with the homogeneous Dirichlet and Neumann boundary conditions. Such equation with the Robin boundary condition was studied in [28]. Using the integral transform methods, including the Laplace transform and the Fourier transform, the exact solutions of problems were obtained.

In this section, we consider the solution of the inhomogeneous fractional heat (diffusion) equation in half-line with all the three homogeneous boundary conditions (Dirichlet, Neumann, Robin). For these cases, we have constructed exact formulas for the solutions. These formulas are obtained by the continuation method. Green's functions of the problems are constructed in terms of the Fox $H-$function. These formulas are convenient for solving direct and inverse problems for a fractional diffusion equation on a half-line. The presentation of this paragraph corresponds to the work of [29].

We consider the following one-dimensional fractional diffusion equation on a semi-infinite straight line:

$$\partial_t^\alpha u - u_{xx} = f(x,t), \quad t > 0, \quad x \in \mathbb{R}_+, \tag{3.76}$$

the solution of which satisfies the initial condition

$$u(x,0) = \varphi(x), \quad x \in \mathbb{R}_+ \cup \{0\}, \tag{3.77}$$

and the boundary condition

$$au_x(0,t) + bu(0,t) = 0, \quad t \geq 0, \tag{3.78}$$

where $\mathbb{R}_+ = \left\{x \in \mathbb{R} \middle| x > 0\right\}$, $|a| + |b| \neq 0$, ∂_t^α is the Caputo fractional differential operator of the order $0 < \alpha \leq 1$ (1.4), $f(x,t)$ and $\varphi(x)$ are given functions.

3.3.1 Cauchy Problem and Auxiliary Lemma

Let $v(x,t)$ be a solution to the Cauchy problem

$$\partial_t^\alpha v - v_{xx} = f(x,t),\ t > 0,\ x \in \mathbb{R}, \tag{3.79}$$

$$v(x,0) = \varphi(x),\ \ x \in \mathbb{R}. \tag{3.80}$$

The solution to the problem (3.79) and (3.80) is determined by the formula (see [10] and also formulas (3.19), (3.19)):

$$v(x,t) = \int\limits_{\mathbb{R}} G_{\alpha,1}(x-\xi,t)\varphi(\xi)d\xi + \int_0^t \int\limits_{\mathbb{R}} G_{\alpha,\alpha}(x-\xi,t-\tau)f(\xi,\tau)d\xi d\tau,\ t > 0,\ x \in \mathbb{R}, \tag{3.81}$$

where

$$G_{\alpha,\beta}(x,t) = \frac{t^{\beta-1}}{\sqrt{\pi}|x|} H^{2,0}_{1,2}\left[\frac{x^2}{4t^\alpha}\left|\begin{matrix}(\beta,\alpha)\\(\frac{1}{2},1),(1,1)\end{matrix}\right.\right] \tag{3.82}$$

is the fundamental solution for $\beta = 1$ and $\beta = \alpha$ of the fractional diffusion operator $\partial_t^\alpha - \frac{\partial^2}{\partial x^2}$, $H^{m,n}_{p,q}\left[z\left|\begin{matrix}(a_j,A_j)_1^p\\(b_j,B_j)_1^q\end{matrix}\right.\right]$ is the generalized hypergeometric $H-$ function defined in Definition 3.2.

The function $G_{\alpha,\beta}$ is infinitely differentiable at $x \neq 0$ for $t > 0$. The regularity of the function $G_{\alpha,\beta}$ and some of its derivatives at $x \neq 0$ is determined by the regularity of the $H-$ function and its derivatives.

From the asymptotic behavior of the $H-$ function for small and large values of the argument and the formulas of differentiation $H-$ functions (see formulas (3.12) and (3.13)), follow the estimates

(1) if $|x|^2 \le t^\alpha$, $x \neq 0$, then

$$\left|\frac{\partial^k}{\partial x^k} G_{\alpha,\beta}(x,t)\right| \le C t^{\beta-1-\frac{(k-1)\alpha}{2}}, \quad k = 0, 1, 2; \tag{3.83}$$

(2) if $|x|^2 \ge t^\alpha$, then

$$\left|\frac{\partial^k}{\partial x^k} G_{\alpha,\beta}(x,t)\right| \le C t^{\beta-1-\frac{(1+k)\alpha}{2}} \exp\left(-\sigma t^{-\frac{\alpha}{2-\alpha}} |x|^{\frac{2}{2-\alpha}}\right) \quad k = 0, 1, 2, \tag{3.84}$$

where $C > 0$, $\sigma > 0$ denote various constants.

Lemma 3.3 *Let the function $\Phi(x)$ be defined on $\mathbb{R}$, have bounded derivatives on it up to the 2-th order inclusive, and a linear combination $\sum_{k=o}^{n} a_k \Phi^{(k)}(x)$, where $a_k = const$, is odd with respect to the point $x = 0$. Then, the function*

$$u(x,t) = \int_{\mathbb{R}} G_{\alpha,\beta}(x-\xi,t)\Phi(\xi)d\xi,$$

$$G_{\alpha,\beta}(x-\xi,t) = \frac{t^{\beta-1}}{\sqrt{\pi}|x-\xi|} H^{2,0}_{1,2}\left[\frac{(x-\xi)^2}{4t^\alpha}\left|\begin{matrix}(\beta,\alpha)\\(\frac{1}{2},1),(1,1)\end{matrix}\right.\right] \tag{3.85}$$

satisfies the condition

$$\sum_{k=0}^{2} a_k \left.\frac{\partial^k u(x,t)}{\partial x^k}\right|_{x=0} = 0. \tag{3.86}$$

To **prove** this assertion, we note that the following equalities are true:

$$\frac{\partial^k G_{\alpha,\beta}}{\partial x^k}(x-\xi,t) = (-1)^k \frac{\partial^k G_{\alpha,\beta}}{\partial \xi^k}(x-\xi,t).$$

Under the conditions of the lemma, one can differentiate (3.85) twice under the integral sign with respect to x. That's why

$$\sum_{k=0}^{2} a_k \frac{\partial^k u(x,t)}{\partial x^k} = \sum_{k=0}^{2} a_k \int_{\mathbb{R}} (-1)^k \frac{\partial^k G}{\partial \xi^k}(x-\xi,t)\Phi(\xi)d\xi.$$

Integrating by parts on the right side of this equality, we "throw over" the derivatives $\frac{\partial^k}{\partial x^k}$ to the function $\Phi(\xi)$. At the same time, taking into account that the non-integral terms vanish in accordance with (3.84), we obtain

$$\sum_{k=0}^{2} a_k \frac{\partial^k u(x,t)}{\partial x^k} = \sum_{k=0}^{2} a_k \int_{\mathbb{R}} G_{\alpha,\beta}(x-\xi,t)\Phi^{(k)}(\xi)d\xi.$$

Under the conditions of lemma the integrand in the last formula is odd with respect to $x = 0$. Therefore, the relation (3.86) is valid.

3.3.2 Continuation Method: Exact Formulas for Solutions

Lemma 3.3 allows us to indicate the method for solving the problem for the homogeneous fractional heat equation

$$\partial_t^\alpha u - u_{xx} = 0,\ t > 0,\ x \in \mathbb{R}_+,$$

with the given initial condition

$$u(x,0) = \varphi(x), \quad x \in \mathbb{R}_+,$$

and a homogeneous boundary condition of the form

$$\sum_{k=0}^{2} a_k \frac{\partial^k u(x,t)}{\partial x^k}\Big|_{x=0} = 0.$$

For this purpose, we continue the function $\varphi(x)$ to the entire real axis, defining a new function $\Phi(x)$, which satisfies conditions

$$\Phi(x) = \begin{cases} \varphi(x) \ \text{ at } \ x \geq 0, \\ \sum\limits_{k=0}^{2} a_k \Phi^{(k)}(x) = -\sum\limits_{k=0}^{2} a_k \varphi^{(k)}(s)\Big|_{s=-x} \quad \text{at } \ x \leq 0 \end{cases} \tag{3.87}$$

and is continuous together with derivatives up to the 2−th order inclusive on $\mathbb{R}$.

We note that in the case of a homogeneous Dirichlet boundary condition, in the formula (3.87) there will be $a_0 = 1$ and $a_{1,2} = 0$. Then, according to Lemma, the function $\Phi(x)$ must be odd, that is, the function $\varphi(x)$ must be extended to the negative semiaxis in an odd way. In the case of a homogeneous Neumann boundary condition, in the formula (3.87) there will be $a_{0,2} = 0$ and $a_1 = 1$ according to Lemma, the function $\Phi'(x)$ must be odd, that is, the function $\Phi(x)$ must be even. Hence the function $\varphi(x)$ must be extended to the negative semiaxis in an even way.

1. Dirichlet problem. We construct a solution of the problem (3.76)–(3.78) with the Dirichlet boundary condition ($a = 0,\ b = 1$). Let us assume that the matching condition $\varphi(0) = 0$ between the initial and boundary conditions is satisfied. We introduce the function $\Phi(x)$, which is an odd continuation of the $\varphi(x)$ function:

$$\Phi(x) = \begin{cases} \varphi(x) & x \geq 0, \\ -\varphi(-x) & x \leq 0 \end{cases} \tag{3.88}$$

and consider Cauchy problem (3.79), (3.80) with $f = 0$ and $\Phi(x)$ instead of $\varphi(x)$.

The solution to the problem (3.79), (3.80), as noted above (see (3.81)), can be written as an analog of Poisson's integral

$$v(x, t) = \int_{\mathbb{R}} G_{\alpha,1}(x - \xi, t)\Phi(\xi)d\xi.$$

We now transform this formula, expressing $\Phi(x)$ through $\varphi(x)$ using (3.88):

$$u(x, t) = v(x, t)\Big|_{x\in\mathbb{R}_+} = \int_0^\infty G_{\alpha,1}(x - \xi, t)\varphi(\xi)d\xi - \int_{-\infty}^0 G_{\alpha,1}(x - \xi, t)\varphi(-\xi)d\xi$$

$$= \int_0^\infty \left[G_{\alpha,1}(x - \xi, t) - G_{\alpha,1}(x + \xi, t)\right] \varphi(\xi)d\xi.$$

The final solution to the problem (3.76)–(3.78) with the boundary condition for $a = 0,\ b = 1$ can be written in the form

$$u(x, t) = \int_0^\infty G^D_{\alpha,1}(x, \xi, t)\varphi(\xi)d\xi. \tag{3.89}$$

where, taking into account (3.82) and (3.89), we get

$$G^D_{\alpha,1}(x,\xi,t) = G_{\alpha,1}(x-\xi,t) - G_{\alpha,1}(x+\xi,t)$$
$$= t^{\beta-1}\Bigg\{\frac{1}{\sqrt{\pi}|x-\xi|}H^{2,0}_{1,2}\left[\frac{(x-\xi)^2}{4t^\alpha}\Bigg|^{(1,\alpha)}_{(\frac{1}{2},1),(1,1)}\right] - \frac{1}{\sqrt{\pi}|x+\xi|}H^{2,0}_{1,2}\left[\frac{(x+\xi)^2}{4t^\alpha}\Bigg|^{(1,\alpha)}_{(\frac{1}{2},1),(1,1)}\right]\Bigg\}.$$

It is natural to call the function $G^D_{\alpha,1}(x,\xi,t)$, by analogy with the classical theory, Green's function of the Dirichlet problem for the fractional heat equation.

Using Green's function $G^D_{\alpha,1}(x,\xi,t)$, one can construct a solution to the problem (3.76)–(3.78) with a non-zero right-hand side and $\varphi = 0$ for $a = 0$, $b = 1$. According to the fractional analog of the Duhamel principle [30], it has the form

$$u(x,t) = \int_0^t\int_0^\infty G^D_{\alpha,\alpha}(x,\xi,t-\tau)f(\xi,\tau)d\xi d\tau. \tag{3.90}$$

where

$$G^D_{\alpha,\alpha}(x,\xi,t-\tau) = (t-\tau)^{\alpha-1}\Bigg\{\frac{1}{\sqrt{\pi}|x-\xi|}H^{2,0}_{1,2}\left[\frac{(x-\xi)^2}{4(t-\tau)^\alpha}\Bigg|^{(\alpha,\alpha)}_{(\frac{1}{2},1),(1,1)}\right] - \frac{1}{\sqrt{\pi}|x+\xi|}H^{2,0}_{1,2}\left[\frac{(x+\xi)^2}{4(t-\tau)^\alpha}\Bigg|^{(\alpha,\alpha)}_{(\frac{1}{2},1),(1,1)}\right]\Bigg\}. \tag{3.91}$$

Summing up the formulas (3.89) and (3.90), we obtain the solution to the problem (3.76)–(3.78) in the closed form

$$u(x,t) = \int_0^\infty G^D_{\alpha,1}(x,\xi,t)\varphi(\xi)d\xi + \int_0^t\int_0^\infty (t-\tau)^{\alpha-1}G^D_{\alpha,\alpha}(x,\xi,t-\tau)f(\xi,\tau)d\xi d\tau.$$

2. Neumann problem. The solution to the problem (3.76)–(3.78) with the Neumann boundary condition ($a = 1,\ b = 0$)) is constructed in a similar way, but in this case the function $\varphi(x)$ can be extended for $x \le 0$ in an even way:

$$\Phi(x) = \begin{cases} \varphi(x) & x \ge 0, \\ \varphi(-x) & x \le 0. \end{cases}$$

Repeating all the arguments carried out in the case of the Dirichlet boundary condition, we finally obtain the solution to the problem (3.76)–(3.78) for $a = 1$, $b = 0$

$$u(x,t)=\int_0^\infty G^N_{\alpha,1}(x,\xi,t)\varphi(\xi)d\xi+\int_0^t\int_0^\infty (t-\tau)^{\alpha-1}G^N_{\alpha,\alpha}(x,\xi,t-\tau)f(\xi,\tau)d\xi d\tau. \tag{3.92}$$

where

$$G^N_{\alpha,\beta}(x,\xi,t-\tau)=(t-\tau)^{\beta-1}\left\{\frac{1}{\sqrt{\pi}|x-\xi|}H^{2,0}_{1,2}\left[\frac{(x-\xi)^2}{4(t-\tau)^\alpha}\left|^{(\beta,\alpha)}_{(\frac{1}{2},1),(1,1)}\right.\right]\right.$$
$$\left.+\frac{1}{\sqrt{\pi}|x+\xi|}H^{2,0}_{1,2}\left[\frac{(x+\xi)^2}{4(t-\tau)^\alpha}\left|^{(\beta,\alpha)}_{(\frac{1}{2},1),(1,1)}\right.\right]\right\}. \tag{3.93}$$

It is also assumed that the matching condition $\varphi'(0)=0$ is satisfied.

3. Robin problem. Using Lemma and assuming $a=1,\ b=-h$ we investigate the more general problem (3.76)–(3.78).

Suppose that the function $\varphi(x)$ satisfies the matching condition $\varphi'(0)-h\varphi(0)=0$. According to Lemma, it is necessary to continue the function $\varphi(x)$ for $x\le 0$ in such a way that the function $\Phi'(x)-h\Phi(x)$ is odd, where $\Phi(x)$ is the continuation of the function $\varphi(x)$ on $\mathbb{R}$. Obviously $\Phi(x)=\varphi(x)$ for $x\in\mathbb{R}_+$. For determining the function $\Phi(x)$ for $x\le 0$ we obtain the Cauchy problem for the ordinary differential equation

$$\begin{cases}\Phi'(x)-h\Phi(x)=g(x), & x<0,\\ \Phi(0)=\varphi(0),\end{cases}$$

where $g(x):=-\varphi'(-x)+h\varphi(-x)$. Solving this problem, we find $\Phi(x)$ for $x\le 0$

$$\Phi(x)=\varphi(-x)+2h\int_0^x e^{h(x-z)}\varphi(-z)dz.$$

Thus, the $\varphi(x)$ function continues as follows:

$$\Phi(x)=\begin{cases}\varphi(x) & x\ge 0,\\ \varphi(-x)+2h\int_0^x e^{h(x-z)}\varphi(-z)dz, & x\le 0.\end{cases} \tag{3.94}$$

Writing the solution to the problem

$$\partial_t^\alpha u-u_{xx}=0,\ t>0,\ x\in\mathbb{R},\quad u(x,0)=\Phi(x),\quad x\in\mathbb{R}$$

in the form of an analog of Poisson's integral, where the function $\Phi(x)$ is determined by the formula (3.94):

$$u(x,t)=\int_{\mathbb{R}} G_{\alpha,1}(x-\xi,t)\Phi(\xi)d\xi=\int_0^\infty G_{\alpha,1}(x-\xi,t)\varphi(\xi)d\xi$$

$$+\int_{-\infty}^0 G_{\alpha,1}(x-\xi,t)\varphi(-\xi)d\xi+2h\int_{-\infty}^0 G_{\alpha,1}(x-\xi,t)\int_0^\xi e^{h(\xi-z)}\varphi(-z)dzd\xi,$$

after transformations we get

$$u(x,t)=\int_0^\infty\Big[G_{\alpha,1}(x-\xi,t)+G_{\alpha,1}(x+\xi,t)-2h\int_0^\infty G_{\alpha,1}(x+\xi+\eta,t)e^{-h\eta}d\eta\Big]\varphi(\xi)d\xi.$$

As a result, we obtain an expression for Green's function of the Robin problem for the fractional heat equation on the half-line:

$$G^R_{\alpha,1}(x,\xi,t)=G_{\alpha,1}(x-\xi,t)+G_{\alpha,1}(x+\xi,t)-2h\int_0^\infty G_{\alpha,1}(x+\xi+\eta,t)e^{-h\eta}d\eta.$$

or, taking into account the formula (3.82)

$$\begin{aligned}G^R_{\alpha,1}(x,\xi,t)=t^{\beta-1}\Bigg\{&\frac{1}{\sqrt{\pi}|x-\xi|}H^{2,0}_{1,2}\left[\frac{(x-\xi)^2}{4t^\alpha}\,\middle|\,\begin{matrix}(1,\alpha)\\(\frac{1}{2},1),(1,1)\end{matrix}\right]\\&-\frac{1}{\sqrt{\pi}|x+\xi|}H^{2,0}_{1,2}\left[\frac{(x+\xi)^2}{4t^\alpha}\,\middle|\,\begin{matrix}(1,\alpha)\\(\frac{1}{2},1),(1,1)\end{matrix}\right]\\-2h\int_0^\infty&\frac{1}{\sqrt{\pi}|x+\xi+\eta|}H^{2,0}_{1,2}\left[\frac{(x+\xi+\eta)^2}{4t^\alpha}\,\middle|\,\begin{matrix}(1,\alpha)\\(\frac{1}{2},1),(1,1)\end{matrix}\right]e^{-h\eta}d\eta\Bigg\}.\end{aligned}$$

With the help of the constructed Green's function the solution to the initial-boundary value problem for an inhomogeneous fractional Eq. (3.76) with the initial condition (3.77) and the homogeneous Robin condition (3.78) for $a=1,\ b=-h$ can be written in the form

$$u(x,t)=\int_0^\infty G^R_{\alpha,1}(x,\xi,t)\varphi(\xi)d\xi+\int_0^t\int_0^\infty G^R_{\alpha,\alpha}(x,\xi,t-\tau)f(\xi,\tau)d\xi d\tau.$$

where

$$G^R_{\alpha,\beta}(x,\xi,t-\tau) = (t-\tau)^{\beta-1}\Bigg\{\frac{1}{\sqrt{\pi}|x-\xi|}H^{2,0}_{1,2}\left[\frac{(x-\xi)^2}{4(t-\tau)^\alpha}\Big|^{(\beta,\alpha)}_{(\frac{1}{2},1),(1,1)}\right]$$
$$-\frac{1}{\sqrt{\pi}|x+\xi|}H^{2,0}_{1,2}\left[\frac{(x+\xi)^2}{4(t-\tau)^\alpha}\Big|^{(\beta,\alpha)}_{(\frac{1}{2},1),(1,1)}\right]$$
$$-2h\int\limits_0^\infty \frac{1}{\sqrt{\pi}|x+\xi+\eta|}H^{2,0}_{1,2}\left[\frac{(x+\xi+\eta)^2}{4(t-\tau)^\alpha}\Big|^{(\beta,\alpha)}_{(\frac{1}{2},1),(1,1)}\right]e^{-h\eta}d\eta\Bigg\}. \tag{3.95}$$

3.3.3 *Particular Cases of Green's Functions*

Representing the $H-$ function by means of a Mellin-Barnes type integral in the following form [10, p. 2]

$$H^{2,0}_{1,2}\left[\frac{x^2}{4t^\alpha}\Big|^{(\beta,\alpha)}_{(\frac{1}{2},1),(1,1)}\right] = \frac{1}{2\pi i}\int_{\sigma-i\infty}^{\sigma+i\infty}\frac{\Gamma\left(\frac{1}{2}+s\right)\Gamma\left(1+s\right)}{\Gamma\left(\beta+\alpha s\right)}\left(\frac{|x|^2}{4t^\alpha}\right)^{-s}ds$$

with $\sigma \in (-\alpha, 1)$, the fundamental solution (3.82) can be rewritten as

$$G_{\alpha,\beta}(x,t) = \frac{t^{\beta-1}}{\sqrt{\pi}|x|}\cdot\frac{1}{2\pi i}\int_{\sigma-i\infty}^{\sigma+i\infty}\frac{\Gamma\left(\frac{1}{2}+s\right)\Gamma\left(1+s\right)}{\Gamma\left(\beta+\alpha s\right)}\left(\frac{|x|^2}{4t^\alpha}\right)^{-s}ds. \tag{3.96}$$

In particular, $\alpha=\beta=1$ (classical heat equation) the representation (3.96) takes the form

$$G_{1,1}(x,t) = \frac{t^{\beta-1}}{\sqrt{\pi}|x|}\cdot\frac{1}{2\pi i}\int_{\sigma-i\infty}^{\sigma+i\infty}\Gamma\left(\frac{1}{2}+s\right)\left(\frac{|x|^2}{4t}\right)^{-s}ds$$
$$= \frac{1}{(2\sqrt{\pi t})}\cdot\frac{1}{2\pi i}\int_{\sigma^*-i\infty}^{\sigma^*+i\infty}\Gamma(s)\left(\frac{|x|^2}{4t}\right)^{-s}ds,\ \sigma^*>0.$$

The contour of integration in the integral of the last formula can be transformed to the loop $L_{-\infty}$, which is started and ended at $-\infty$, encircling all poles $s_j=-j,\ j=0,1,2,\cdots$ of the function $\Gamma(s)$. In view of the Jordan lemma, the Cauchy residue theorem and the formula $res_{s=-j}\Gamma(s) = (-1)^j/(j!),\ j=0,1,2,\cdots$, we get the following equality

$$G_{1,1}(x,t) = \frac{1}{2\sqrt{\pi t}}\cdot\frac{1}{2\pi i}\int_{\sigma^*-i\infty}^{\sigma^*+i\infty}\Gamma\left(s\right)\left(\frac{x^2}{4t}\right)^{-s}ds = \frac{1}{2\sqrt{\pi t}}\sum_{j=0}^{\infty}\frac{(-1)^j}{j!}\left(\frac{x^2}{4t}\right)^{-j}.$$

Thus the fundamental solution $G_{1,1}$ to the one-dimensional heat equation takes its classical form

$$G_{1,1}(x,t) = \frac{1}{2\sqrt{\pi t}} \sum_{j=0}^{\infty} \frac{(-1)^j}{j!} \left(\frac{|x|^2}{4t}\right)^{-j} = \frac{1}{2\sqrt{\pi t}} \exp\left(-\frac{x^2}{4t}\right).$$

Then, as follow from the formulas (3.91), (3.93), (3.95) ($\alpha = \beta$) the Green functions of the problem (3.76)–(3.78) in the case of $\alpha = 1$ (i.e., (1) is a one-dimensional classical inhomogeneous heat equation) have the following forms:

(1) in the case of the Dirichlet boundary condition

$$G_{1,1}^D(x,\xi,t-\tau) = \frac{1}{2\sqrt{\pi(t-\tau)}}\left[\exp\left(-\frac{(x-\xi)^2}{4(t-\tau)}\right) - \exp\left(-\frac{(x+\xi)^2}{4(t-\tau)}\right)\right];$$

(2) in the case of the Neumann boundary condition

$$G_{1,1}^N(x,\xi,t-\tau) = \frac{1}{2\sqrt{\pi(t-\tau)}}\left[\exp\left(-\frac{(x-\xi)^2}{4(t-\tau)}\right) + \exp\left(-\frac{(x+\xi)^2}{4(t-\tau)}\right)\right];$$

(3) in the case of the Robin boundary condition

$$G_{1,1}^R(x,\xi,t-\tau) = \frac{1}{2\sqrt{\pi(t-\tau)}}\left[\exp\left(-\frac{(x-\xi)^2}{4(t-\tau)}\right) + \exp\left(-\frac{(x+\xi)^2}{4(t-\tau)}\right) - 2h\int_0^{\infty} \exp\left(-\frac{(x+\xi+\eta)^2}{4(t-\tau)} - h\eta\right) d\eta\right].$$

3.4 Conclusion

In Sects. 3.1 and 3.2, we established the existence of solutions on the interval $[0, T^*]$, proved the global uniqueness, and derived stability estimates for the inverse problems associated with fractional diffusion equations involving Caputo and Hilfer derivatives, where the direct problems are formulated as Cauchy problems. These theorems are constructive, as they provide an explicit algorithm for determining T^*.

By solving the system of integral equations equivalent to the inverse problems, for instance, using the method of successive approximations (whose convergence to the solution is ensured by the proved theorems), we uniquely determine the unknown functions $\vartheta(x,t)$ and $q(t)$ for $T \in [0, T^*]$ in the domain D_T. Consequently, we define the solution to the inverse problem, specifically the function $q(t)$ on the segment $[0, T^*]$.

In Sect. 3.3, we applied the continuation method to extend the solution from the infinite axis, deriving exact solutions for the first, second, and third initial-boundary value problems for the one-dimensional time-fractional diffusion equation with a Caputo fractional derivative of order α $(0 < \alpha < 1)$. These solutions are expressed in terms of Green's functions corresponding to Dirichlet, Neumann, and Robin boundary conditions. Green's functions are formulated using the Fox H-function, a widely used tool in fractional calculus. Additionally, we demonstrated that the derived formulas coincide with the well-known solutions of the corresponding problems in the limiting case $\alpha \to 1^-$.

The constructed formulas from Sect. 3.3 allow for further study of inverse problems concerning the determination of the zero coefficient in the fractional diffusion equation under Dirichlet, Neumann, or Robin boundary conditions, along with an overdetermination condition at the boundary of the spatial domain. The existence and uniqueness of solutions for such inverse problems can be established using an integral equation approach based on Green's functions, as applied in Sects. 3.1 and 3.2.

References

1. Schneider, W.R., Wyss, W.: Fractional diffusion and wave equations. J. Math. Phys. **30**, 134–144 (1989)
2. Kochubei, A.N.: Fractional-order diffusion. Differential Equations **26**, 485–492 (1990)
3. Voroshilov, A.A., Kilbas, A.A.: Conditions for the existence of a classic solution of the Cauchy problem for the diffusion-wave equation with Caputo partial derivative. Dokl. Akad. Nauk **414**(4), 1–4 (2007)
4. Eidelman, S.D., Kochubei, A.N.: Cauchy problem for fractional diffusion equations. J. Differ. Equ. **199**, 211–255 (2004)
5. Durdiev, D.K.: Inverse Coefficient Problem For the Time-fractional Diffusion Equation. Eurasian Journal of Mathematical and Computer Applications **9**(1), 44–54 (2021)
6. Durdiev, D.K., Rahmonov, A.A., Bozorov, Z.R.: A two-dimensional diffusion coefficient determination problem for the time-fractional equation. Math. Methods Appl. Sci. **44**(13), 10753–10761 (2021)
7. Durdiev, U.D.: Problem of determining the reaction coefficient in a fractional diffusion equation. Differential Equations **57**(9), 1195–1204 (2021)
8. Durdiev, D.K., Akylbayev, M., Maxumova, Z., Iskakova, A.: A 2d convolution kernel determination problem for the time-fractional diffusion equation. Lobachevskii J. Math. **45**(3), 1044–1058 (2024)
9. Durdiev, D.K., Rahmonov, A.A.: A multidimensional diffusion coefficient determination problem for the time-fractional equation. Turkish J. Math. **46**(6), 2250–2263 (2022)
10. Mathai, A.M., Saxena, R.K., Haubold, H.J.: The H-function: Theory and application. Springer, Berlin/Heidelberg (2010)
11. Prudnikov, A.P., Brychkov, Y.A., Marichev, O.I.: Integrals and series, vol. 3. More Special Functions, Gordon and Breach, New York (1990)
12. Srivastava, H.M., Gupta, K.C., Goyal, S.P.: The H-functions of one and two variables with applications. South Asian publishers, New York (1982)

13. Yang, X. J. 2021. The Y-function applied in the study of an anomalous diffusion. *Thermal Sci.* 25(6)B: 4465–4475
14. Yang, X. J., Yu, X. J. 2022. New generalization of the Y-function applied to model the anomalous diffusion. *Thermal Sci.* 26(2)A: 1069–1079
15. Tomovski, Z., Hilfer, R., Srivastava, N.M.: Fractional and operational calculus with generalized fractional derivatives operators and Mittag-Leffler type functions. Integr. Transform. Spec. Funct. **21**, 797–814 (2010)
16. Samko, S.G., Kilbas, A.A., Marichev, O.I.: Fractional integrals and derivatives. Gordon and Breach Science Publishers, Yveron, Theory and applications (1993)
17. Kilbas, A.A., Srivastava, H.M., Trujillo, J.J.: Theory and application of fractional differetial equations. Elsevier, North-Holland Mathematical Studies, Amsterdam (2006)
18. Henry, D. 1981. *Geometric theory of semi linear parabolic equations.* Berlin Germany
19. Hilfer, R.: Applications of Fractional Calculus in Physics. World Scientific, Singapore (2000)
20. Lukashchuk, S.Y.: Conservation laws for time-fractional subdiffusion and diffusion-wave equations. Nonlinear Dyn **80**, 791–802 (2015)
21. Hilfer, R.: Experimental evidence for fractional time evolution in glass forming materials. Chemical Phys **284**, 399–408 (2002)
22. Durdiev, D.K.: Inverse coefficient problem for the time-fractional diffusion equation with Hilfer operator. Math. Meth. Appl. Sci. **46**(16), 1–16 (2023)
23. Hilfer, R., Luchko, Y., Tomovski, Z.: Operational method for the solution of fractional differential equations with generalized Riemann-Liouville fractional derivatives. Frac. Calc. Appl. Anal. **12**(3), 299–318 (2009)
24. Boyadjiev, L., Luchko, Y.: Mellin Integral Transform Approach to Analyze the Multidimensional Diffusion-Wave Equations. Chaos, Solitons and Fractals **102**, 127–134 (2017)
25. Mainardi, F.: Fractional relaxation-oscillation and fractional diffusion-wave phenomena. Chaos Sol. Frac. **7**(9), 1461–1477 (1996)
26. Mainardi, F.: The fundamental solutions for the fractional diffusion-wave equation. Appl. Math. Lett. **9**(6), 23–28 (1996)
27. Metzler, R., Klafter, J.: Boundary value problems for fractional diffusion equations. Phys. A. **278**, 107–125 (2000)
28. Duan, J.S., Wang, Z., Fu, S.Z.: Fractional diffusion equation in half-space with Robin boundary condition. Cent. Eur. J. Phys. **11**(6), 799–805 (2013)
29. Durdiev, D.K.: Initial boundary value and inverse coefficient problems for one-dimensional fractional diffusion equation in a half-line. Lobachevskii J. Math. **45**(7), 3265–3279 (2024)
30. Umarov, S., Saydamatov, E.: A fractional analog of the Duhamel principle. Fract. Calc. Appl. Anal. **9**(1), 57–69 (2006)

Chapter 4
Carleman Estimate Method in Inverse Problems

This chapter discusses the inverse source and inverse coefficient problems for a first- and half-order time-fractional diffusion equation by the Carleman estimate method.

In Sect. 4.1, the first- and half-order time-fractional equation with the zero initial condition is considered. An inverse source problem of determining the time-independent source factor by the spatial data at an arbitrarily fixed time and the conditional stability estimate of Hölder type in this inverse problem are investigated. A method is based on the Bukhgeim-Klibanov method [1] by means of the Carleman estimate. Also the Carleman estimate for the first- and half-order time-fractional diffusion equation is derived.

In Sect. 4.2, Lipschitz stability in inverse source and inverse coefficient problems for a first- and half-order time-fractional diffusion equation are established by the Carleman estimate method.

In writing this chapter, the articles [2, 3] were used, in which the method of the Carleman estimates was applied to the study of inverse problems for the first- and half-order time-fractional equation.

4.1 Hölder Stability Estimate in an Inverse Source Problem

A Carleman estimate is a weighted L^2 inequality for a solution of a partial differential equation and is introduced by Carleman [4] to prove the uniqueness in the Cauchy problem for a first-order elliptic system in $\mathbb{R}^2$ with non-analytic coefficients. Since Carleman's paper, there have been a lot of work for Carleman estimates and their applications (see, e.g., monographs [5–7] and references therein). In this section we establish the local Carleman estimate for a first- and half-order time-fractional diffusion equation with the zero initial condition by using the local Carleman estimate for parabolic equations. The main idea of the proof is a transformation from a first- and

D. K. Durdiev, *Inverse Problems for Fractional Diffusion Equations*,
Industrial and Applied Mathematics, https://doi.org/10.1007/978-981-96-5338-6_4

half-order time-fractional diffusion equation to an integer-order partial differential equation.

As the most relevant work to problems considered in this paragraph, we may refer to [8, 9]. In paper [9], it was established the conditional Hölder type stability estimate in determining the time-independent source factor of a single-term time FDE from the additional data at an arbitrarily fixed time by using the local Carleman estimate derived in [10]. In [8], the Lipschitz type stability estimates by the additional data at an arbitrarily fixed time and boundary/interior data in inverse source problems for a single term time FDE with suitable boundary conditions and the zero initial condition by using the Carleman estimates with boundary/interior data were obtained.

Let $T > 0$, $\Omega \subset \mathbb{R}^n$ be a bounded domain with sufficiently smooth boundary $\partial\Omega$. We set $Q = \Omega \times (0, T)$. We consider the following a first- and half-order time-fractional diffusion equation and the initial condition:

$$\left(\rho_1 \partial_t + \rho_2 \partial_t^{\frac{1}{2}} - L\right) u(x,t) = g(x,t), \quad (x,t) \in Q, \tag{4.1}$$

$$u(x,0) = 0, \quad x \in \Omega, \tag{4.2}$$

where $\rho_1 > 0$, $\rho_2 \neq 0$ are constants, $\partial_t^{\frac{1}{2}}$ is a Caputo type fractional derivative of half order:

$$\partial_t^{\frac{1}{2}} u(x,t) := \frac{1}{\Gamma\left(\frac{1}{2}\right)} \int_0^t \frac{\partial_t u(x,\tau)}{(t-\tau)^{\frac{1}{2}}} d\tau, \quad (x,t) \in Q,$$

and L is a symmetric uniformly elliptic operator:

$$Lu(x,t) := \sum_{i,j=1}^{n} \partial_i \left(a_{ij}(x)\partial_j u(x,t)\right)$$

$$+ \sum_{j=1}^{n} b_j(x)\partial_j u(x,t) + c(x)u(x,t), \quad (x,t) \in Q. \tag{4.3}$$

Throughout this section, we assume that $a_{ij} \in C^3(\overline{\Omega})$, $a_{ij} = a_{ji} (1 \le i, j \le n)$, and that there exists a constant $m > 0$ such that

$$\frac{1}{m}|\xi|^2 \le \sum_{i,j=1}^{n} a_{ij}(x)\xi_i\xi_j \le m|\xi|^2, \quad \xi = (\xi_1, \dots, \xi_n) \in \mathbb{R}^n, x \in \overline{\Omega}, \tag{4.4}$$

and $b_j \in C^2\left(\overline{\Omega}\right)$ $(1 \le j \le n)$, $c \in C^2\left(\overline{\Omega}\right)$. Here and henceforth we use notations $\partial_t = \frac{\partial}{\partial t}$, $\partial_i = \frac{\partial}{\partial x_i} (i = 1, 2, \dots, n)$. We also use the multi index $\alpha = (\alpha_1, \alpha_2, \dots, \alpha_n)$ with $\alpha_j \in \mathbb{N} \cup \{0\} (j = 1, 2, \dots, n)$, $\partial_x^\alpha = \partial_1^{\alpha_1} \partial_2^{\alpha_2} \cdots \partial_n^{\alpha_n}$, $|\alpha| = \alpha_1 + \alpha_2 + \cdots + \alpha_n$. Let $\nu = \nu(x)$ be the outwards unit normal vector to $\partial\Omega$ at x and let $\partial_\nu = \nu \cdot \nabla$.

In this section we consider an inverse problem of determining the time-independent factor of the source term g of (4.1) by the additional data $u\,(x, t_0)\,, x \in \Omega$ where $t_0 \in (0, T)$ is an arbitrarily fixed time. We establish the conditional Hölder type stability estimate in our inverse source problem.

4.1.1 Hölder Stability Estimate

Let $t_0 \in (0, T)$ be an arbitrarily fixed time. Let γ be an arbitrarily fixed open connected sub-boundary of $\partial\Omega$ and let $\omega \Subset \Omega \cup \gamma$ be a sub-domain with the smooth boundary $\partial\omega$. We set $\Gamma = \gamma \times (0, T)$. We assume that

$$\begin{cases} R \in C^2\left([0, T]; C^2(\overline{\Omega})\right) \cap C^3([0, T]; C(\overline{\Omega})), \quad \partial_t^{\frac{1}{2}} R \in C^2([0, T]; C(\overline{\Omega})), \\ \text{and } |R\,(x, t_0)| > 0, \quad x \in \overline{\Omega}. \end{cases} \tag{4.5}$$

We consider the following equation and the initial condition:

$$\left(\rho_1 \partial_t + \rho_2 \partial_t^{\frac{1}{2}} - L\right) u(x, t) = f(x) R(x, t), \ (x, t) \in Q, \tag{4.6}$$

$$u(x, 0) = 0, \ x \in \Omega, \tag{4.7}$$

$$u(x, t) = \partial_\nu u(x, t) = 0, \ (x, t) \in \Gamma, \tag{4.8}$$

and we investigate the following inverse problem.
Inverse source problem: Determine $f(x), x \in \omega$ by data $u\,(x, t_0)\,, x \in \Omega$.

Theorem 4.1 *Let $u, \partial_t u, \partial_t^2 u \in L^2\left(0, T; H^4(\Omega)\right) \cap H^1\left(0, T; H^2(\Omega)\right) \cap H^2\left(0, T; L^2(\Omega)\right)$ and u satisfies (4.6)-(4.8) and the additional boundary condition $\partial_x^\alpha u(x, t) = 0, (x, t) \in \Gamma, |\alpha| = 2$. We suppose that $f \in H^2(\Omega)$ with $f = \partial_\nu f = 0$ on γ and R satisfies (4.5), Moreover we assume that there exists $M > 0$ such that*

$$\begin{cases} \left\|\partial_t^k u\right\|_{L^2(0,T;H^4(\Omega))} + \left\|\partial_t^k u\right\|_{H^1(0,T;H^2(\Omega))} \\ \qquad + \left\|\partial_t^k u\right\|_{H^2(0,T;L^2(\Omega))} \le M, \quad k = 0, 1, 2, \\ \|f\|_{H^2(\Omega)} \le M. \end{cases} \tag{4.9}$$

Then there exist constants $C > 0$ and $\kappa \in (0, 1)$ such that

$$\|f\|_{H^2(\omega)} \le C\, \|u\,(\cdot, t_0)\|^\kappa_{H^4(\Omega)}\,. \tag{4.10}$$

Remark 4.1 For simplicity of the statement of Theorem 4.1, we observed the additional data $u\,(\cdot, t_0)$ on the whole domain Ω. More precisely, it is sufficient to observe the data $u\,(\cdot, t_0)$ on the sub-domain Ω_2 of Ω such that $\omega \subset \Omega_2$, where Ω_2 is defined in the proof of Theorem 4.1 and the domain depends on the weight function of the Carleman estimate.

Remark 4.2 It is required to assume that the additional boundary condition $\partial_x^\alpha u(x,t) = 0$, $(x,t) \in \Gamma$, $|\alpha| = 2$ when we estimate the boundary term of the Carleman estimate in the proof of our Theorem 4.1. In the case of $n = 1$, that is, $\Omega \subset \mathbb{R}$ is the one-dimensional case in space, we may relax this additional boundary condition. Indeed,by $f = 0$ on γ and $u = \partial_1 u = 0$ on Γ,(4.6) implies that $\partial_1^2 u = 0$ on Γ. In the multidimensional case in space, taking a small sub-domain $\tilde{\omega} \subset \Omega$ such that $\Gamma \subset \partial\tilde{\omega}$, we may also assume that $u(x,t) = 0$, $(x,t) \in \tilde{\omega}$ instead of boundary conditions.

Remark 4.3 Comparing our result (Theorem 4.1) with the result by Yamamoto and Zhang [9], they proved the conditional stability estimate of Holder type in the case of $\rho_1 = 0$, $\rho_2 = 1$ in (4.1) and $n = 1$. On the other hand, we assume that $\rho_1 > 0$, $\rho_2 \neq 0$ in (4.1) and $n \in \mathbb{N}$ in our result.

4.1.2 Reduction of the First- and Half-Order Fractional Equation

We reduce (4.1) to the following equation:

$$\rho_2^2 \partial_t u(x,t) - (\rho_1 \partial_t - L)^2 u(x,t) = G(x,t), \quad (x,t) \in Q$$

and derive the Carleman estimate for the above equation. Let us justify the above reduction from (4.1). Using the idea of Xu, Cheng and Yamamoto [10], we may prove the following Lemma.

Lemma 4.1 *If* $u \in C\left([0,T]; H^4(\Omega)\right) \cap C^1\left([0,T]; H^2(\Omega)\right) \cap C^2\left([0,T]; L^2(\Omega)\right)$ *satisfies (4.1) and (4.2), then* u *satisfies*

$$\rho_2^2 \partial_t u(x,t) - (\rho_1 \partial_t - L)^2 u(x,t) = G(x,t), \quad (x,t) \in Q, \tag{4.11}$$

where

$$G(x,t) = \left[\rho_2 \partial_t^{\frac{1}{2}} - (\rho_1 \partial_t - L)\right] g(x,t) + \frac{\rho_2 g(x,0)}{\sqrt{\pi t}}, \quad (x,t) \in Q. \tag{4.12}$$

Proof of Lemma. Applying $\left[\rho_2 \partial_t^{\frac{1}{2}} - (\rho_1 \partial_t - L)\right]$ to (4.1), we have

$$\left[\rho_2 \partial_t^{\frac{1}{2}} - (\rho_1 \partial_t - L)\right] \left(\rho_2 \partial_t^{\frac{1}{2}} + \rho_1 \partial_t - L\right) u = \left[\rho_2 \partial_t^{\frac{1}{2}} - (\rho_1 \partial_t - L)\right] g.$$

Expanding the left-hand side of the above equation, we get

$$\rho_2^2 \partial_t^{\frac{1}{2}} \partial_t^{\frac{1}{2}} u + \rho_1 \rho_2 \partial_t^{\frac{1}{2}} \partial_t u - \rho_1 \rho_2 \partial_t \partial_t^{\frac{1}{2}} u - \rho_1^2 \partial_t^2 u + 2\rho_1 \partial_t L u - L^2 u$$

$$= \left[\rho_2 \partial_t^{\frac{1}{2}} - (\rho_1 \partial_t - L)\right] g.$$

Since $(\rho_1 \partial_t - L)^2 = \rho_1^2 \partial_t^2 - 2\rho_1 \partial_t L + L^2$, we have

$$\rho_2^2 \partial_t^{\frac{1}{2}} \partial_t^{\frac{1}{2}} u + \rho_1 \rho_2 \partial_t^{\frac{1}{2}} \partial_t u - \rho_1 \rho_2 \partial_t \partial_t^{\frac{1}{2}} u - (\rho_1 \partial_t - L)^2 u = \left[\rho_2 \partial_t^{\frac{1}{2}} - (\rho_1 \partial_t - L)\right] g. \tag{4.13}$$

Henceforth we calculate $\partial_t^{\frac{1}{2}} \partial_t^{\frac{1}{2}} u$, $\partial_t^{\frac{1}{2}} \partial_t u - \partial_t \partial_t^{\frac{1}{2}} u$.

Let $\frac{\partial^\alpha}{\partial t^\alpha}$ denote the Riemann-Liouville fractional derivative. By the relations between the Riemann-Liouville and the Caputo fractional derivative (see Sect. 1.1.1, formula (1.7)), we have

$$\partial_t^{\frac{1}{2}} u(x,t) = \frac{\partial^{\frac{1}{2}}}{\partial t^{\frac{1}{2}}} u(x,t) - \frac{u(x,0)}{\sqrt{\pi t}}, \quad (x,t) \in Q \tag{4.14}$$

and

$$\frac{\partial^{\frac{1}{2}}}{\partial t^{\frac{1}{2}}} \frac{\partial^{\frac{1}{2}}}{\partial t^{\frac{1}{2}}} u(x,t) = \partial_t u(x,t), \quad (x,t) \in Q. \tag{4.15}$$

Setting $v = \partial_t^{\frac{1}{2}} u$, (4.14) gives us

$$\partial_t^{\frac{1}{2}} v(x,t) = \frac{\partial^{\frac{1}{2}}}{\partial t^{\frac{1}{2}}} v(x,t) - \frac{v(x,0)}{\sqrt{\pi t}}, \quad (x,t) \in Q. \tag{4.16}$$

By (4.15) and (4.2), we see that

$$\frac{\partial^{\frac{1}{2}}}{\partial t^{\frac{1}{2}}} v(x,t) = \frac{\partial^{\frac{1}{2}}}{\partial t^{\frac{1}{2}}} \partial_t^{\frac{1}{2}} u(x,t) = \frac{\partial^{\frac{1}{2}}}{\partial t^{\frac{1}{2}}} \frac{\partial^{\frac{1}{2}}}{\partial t^{\frac{1}{2}}} u(x,t) = \partial_t u(x,t), \quad (x,t) \in Q.$$

Together this with (4.16), we have

$$\partial_t^{\frac{1}{2}} \partial_t^{\frac{1}{2}} u(x,t) = \partial_t u(x,t) - \frac{\partial_t^{\frac{1}{2}} u(x,0)}{\sqrt{\pi t}}, \quad (x,t) \in Q. \tag{4.17}$$

By the definition of the Caputo derivative,

$$\partial_t^{\frac{1}{2}} \partial_t u(x,t) = \frac{1}{\Gamma\left(\frac{1}{2}\right)} \int_0^t \frac{\partial_t^2 u(x,\tau)}{(t-\tau)^{\frac{1}{2}}} d\tau, \quad (x,t) \in Q \tag{4.18}$$

and

$$\partial_t^{\frac{1}{2}} u(x,t) = \frac{1}{\Gamma\left(\frac{1}{2}\right)} \int_0^t \frac{\partial_t u(x,\tau)}{(t-\tau)^{\frac{1}{2}}} d\tau$$
$$= \frac{1}{\Gamma\left(\frac{1}{2}\right)} \left(2t^{\frac{1}{2}} \partial_t u(x,0) + 2\int_0^t (t-\tau)^{\frac{1}{2}} \partial_t^2 u(x,\tau) d\tau \right), \quad (x,t) \in Q.$$

Therefore

$$\partial_t \partial_t^{\frac{1}{2}} u(x,t) = \frac{1}{\Gamma\left(\frac{1}{2}\right)} \left(\frac{1}{t^{\frac{1}{2}}} \partial_t u(x,0) + \int_0^t \frac{\partial_t^2 u(x,\tau)}{(t-\tau)^{\frac{1}{2}}} d\tau \right), \quad (x,t) \in Q. \tag{4.19}$$

By (4.18) and (4.19), we obtain

$$\partial_t^{\frac{1}{2}} \partial_t u(x,t) - \partial_t \partial_t^{\frac{1}{2}} u(x,t) = -\frac{\partial_t u(x,0)}{\sqrt{\pi t}}, \quad (x,t) \in Q. \tag{4.20}$$

Here we note that $\Gamma\left(\frac{1}{2}\right) = \sqrt{\pi}$. (4.17) and (4.20) yield

$$\rho_2^2 \partial_t^{\frac{1}{2}} \partial_t^{\frac{1}{2}} u(x,t) + \rho_1\rho_2 \partial_t^{\frac{1}{2}} \partial_t u(x,t) - \rho_1\rho_2 \partial_t \partial_t^{\frac{1}{2}} u(x,t)$$
$$= \rho_2^2 \partial_t u(x,t) - \frac{\rho_2^2 \partial_t^{\frac{1}{2}} u(x,0)}{\sqrt{\pi t}} - \frac{\rho_1\rho_2 \partial_t u(x,0)}{\sqrt{\pi t}}$$
$$= \rho_2^2 \partial_t u(x,t) - \frac{\rho_2}{\sqrt{\pi t}} \left(\rho_2 \partial_t^{\frac{1}{2}} u(x,0) + \rho_1 \partial_t u(x,0) \right), \quad (x,t) \in Q. \tag{4.21}$$

By (4.1) and (4.2), we see that

$$\rho_2 \partial_t^{\frac{1}{2}} u(x,0) + \rho_1 \partial_t u(x,0) = g(x,0), \quad x \in \Omega.$$

Combining this with (4.21), we have

$$\rho_2^2 \partial_t^{\frac{1}{2}} \partial_t^{\frac{1}{2}} u(x,t) + \rho_1\rho_2 \partial_t^{\frac{1}{2}} \partial_t u(x,t) - \rho_1\rho_2 \partial_t \partial_t^{\frac{1}{2}} u(x,t)$$
$$= \rho_2^2 \partial_t u(x,t) - \frac{\rho_2 g(x,0)}{\sqrt{\pi t}}, \quad (x,t) \in Q. \tag{4.22}$$

By (4.13) and (4.22), we conclude that

$$\rho_2^2 \partial_t u(x,t) - (\rho_1 \partial_t - L)^2 u(x,t)$$
$$= \left[\rho_2 \partial_t^{\frac{1}{2}} - (\rho_1 \partial_t - L) \right] g(x,t) + \frac{\rho_2 g(x,0)}{\sqrt{\pi t}}, \quad (x,t) \in Q.$$

Thus we complete the proof.

4.1.3 Carleman Estimate for a Fractional Equation with Uniform Elliptic Operator

Let us introduce the weight function for the Carleman estimate. Let $\beta > 0$, $t_0 \in (0, T)$. Let $d \in C^2(\overline{\Omega})$ satisfy $|\nabla d| \neq 0$ in $\overline{\Omega}$. We set the weight function:

$$\varphi(x,t) = e^{\lambda\psi(x,t)}, \quad \psi(x,t) = d(x) - \beta\,(t - t_0)^2, \quad (x,t) \in Q.$$

Let $V \subset Q$ be a sub-domain with the smooth boundary ∂V. And we set $w(x,t) = \rho_1 \partial_t u(x,t) - Lu(x,t)$, $(x,t) \in Q$.

Now we are ready to state the Carleman estimate for (4.1).

Theorem 4.2 *There exists $\lambda_0 > 0$ such that for any $\lambda > \lambda_0$, we can choose $s_0(\lambda) > 0$ satisfying: there exists $C = C\,(s_0, \lambda_0) > 0$ such that*

$$\begin{aligned}
&\int_V \left[\frac{1}{s^2}\left(\left|\partial_t^2 u\right|^2 + \sum_{i,j=1}^{n} \left|\partial_t\partial_i\partial_j u\right|^2 \right) + \lambda^2 \left|\nabla\partial_t u\right|^2 \right. \\
&\left. + s^2\lambda^4 \left(|\partial_t u|^2 + \sum_{i,j=1}^{n} \left|\partial_i\partial_j u\right|^2 \right) + s^4\lambda^6 |\nabla u|^2 + s^6\lambda^8 |u|^2 \right] e^{2s\varphi} dx dt \\
&\le C \int_V \left|\left[\rho_2^2\partial_t - (\rho_1\partial_t - L)^2\right] u\right|^2 e^{2s\varphi} dx dt + C e^{Cs} B
\end{aligned} \tag{4.23}$$

for all $s > s_0$ and all $u \in L^2\left(0, T; H^4(\Omega)\right) \cap H^1\left(0, T; H^2(\Omega)\right) \cap H^2\left(0, T; L^2(\Omega)\right)$, where B is the following boundary term:

$$B = \int_{\partial V} \left(|\nabla w|^2 + |\partial_t w|^2 + |w|^2 + |\nabla\partial_t u|^2 + |\nabla u|^2 + \left|\partial_t^2 u\right|^2 + |\partial_t u|^2 + |u|^2 \right) ds dt.$$

As Lemmas 4.2 and 4.3, we introduce previous results of Carleman estimates with two large parameters for the second-order partial differential equations. As for the proofs of two lemmas, see, e.g., [12–15]. Eller and Isakov [12] established Carleman estimates by using differential quadratic forms, an approach of Hörmander [5]. On the other hand, we see the direct derivation of Carleman estimates by using integrating parts in Yamamoto [15].

To prove Theorem 4.2, we use the following Carleman estimate for parabolic equations.

Lemma 4.2 *There exists $\lambda_0 > 0$ such that for any $\lambda > \lambda_0$, we can choose $s_0(\lambda) > 0$ satisfying: there exists $C = C\,(s_0, \lambda_0) > 0$ such that*

$$\int_V \left[\frac{1}{s} \left(|\partial_t v|^2 + \sum_{i,j=1}^n \left|\partial_i \partial_j v\right|^2 \right) + s\lambda^2 |\nabla v|^2 + s^3\lambda^4 |v|^2 \right] e^{2s\varphi} dxdt$$

$$\le C \int_V |(\rho_1 \partial_t - L)\, v|^2 \, e^{2s\varphi} dxdt + Ce^{Cs} \int_{\partial V} \left(|\nabla v|^2 + |\partial_t v|^2 + |v|^2 \right) dsdt$$

for all $s > s_0$ *and all* $v \in L^2\left(0, T; H^2(\Omega)\right) \cap H^1\left(0, T; L^2(\Omega)\right)$.

Moreover we state the Carleman estimate for the elliptic equation which we use in the proof of the stability estimate in our inverse problem.

Let $\tilde{L}$ be a symmetric uniformly elliptic operator:

$$\widetilde{L}\widetilde{v}(x) := \sum_{i,j=1}^n \partial_i \left(\widetilde{a}_{ij}(x) \partial_j \widetilde{v}(x)\right) + \sum_{j=1}^n \widetilde{b}_j(x) \partial_j \widetilde{v}(x) + \widetilde{c}(x)\widetilde{v}(x),\ \widetilde{v} \in \Omega.$$

We assume that $\tilde{a}_{ij} \in C^1(\overline{\Omega}), \tilde{a}_{ij} = \tilde{a}_{ji} (1 \le i, j \le n)$ and that there exists a constant $\tilde{m} > 0$ such that

$$\frac{1}{\tilde{m}} |\xi|^2 \le \sum_{i,j=1}^n \tilde{a}_{ij}(x) \xi_i \xi_j \le \tilde{m} |\xi|^2, \quad \xi = (\xi_1, \ldots, \xi_n) \in \mathbb{R}^n, x \in \overline{\Omega},$$

and $\widetilde{b}_j \in C(\overline{\Omega}) (1 \le j \le n), \quad \widetilde{c} \in C(\overline{\Omega})$.

Set $\varphi_0(x) := \varphi\,(x, t_0) = e^{\lambda d(x)}, x \in \Omega$. Let $D \subset \Omega$ be a sub-domain with the smooth boundary ∂D. Then we have the following Lemma.

Lemma 4.3 *There exists* $\lambda_0 > 0$ *such that for any* $\lambda > \lambda_0$, *we can choose* $s_0(\lambda) > 0$ *satisfying: there exists* $C = C\,(s_0, \lambda_0) > 0$ *such that*

$$\int_D \left(\frac{1}{s} \sum_{i,j=1}^n \left|\partial_i \partial_j \tilde{v}\right|^2 + s\lambda^2 |\nabla \tilde{v}|^2 + s^3 \lambda^4 |\tilde{v}|^2 \right) e^{2s\varphi_0} dx$$

$$\le C \int_D |\tilde{L}\tilde{v}|^2 e^{2s\varphi_0} dx + Ce^{Cs} \int_{\partial D} \left(|\nabla \tilde{v}|^2 + |\tilde{v}|^2 \right) dsdt$$

for all $s > s_0$ *and all* $\tilde{v} \in H^2(\Omega)$.

Proof of Theorem 4.2. Since $w = \rho_1 \partial_t u - Lu$ in Q, we have

$$\rho_2^2 \partial_t u(x,t) - (\rho_1 \partial_t - L)\, w(x,t) = G(x,t), \quad (x,t) \in Q$$

by (4.11). So we obtain two parabolic equations with respect to u and w :

$$\rho_1 \partial_t w(x,t) - Lw(x,t) = \rho_2^2 \partial_t u(x,t) - G(x,t), \quad (x,t) \in Q, \tag{4.24}$$

$$\rho_1 \partial_t u(x,t) - Lu(x,t) = w(x,t), \quad (x,t) \in Q. \tag{4.25}$$

Applying Lemma 4.2 to (4.24) and (4.25), we have

$$\begin{aligned}
&\int_V \left(\frac{1}{s} |\partial_t w|^2 + s\lambda^2 |\nabla w|^2 + s^3\lambda^4 |w|^2 \right) e^{2s\varphi} dxdt \\
&\qquad \le C \int_V |\partial_t u|^2 e^{2s\varphi} dxdt + C \int_V |G|^2 e^{2s\varphi} dxdt \\
&\qquad + Ce^{Cs} \int_{\partial V} \left(|\nabla w|^2 + |\partial_t w|^2 + |w|^2 \right) dsdt,
\end{aligned} \tag{4.26}$$

and

$$\begin{aligned}
&\int_V \left[\frac{1}{s} \left(|\partial_t u|^2 + \sum_{i,j=1}^n |\partial_i \partial_j u|^2 \right) + s\lambda^2 |\nabla u|^2 + s^3\lambda^4 |u|^2 \right] e^{2s\varphi} dxdt \\
&\le C \int_V |w|^2 e^{2s\varphi} dxdt + Ce^{Cs} \int_{\partial V} \left(|\nabla u|^2 + |\partial_t u|^2 + |u|^2 \right) dsdt.
\end{aligned} \tag{4.27}$$

Substituting the estimate of $|\partial_t u|^2$ in (4.27) into the right-hand side of (4.26), we obtain

$$\begin{aligned}
&\int_V \left(\frac{1}{s} |\partial_t w|^2 + s\lambda^2 |\nabla w|^2 + s^3\lambda^4 |w|^2 \right) e^{2s\varphi} dxdt \\
&\qquad \le C \int_V s|w|^2 e^{2s\varphi} dxdt + C \int_V |G|^2 e^{2s\varphi} dxdt + Ce^{Cs} B_1,
\end{aligned}$$

where

$$B_1 = \int_{\partial V} \left(|\nabla w|^2 + |\partial_t w|^2 + |w|^2 + |\nabla u|^2 + |\partial_t u|^2 + |u|^2 \right) dsdt.$$

Taking sufficient large $s > 0$, we can absorb the first term on the right-hand side into the left-hand side, we have

$$\begin{aligned}
&\int_V \left(\frac{1}{s} |\partial_t w|^2 + s\lambda^2 |\nabla w|^2 + s^3\lambda^4 |w|^2 \right) e^{2s\varphi} dxdt \\
&\qquad \le C \int_V |G|^2 e^{2s\varphi} dxdt + Ce^{Cs} B_1.
\end{aligned} \tag{4.28}$$

Combining this with (4.27), we get

$$\int_V \left[s^2\lambda^4 \left(|\partial_t u|^2 + \sum_{i,j=1}^{n} \left|\partial_i\partial_j u\right|^2 \right) + s^4\lambda^6|\nabla u|^2 + s^6\lambda^8|u|^2 \right] e^{2s\varphi} dxdt$$

$$\le C \int_V |G|^2 e^{2s\varphi} dxdt + Ce^{Cs} B_1. \tag{4.29}$$

By (4.25) and (4.28), we have

$$\int_V \frac{1}{s} \left|\partial_t\left(\rho_1 \partial_t u - Lu\right)\right|^2 e^{2s\varphi} dxdt \le C \int_V |G|^2 e^{2s\varphi} dxdt + Ce^{Cs} B_1. \tag{4.30}$$

Setting $u_0 = \partial_t u$, the left-hand side of (4.30) gives us

$$\int_V \frac{1}{s} \left|\rho_1 \partial_t u_0 - Lu_0\right|^2 e^{2s\varphi} dxdt \le C \int_V |G|^2 e^{2s\varphi} dxdt + Ce^{Cs} B_1. \tag{4.31}$$

By Lemma 4.2, we may estimate the left-hand side of (4.31) from below and we get

$$\int_V \left[\frac{1}{s^2} \left(|\partial_t u_0|^2 + \sum_{i,j=1}^{n} \left|\partial_i\partial_j u_0\right|^2 \right) + \lambda^2 \left|\nabla u_0\right|^2 + s^2\lambda^4 \left|u_0\right|^2 \right] e^{2s\varphi} dxdt$$

$$\le C \int_V |G|^2 e^{2s\varphi} dxdt + Ce^{Cs} \int_{\partial V} \left(|\nabla u_0|^2 + |\partial_t u_0|^2 + |u_0|^2\right) dsdt + Ce^{Cs} B_1,$$

that is,

$$\int_V \left[\frac{1}{s^2} \left(\left|\partial_t^2 u\right|^2 + \sum_{i,j=1}^{n} \left|\partial_t\partial_i\partial_j u\right|^2 \right) + \lambda^2 \left|\nabla \partial_t u\right|^2 + s^2\lambda^4 \left|\partial_t u\right|^2 \right] e^{2s\varphi} dxdt$$

$$\le C \int_V |G|^2 e^{2s\varphi} dxdt + Ce^{Cs} B. \tag{4.32}$$

Together this with (4.29), we have (4.23).

4.1.4 Proof of the Hölder Stability Estimate

We choose a suitable weight functions φ, that is, $\beta > 0$ and a distance function d. Take $\delta > 0$ such that

$$0 < t_0 - 2\delta < t_0 < t_0 + 2\delta < T.$$

Taking a bounded domain Ω_0 with the smooth boundary $\partial\Omega_0$ such that

$$\Omega \subset \Omega_0, \quad \Omega \neq \Omega_0, \quad \bar{\gamma} = \overline{\partial\Omega \cap \Omega_0}, \quad \partial\Omega\backslash\gamma \subset \partial\Omega_0,$$

we choose an open subset $\omega_0 \Subset \Omega_0\backslash\overline{\Omega}$. There exists a distance function $d \in C^2(\mathbb{R}^n)$ such that

$$d(x) > 0, x \in \Omega_0, \quad d(x) = 0, x \in \partial\Omega_0, \quad |\nabla d(x)| > 0, x \in \overline{\Omega}.$$

The existence of such a function is proved in [16] (see the proof of Lemma 1.1 in [16]). We take $\varepsilon_0 \in (0, 1)$ such that

$$\omega \subset \left\{x \in \Omega_0 \mid d(x) > \varepsilon\|d\|_{C(\overline{\Omega_0})}\right\} \cap \overline{\Omega}, \quad \varepsilon \in (0, \varepsilon_0].$$

Take $\beta > 0$ such that

$$\frac{\|d\|_{C(\overline{\Omega_0})}}{4\delta^2} < \beta < \frac{\|d\|_{C(\overline{\Omega_0})}}{3\delta^2}.$$

Fixing $\varepsilon \in (0, \varepsilon_0]$, we set

$$\mu_k = \varepsilon\left(\frac{k}{3}\|d\|_{C(\overline{\Omega_0})} - \beta\delta^2\right) > 0, \quad k = 1, 2, 3.$$

We set $Q_k = \{(x,t) \in Q \mid \psi(x,t) > \mu_k\}$, $Q_k^- = Q_k \cap \{(x,t) \in Q \mid t < t_0\}$, $\Omega_k = Q_k \cap \{(x,t) \in Q \mid t = t_0\}$ for $k = 1, 2, 3$. Since $0 < \mu_1 < \mu_2 < \mu_3$, we have

$$\omega \times \left(t_0 - \sqrt{\varepsilon}\delta, t_0 + \sqrt{\varepsilon}\delta\right) \subset Q_3 \subset Q_2 \subset Q_1 \subset \overline{\Omega} \times (t_0 - 2\delta, t_0 + 2\delta) \subset \bar{Q}$$

and $\omega \subset \Omega_3 \subset \Omega_2 \subset \Omega_1 \subset \overline{\Omega}$. For this choice of the weight function when we are given the interior domain ω and the sub-boundary γ, we may refer to §5 in Yamamoto [15].

By Lemma 4.1, (4.6) gives us the following equation

$$\rho_2^2\partial_t u(x,t) - (\rho_1\partial_t - L)^2 u(x,t) = F(x,t), \quad (x,t) \in Q, \tag{4.33}$$

where

$$F(x,t) = \left[\rho_2\partial_t^{\frac{1}{2}} - (\rho_1\partial_t - L)\right](f(x)R(x,t)) + \rho_2 f(x)\frac{R(x,0)}{\sqrt{\pi t}} \tag{4.34}$$

$$= R(x,t)\sum_{i,j=1}^{n} \partial_i\left(a_{ij}(x)\partial_j f(x)\right)$$

$$+\sum_{j=1}^{n}\left(2\sum_{i=1}^{n}a_{ij}(x)\partial_i R(x,t)+b_j(x)R(x,t)\right)\partial_j f(x)$$

$$+\left[\rho_2\partial_t^{\frac{1}{2}}R(x,t)-\rho_1\partial_t R(x,t)+\sum_{i,j=1}^{n}\partial_i\left(a_{ij}(x)\partial_j R(x,t)\right)\right.$$

$$\left.+\sum_{j=1}^{n}b_j(x)\partial_j R(x,t)+c(x)R(x,t)+\frac{\rho_2 R(x,0)}{\sqrt{\pi t}}\right]f(x),\,(x,t)\in Q.$$

Expanding the left-hand side of (4.33), we get the following equation:

$$\rho_2^2\partial_t u(x,t)-\rho_1^2\partial_t^2 u(x,t)+2\rho_1\partial_t Lu(x,t)-L^2u(x,t)=F(x,t),\ (x,t)\in Q. \tag{4.35}$$

By Eq. (4.35) at $t=t_0$, we have

$$\rho_2^2\partial_t u\,(x,t_0)-\rho_1^2\partial_t^2 u\,(x,t_0)+2\rho_1\partial_t Lu\,(x,t_0)-L^2u\,(x,t_0)=F\,(x,t_0)\,,\ x\in\Omega. \tag{4.36}$$

Taking the weighted L^2 norm of (4.36) in Ω_2, we obtain

$$\int_{\Omega_2}|F\,(x,t_0)|^2\,e^{2s\varphi(x,t_0)}dx\le C\int_{\Omega_2}|\partial_t u\,(x,t_0)|^2\,e^{2s\varphi(x,t_0)}dx$$

$$+C\int_{\Omega_2}\left|\partial_t^2 u\,(x,t_0)\right|^2e^{2s\varphi(x,t_0)}dx+C\int_{\Omega_2}|\partial_t Lu\,(x,t_0)|^2\,e^{2s\varphi(x,t_0)}dx$$

$$+C\int_{\Omega_2}\sum_{|\alpha|\le 4}\left|\partial_x^\alpha u\,(x,t_0)\right|^2e^{2s\varphi(x,t_0)}dx. \tag{4.37}$$

Henceforth we estimate from the first term to the third term on the right-hand side of (4.37) by the Carleman estimate (Theorem 4.2).

To use the Carleman estimate, we introduce a cut-off function $\chi\in C^\infty\left(\mathbb{R}^{n+1}\right)$ such that $0\le\chi\le 1$ in $\mathbb{R}^{n+1}$, $\operatorname{supp}\chi\subset\left\{(x,t)\in R^{n+1}\mid\psi(x,t)>\mu_1\right\}$ and $\chi\equiv 1$ in $\left\{(x,t)\in R^{n+1}\mid\psi(x,t)>\mu_2\right\}$.

Set $y=\chi\partial_t u$, $z=\chi\partial_t^2 u$. By (4.33), we have

$$\rho_2^2\partial_t y(x,t)-\rho_1^2\partial_t^2 y(x,t)+2\rho_1\partial_t Ly(x,t)-L^2y(x,t)$$

$$=\chi\partial_t F(x,t)+h_1(x,t),\quad (x,t)\in Q, \tag{4.38}$$

$$\rho_2^2\partial_t z(x,t)-\rho_1^2\partial_t^2 z(x,t)+2\rho_1\partial_t Lz(x,t)-L^2z(x,t)$$

$$=\chi\partial_t^2 F(x,t)+h_2(x,t),\quad (x,t)\in Q, \tag{4.39}$$

where h_1 and h_2 are linear combinations of $\partial_x^\alpha \partial_t u$ $(|\alpha| \le 3)$, $\partial_x^\alpha \partial_t^2 u$ $(|\alpha| \le 1)$ and $\partial_x^\alpha \partial_t^2 u$ $(|\alpha| \le 3)$, $\partial_x^\alpha \partial_t^3 u$ $(|\alpha| \le 1)$, respectively, with coefficients containing $\partial_x^\alpha \chi$ $(1 \le |\alpha| \le 4)$, $\partial_x^\alpha \partial_t \chi$ $(|\alpha| \le 2)$.

Fixing $\lambda > 0$ and applying Theorem 4.2 to (4.38) and (4.39) in Q_1, we have

$$\int_{Q_1} \Bigg[s^2 \left(|\partial_t y|^2 + |\partial_t z|^2\right) + s^2 \left(\sum_{i,j=1}^n \left|\partial_i \partial_j y\right|^2 + \sum_{i,j=1}^n \left|\partial_i \partial_j z\right|^2 \right) \tag{4.40}$$

$$+ s^4 \left(|\nabla y|^2 + |\nabla z|^2\right) + s^6 \left(|y|^2 + |z|^2\right)\Bigg] e^{2s\varphi} dxdt$$

$$\le C \int_{Q_1} \chi^2 \left(|\partial_t F|^2 + \left|\partial_t^2 F\right|^2\right) e^{2s\varphi} dxdt + C \int_{Q_1} \left(|h_1|^2 + |h_2|^2\right) e^{2s\varphi} dxdt.$$

Here we note that the boundary term on ∂Q_1 of the Carleman estimate vanishes. Indeed, by the choice of χ, the boundary term on $\partial Q_1 \backslash \Gamma$ becomes 0. Since $f = \partial_\nu f = 0$ on γ,(4.8) and $\partial_x^\alpha u = 0$ on Γ, $|\alpha| = 2$, moreover, the boundary term on $\partial Q_1 \cap \Gamma$ vanishes.

Noting that $h_1(x, t) = h_2(x, t) = 0, \quad (x, t) \in Q_2$,

$$1 \le e^{2s\varphi(x,t)} \le e^{2s\exp(\lambda\mu_2)}, \quad (x, t) \in Q_1 \backslash Q_2, \tag{4.41}$$

and that $h_1, h_2 \in L^2(Q_1)$ have an upper bound depending on M by (4.9), we obtain

$$\int_{Q_1} \left(|h_1|^2 + |h_2|^2\right) e^{2s\varphi} dxdt = \int_{Q_1 \backslash Q_2} \left(|h_1|^2 + |h_2|^2\right) e^{2s\varphi} dxdt$$

$$\le C e^{2s\exp(\lambda\mu_2)}. \tag{4.42}$$

By (4.5) and (4.34), we get

$$\int_{Q_1} \chi^2 \left(|\partial_t F|^2 + \left|\partial_t^2 F\right|^2\right) e^{2s\varphi} dxdt$$

$$\le C \int_{Q_2} \sum_{|\alpha| \le 2} \left|\partial_x^\alpha f\right|^2 e^{2s\varphi} dxdt + C \int_{Q_1 \backslash Q_2} \sum_{|\alpha| \le 2} \left|\partial_x^\alpha f\right|^2 e^{2s\varphi} dxdt$$

$$\le C \int_{Q_2} \sum_{|\alpha| \le 2} \left|\partial_x^\alpha f\right|^2 e^{2s\varphi} dxdt + C e^{2s\exp(\lambda\mu_2)}. \tag{4.43}$$

In the last in equality of (4.43), we used the following inequality

$$\int_{Q_1\setminus Q_2}\sum_{|\alpha|\le 2}\left|\partial_x^\alpha f\right|^2 e^{2s\varphi}dxdt \le Ce^{2s\exp(\lambda\mu_2)}.$$

which is obtained by (4.9) and (4.41). Combining (4.40) with (4.42) and (4.43), we have

$$\int_{Q_1}\left[s^2\left(|\partial_t y|^2+|\partial_t z|^2\right)+s^2\left(\sum_{i,j=1}^n\left|\partial_i\partial_j y\right|^2+\sum_{i,j=1}^n\left|\partial_i\partial_j z\right|^2\right)\right.$$

$$\left.+s^4\left(|\nabla y|^2+|\nabla z|^2\right)+s^6\left(|y|^2+|z|^2\right)\right]e^{2s\varphi}dxdt$$

$$\le C\int_{Q_2}\sum_{|\alpha|\le 2}\left|\partial_x^\alpha f\right|^2 e^{2s\varphi}dxdt + Ce^{2s\exp(\lambda\mu_2)}.$$

Since $y=\partial_t u$, $z=\partial_t^2 u$ in Q_2 and $Q_2\subset Q_1$, we obtain

$$\int_{Q_2}\left[s^2\left(\left|\partial_t^2 u\right|^2+\left|\partial_t^3 u\right|^2\right)+s^2\left(\sum_{i,j=1}^n\left|\partial_i\partial_j\partial_t u\right|^2+\sum_{i,j=1}^n\left|\partial_i\partial_j\partial_t^2 u\right|^2\right)\right.$$

$$\left.+s^4\left(|\nabla\partial_t u|^2+\left|\nabla\partial_t^2 u\right|^2\right)+s^6\left(|\partial_t u|^2+\left|\partial_t^2 u\right|^2\right)\right]e^{2s\varphi}dxdt$$

$$\le C\int_{Q_2}\sum_{|\alpha|\le 2}\left|\partial_x^\alpha f\right|^2 e^{2s\varphi}dxdt + Ce^{2s\exp(\lambda\mu_2)}. \tag{4.44}$$

Since $\chi=0$ in $\Omega\times(0,t_0)\setminus Q_1^-$, we have

$$\int_{\Omega_2}|\partial_t u(x,t_0)|^2 e^{2s\varphi(x,t_0)}dx \le \int_{\Omega_1}|\chi(x,t_0)\,\partial_t u(x,t_0)|^2 e^{2s\varphi(x,t_0)}dx$$

$$=\int_0^{t_0}\int_{\Omega_1}\partial_t\left(|\chi\,\partial_t u|^2 e^{2s\varphi}\right)dxdt$$

$$\le C\int_0^{t_0}\int_{\Omega_1}\left[\chi^2|\partial_t u|\left|\partial_t^2 u\right|+\left(\chi\,|\partial_t\chi|+s\chi^2\right)|\partial_t u|^2\right]e^{2s\varphi}dxdt$$

$$=C\int_{Q_1^-}\left[\chi^2|\partial_t u|\left|\partial_t^2 u\right|+\left(\chi\,|\partial_t\chi|+s\chi^2\right)|\partial_t u|^2\right]e^{2s\varphi}dxdt$$

$$\le C\int_{Q_1}s\left(|\partial_t u|^2+\left|\partial_t^2 u\right|^2\right)e^{2s\varphi}dxdt$$

$$\leq C \int_{Q_2} s \left(|\partial_t u|^2 + \left|\partial_t^2 u\right|^2 \right) e^{2s\varphi} dxdt + Ce^{2s\exp(\lambda\mu_2)}. \tag{4.45}$$

Together this with (4.44), we see that

$$\int_{\Omega_2} |\partial_t u(x, t_0)|^2 e^{2s\varphi(x,t_0)} dx \leq \frac{C}{s^5} \int_{Q_2} \sum_{|\alpha| \leq 2} \left|\partial_x^\alpha f\right|^2 e^{2s\varphi} dxdt + Ce^{2s\exp(\lambda\mu_2)}. \tag{4.46}$$

Similarly, we may obtain

$$\int_{\Omega_2} \left|\partial_t^2 u(x, t_0)\right|^2 e^{2s\varphi(x,t_0)} dx$$

$$\leq \frac{C}{s} \int_{Q_2} \sum_{|\alpha| \leq 2} \left|\partial_x^\alpha f\right|^2 e^{2s\varphi} dxdt + Ce^{2s\exp(\lambda\mu_2)}, \tag{4.47}$$

$$\int_{\Omega_2} |\partial_t Lu(x, t_0)|^2 e^{2s\varphi(x,t_0)} dx$$

$$\leq \frac{C}{s} \int_{Q_2} \sum_{|\alpha| \leq 2} \left|\partial_x^\alpha f\right|^2 e^{2s\varphi} dxdt + Ce^{2s\exp(\lambda\mu_2)}. \tag{4.48}$$

By (4.37), (4.46)–(4.48), we have

$$\int_{\Omega_2} |F(x, t_0)|^2 e^{2s\varphi(x,t_0)} dx \leq \frac{C}{s} \int_{Q_2} \sum_{|\alpha| \leq 2} \left|\partial_x^\alpha f\right|^2 e^{2s\varphi} dxdt$$

$$+ C \int_{\Omega_2} \sum_{|\alpha| \leq 4} \left|\partial_x^\alpha u(x, t_0)\right|^2 e^{2s\varphi(x,t_0)} dx + Ce^{2s\exp(\lambda\mu_2)}. \tag{4.49}$$

Next we estimate

$$\int_{\Omega_2} |F(x, t_0)|^2 e^{2s\varphi(x,t_0)} dx$$

from below. To apply the Carleman estimate for elliptic equation, we introduce a cut-off function $\tilde{\chi} \in C^\infty(\mathbb{R}^n)$ such that $0 \leq \tilde{\chi} \leq 1$ in $\mathbb{R}^n$, $\operatorname{supp} \tilde{\chi} \subset \{x \in R^n \mid d(x) > \mu_1\}$ and $\tilde{\chi} \equiv 1$ in $\{x \in R^n \mid d(x) > \mu_2\}$.

Set $\tilde{f} = \tilde{\chi} f$. By (4.34) at $t = t_0$, we have

$$\sum_{i,j=1}^{n} \partial_i \left(a_{ij}(x) \partial_j \tilde{f}(x) \right)$$

$$+\frac{1}{R(x,t_0)}\sum_{j=1}^{n}\left(2\sum_{i=1}^{n}a_{ij}(x)\partial_i R(x,t_0)+b_j(x)R(x,t_0)\right)\partial_j\tilde{f}(x)$$

$$+\frac{1}{R(x,t_0)}\left[\rho_2\partial_t^{\frac{1}{2}}R(x,t_0)-\rho_1\partial_t R(x,t_0)+\sum_{i,j=1}^{n}\partial_i\left(a_{ij}(x)\partial_j R(x,t_0)\right)\right.$$

$$\left.+\sum_{j=1}^{n}b_j(x)\partial_j R(x,t_0)+c(x)R(x,t_0)+\frac{\rho_2 R(x,0)}{\sqrt{\pi t_0}}\right]\tilde{f}(x)$$

$$=\frac{\tilde{\chi}(x)F(x,t_0)}{R(x,t_0)}+h_3(x),\quad x\in\Omega, \tag{4.50}$$

where

$$h_3(x)=2\sum_{i,j=1}^{n}a_{ij}\partial_i\tilde{\chi}(x)\partial_j f(x)+\left[\sum_{i,j=1}^{n}\partial_i\left(a_{ij}\partial_j\tilde{\chi}(x)\right)\right.$$

$$\left.+\frac{1}{R(x,t_0)}\sum_{j=1}^{n}\left(2\sum_{i=1}^{n}a_{ij}(x)\partial_i R(x,t_0)+b_j(x)R(x,t_0)\right)\left(\partial_j\tilde{\chi}\right)\right]f(x),\ x\in\Omega.$$

Here we see that $h_3(x)=0,\ x\in\Omega_2$ and $\|h_3\|_{L^2(\Omega)}\le C$ by (4.9). Applying Lemma 4.3 to (4.50) in Ω_1 and noting that $\tilde{f}(x)=f(x),\ x\in\Omega_2$, (4.5), (4.9) and (4.41), we obtain

$$\frac{1}{s}\int_{\Omega_2}\sum_{|\alpha|\le 2}\left|\partial_x^\alpha f(x)\right|^2 e^{2s\varphi(x,t_0)}dx$$

$$\le C\int_{\Omega_2}\left(\frac{1}{s}\sum_{i,j=1}^{n}\left|\partial_i\partial_j f\right|^2+s|\nabla f|^2+s^3|f|^2\right)e^{2s\varphi(x,t_0)}dx$$

$$\le C\int_{\Omega_1}\left(\frac{1}{s}\sum_{i,j=1}^{n}\left|\partial_i\partial_j\tilde{f}\right|^2+s|\nabla\tilde{f}|^2+s^3|\tilde{f}|^2\right)e^{2s\varphi(x,t_0)}dx$$

$$\le C\int_{\Omega_1}\left|\frac{\tilde{\chi}(x)F(x,t_0)}{R(x,t_0)}\right|^2 e^{2s\varphi(x,t_0)}dx+C\int_{\Omega_1\setminus\Omega_2}|h_3(x)|^2e^{2s\varphi(x,t_0)}dx$$

$$\le C\int_{\Omega_2}|F(x,t_0)|^2e^{2s\varphi(x,t_0)}dx+C\int_{\Omega_1\setminus\Omega_2}|F(x,t_0)|^2e^{2s\varphi(x,t_0)}dx+$$

$$+ Ce^{2s} \exp(\lambda \mu_2) \le C \int_{\Omega_2} |F(x, t_0)|^2 e^{2s\varphi(x,t_0)} dx + Ce^{2s\exp(\lambda\mu_2)}. \tag{4.51}$$

Here the boundary term on $\partial\Omega_1$ of the Carleman estimate vanishes. Indeed, by $f = \partial_\nu f = 0$ on γ and the choice of $\tilde{\chi}$, we see that $\tilde{f} = \partial_\nu \tilde{f} = 0$ on $\partial\Omega_1$.

By (4.49) and (4.51), we obtain

$$\frac{1}{s} \int_{\Omega_2} \sum_{|\alpha|\le 2} \left|\partial_x^\alpha f(x)\right|^2 e^{2s\varphi(x,t_0)} dx \le \frac{C}{s} \int_{Q_2} \sum_{|\alpha|\le 2} \left|\partial_x^\alpha f\right|^2 e^{2s\varphi} dx dt$$

$$+ C \int_{\Omega_2} \sum_{|\alpha|\le 4} \left|\partial_x^\alpha u(x, t_0)\right|^2 e^{2s\varphi(x,t_0)} dx + Ce^{2s\exp(\lambda\mu_2)}. \tag{4.52}$$

Let us estimate the first integral term on the right-hand side of (4.52).

$$\int_{Q_2} \sum_{|\alpha|\le 2} \left|\partial_x^\alpha f\right|^2 e^{2s\varphi} dx dt \le$$

$$\int_{\Omega_2} \sum_{|\alpha|\le 2} \left|\partial_x^\alpha f\right|^2 e^{2s\varphi} e^{2s\varphi(x,t_0)} \left(\int_0^T e^{-2s(\varphi(x,t_0)-\varphi(x,t))} dt\right) dx.$$

Noting that $\varphi(x, t_0) - \varphi(x, t) = e^{\lambda d(x)} \left(1 - e^{-\beta(t-t_0)^2}\right) \ge 0, (x, t) \in Q$, by Lebesgue's dominated convergence theorem, $\int_0^T e^{-2s(\varphi(x,t_0)-\varphi(x,t))} dt$ converges pointwise to 0 in Ω as s tends to ∞. By Dini's theorem, moreover, we may see that $\int_0^T e^{-2s(\varphi(x,t_0)-\varphi(x,t))} dt$ converges uniformly to 0 in Ω_2 as s goes to ∞. Hence taking sufficient large $s > 0$, we may absorb the first term on the right-hand side of (4.52) into the left-hand side, that is, there exists $s_1 > 0$ such that

$$\int_{\Omega_2} \sum_{|\alpha|\le 2} \left|\partial_x^\alpha f(x)\right|^2 e^{2s\varphi(x,t_0)} dx$$

$$\le C \int_{\Omega_2} \sum_{|\alpha|\le 4} \left|\partial_x^\alpha u(x, t_0)\right|^2 e^{2s\varphi(x,t_0)} dx + Ce^{2s\exp(\lambda\mu_2)} \tag{4.53}$$

for all $s > s_1$.

By $e^{2s\varphi(x,t_0)} \ge e^{2s\exp(\lambda\mu_3)}$, $x \in \Omega_3$, we estimate the left-hand side of (4.53) from below.

$$\int_{\Omega_2} \sum_{|\alpha|\le 2} \left|\partial_x^\alpha f(x)\right|^2 e^{2s\varphi(x,t_0)} dx \ge \int_{\Omega_3} \sum_{|\alpha|\le 2} \left|\partial_x^\alpha f(x)\right|^2 e^{2s\varphi(x,t_0)} dx$$

$$\geq e^{2s\exp(\lambda\mu_3)} \int_{\Omega_3} \sum_{|\alpha|\leq 2} \left|\partial_x^\alpha f(x)\right|^2 dx.$$

Combining this with (4.53), we have

$$e^{2s\exp(\lambda\mu_3)} \int_{\Omega_3} \sum_{|\alpha|\leq 2} \left|\partial_x^\alpha f(x)\right|^2 dx \tag{4.54}$$

$$\leq C \int_{\Omega_2} \sum_{|\alpha|\leq 4} \left|\partial_x^\alpha u(x, t_0)\right|^2 e^{2s\varphi(x,t_0)} dx + Ce^{2s\exp(\lambda\mu_2)},$$

for all $s > s_1$. Dividing the both side of (4.54) by $e^{2s\exp(\lambda\mu_3)}$, we get

$$\int_{\Omega_3} \sum_{|\alpha|\leq 2} \left|\partial_x^\alpha f(x)\right|^2 dx \tag{4.55}$$

$$\leq C \int_{\Omega_2} \sum_{|\alpha|\leq 4} \left|\partial_x^\alpha u(x, t_0)\right|^2 e^{2s[\varphi(x,t_0)-\exp(\lambda\mu_3)]} dx + Ce^{2s[\exp(\lambda\mu_2)-\exp(\lambda\mu_3)]},$$

for all $s > s_1$. Noting that $\varphi(x, t_0) - \exp(\lambda\mu_3) > 0$, $x \in \Omega_2$ and $\exp(\lambda\mu_2) - \exp(\lambda\mu_3) < 0$, there exist $C_1, C_2, D_1, D_2 > 0$ such that

$$\|f\|^2_{H^2(\Omega_3)} \leq C_1 \|u(\cdot, t_0)\|^2_{H^4(\Omega_2)} e^{D_1 s} + C_2 e^{-D_2 s},$$

for all $s > s_1$. Taking $C_3 > 0$ such that $C_1 \leq C_3 e^{-D_1 s_1}$, $C_2 \leq C_3 e^{D_2 s_1}$ and setting $\sigma = s - s_1$, we have

$$\|f\|^2_{H^2(\Omega_3)} \leq C_3 \left(\|u(\cdot, t_0)\|^2_{H^4(\Omega_2)} e^{D_1\sigma} + e^{-D_2\sigma}\right), \tag{4.56}$$

for all $\sigma > 0$. If $\|u(\cdot, t_0)\|_{H^4(\Omega_2)} \geq 1$, by $\|f\|_{H^2(\Omega)} \leq M$, we may obtain the stability estimate immediately. Moreover, provided that $\|u(\cdot, t_0)\|_{H^4(\Omega_2)} = 0$, we have

$$\|f\|^2_{H^2(\Omega_3)} \leq C_3 e^{-D_2\sigma}, \tag{4.57}$$

by (4.56). As σ goes to ∞, the right-hand side of (4.57) tends to 0. Hence we get $\|f\|_{H^2(\Omega_3)} = 0$. So, it is sufficient to assume that $0 < \|u(\cdot, t_0)\|_{H^4(\Omega_2)} < 1$. Taking

$$\sigma = -\frac{\log\left(\|u(\cdot, t_0)\|^2_{H^4(\Omega_2)}\right)}{D_1 + D_2} > 0,$$

which minimize the right-hand side of (4.56), we see that

$$\|u(\cdot, t_0)\|^2_{H^4(\Omega_2)} e^{D_1\sigma} + e^{-D_2\sigma} = 2\|u(\cdot, t_0)\|^{2\kappa}_{H^4(\Omega_2)}$$

with $\kappa = \frac{D_2}{D_1+D_2} \in (0, 1)$. Together this with (4.56), we have

$$\|f\|_{H^2(\Omega_3)} \le C\|u(\cdot, t_0)\|^{\kappa}_{H^4(\Omega_2)}.$$

Since $\omega \subset \Omega_3$ and $\Omega_2 \subset \Omega$, we get (4.10). Thus we complete the proof.

4.2 Lipschitz Stability in Inverse Source and Inverse Coefficient Problems

In Sect. 4.1, the Hölder stability is proven for the inverse source problem of (4.1). In this section, we further prove the Lipshitz stability not only for the inverse source problem but also for the inverse coefficient problem for (4.1).

The Carleman estimate is a weighted L^2 inequality for a solution of a partial differential equation. In the case of parabolic equations with one first-order time derivative, the global Lipschitz stability was proven by using this method of Carleman estimates [17]. In this section we make use of the Carleman estimate for parabolic equations. The Carleman estimates have been used for differential equations with a single-term time-fractional derivative [18]-[20].

This section is organized as follows. In Sect. 4.2.1, inverse source problems are considered. In Sect. 4.2.2, inverse coefficient problems are considered. In Sect. 4.2.3, the Carleman estimate necessary for this section is established. Finally, proofs of the main theorems are given in Sect. 4.2.4.

4.2.1 Inverse Source Problem Formulation and Its Investigation

We consider the first- and half-order time-fractional diffusion equation (4.1) with the following boundary

$$u(x, t) = h_1(x, t),\ (x, t) \in \partial Q \times (0, T) \tag{4.58}$$

and initial conditions

$$u(x, 0) = h_2(x),\ x \in \Omega, \tag{4.59}$$

where $\rho_1 > 0$, $\rho_2 \neq 0$ are constants, and the operator L was determined in Sect. 4.1.

In this subsection, we study the inverse problems of determining the time-independent source factor of (4.1) from spatial data and two types of observations. One is the boundary observation and the other is the interior observation.

Let $t_0 \in (0, T)$ be an arbitrarily fixed time. Let γ be an arbitrarily fixed open connected sub-boundary of $\partial\Omega$ and let ω be an arbitrarily fixed sub-domain of Ω such that $\omega \subseteq \Omega$. We set $\sum = \gamma \times (0, T)$ and $Q_\omega = \omega \times (0, T)$.

Moreover we choose $\delta > 0$ such that

$$0 < t_0 - \delta < t_0 < t_0 + \delta < T,$$

and we set $Q_\delta = \Omega \times (t_0 - \delta, t_0 + \delta)$, $\sum_\delta = \gamma \times (t_0 - \delta, t_0 + \delta)$, $Q_{\omega,\delta} = \omega \times (t_0 - \delta, t_0 + \delta)$

We assume that

$$\begin{cases} R \in C([0, T); C(\overline{\Omega}) \cap C^2((0, T); C^2(\overline{\Omega})) \cap C^3((0, T); C(\overline{\Omega})) \\ \partial_t^{\frac{1}{2}} R \in C^2((0, T); C(\overline{\Omega}))\ and\ |R(x, t_0)| > 0,\ x \in \overline{\omega}. \end{cases} \tag{4.60}$$

Furthermore we define

$$U = L^2((0, T); H^4(\Omega)) \cap H^1((0, T); H^2(\Omega)) \cap H^2((0, T); L^2(\Omega)).$$

Let us assume that $g(x, t)$ in (4.1) has the form

$$g(x, t) = f(x)R(x, t) \tag{4.61}$$

and set $h_1 = 0$ in Q, $h_2 = 0$ in Ω.

We consider

$$\left(\rho_1 \partial_t + \rho_2 \partial_t^{\frac{1}{2}} - L\right) u(x, t) = f(x)R(x, t),\ (x, t) \in Q, \tag{4.62}$$

$$u(x, t) = 0,\ (x, t) \in \partial\Omega \times (0, T), \tag{4.63}$$

$$u(x, 0) = 0,\ x \in \Omega, \tag{4.64}$$

and investigate two kinds of inverse problems depending on the type of observation.

In the inverse source problem via boundary observation, we determine $f(x)$, $x \in \Omega$, by spatial data $u(x, t_0)$, $x \in \Omega$, and boundary data on Σ. In the inverse source problem via interior observation, we determine $f(x)$, $x \in \Omega$, by spatial data $u(x, t_0)$, $x \in \Omega$, and interior data in Q_ω. The main theorems, Theorems 4.3 and 4.4 are stated as follows.

Theorem 4.3 *Let us assume that $u, \partial_t u, \partial_t^2 u \in U$ and u satisfies (4.62)–(4.64). We suppose that $f \in H^2(\Omega)$ with $f = 0$ on $\partial\Omega$ and $\nabla f = 0$ on γ and R satisfies (4.60). Then there exist constants $C > 0$ such that*

$$\|f\|_{H^2(\Omega)} \le C\|u(\cdot, t_0)\|_{H^4(\Omega)} + CB, \tag{4.65}$$

where

$$B = \|\nabla\partial_t^3 u\|_{L^2(\Sigma_\delta)} + \|\nabla\partial_t^2 u\|_{L^2(\Sigma_\delta)} + \|\nabla\partial_t u\|_{L^2(\Sigma_\delta)}.$$

Theorem 4.4 *Let us assume that* $u, \partial_t u, \partial_t^2 u \in U$ *and satisfies (4.62)–(4.64). We suppose that* $f \in H^2(\Omega)$ *with* $f = 0$ *on* $\partial\Omega$ *and* $f = 0$ *on* ω *and* R *satisfies (4.60). Then there exist constants* $C > 0$ *such that*

$$\|f\|_{H^2(\Omega)} \le C\|u(\cdot, t_0)\|_{H^4(\Omega)} + CI, \tag{4.66}$$

where

$$I = \|\nabla\partial_t^3 u\|_{L^2(Q_{\omega,\delta})} + \|\nabla\partial_t^2 u\|_{L^2(Q_{\omega,\delta})} + \|\nabla\partial_t u\|_{L^2(Q_{\omega,\delta})}.$$

Remark 4.4 There is another approach to obtain the Lipschitz stability in inverse source problems by final observation data. In [21], Sakamoto and Yamamoto considered the perturbation of the single-term time-fractional diffusion equations with a parameter as the diffusion coefficient and they obtained the stability estimate by means of the analytic perturbation theory under the appropriate assumptions on the parameter. In our case, however, we may not adopt their methodology directly since we consider the diffusion coefficient without the perturbation.

4.2.2 Inverse Coefficient Problems: Lipschitz Stability Estimates

Determination of the zeroth-order coefficient. Let us consider the inverse problem of determining the zeroth-order coefficient. In (4.1), we consider two coefficients $c^{(k)}(x),\ x \in \Omega\ (k = 1, 2)$, where $c^{(k)} \in C^2(\overline{\Omega}),\ c^{(k)}(x) \ge 0,\ x \in \Omega\ (k = 1, 2)$. We assume that there exists a constant $M_1 > 0$ such that

$$\|c^{(k)}\|_{C^2(\overline{\Omega})} \le M_1 \tag{4.67}$$

for $k = 1, 2$. Let $u^{(k)}(x, t)$ be the corresponding solutions. We write L as

$$Lu^{(k)}(x, t) = L_0^{(k)} u^{(k)}(x, t), \tag{4.68}$$

where $L_0^{(k)}$ is defined as

$$L_0^{(k)} u(x, t) = \sum_{i,j=1}^{n} \partial_i(a_{ij}(x)\partial_j u^{(k)}(x, t))$$
$$- \sum_{j=1}^{n} b_j(x)\partial_j u^{(k)}(x, t) - c^{(k)}(x)u^{(k)}(x, t), \quad (x, t) \in Q$$

for $k = 1, 2$.

By subtraction we obtain

$$\begin{cases} (\rho_1 \partial_t + \rho_2 \partial_t^{\frac{1}{2}} - L_0^1) u(x,t) = f(x) R(x,t), & (x,t) \in Q, \\ u(x,t) = 0, & (x,t) \in \partial\Omega \times (0,T), \\ u(x,0) = 0, & x \in \Omega, \end{cases}$$

where

$$f(x) = c^{(1)}(x) - c^{(2)}(x), u(x,t) = u^{(1)}(x,t) - u^{(2)}(x,t), R(x,t) = -u^{(2)}(x,t)$$

for $(x,t) \in Q$. Thus we arrive at the following two theorems as a direct consequence of the inverse source problems. In both cases the Lipschitz stability is obtained. Theorem 4.5 is proved using the inverse source problem via boundary observation stated in Theorem 4.3. Theorem 4.6 is proved using the inverse source problem via interior observation stated in Theorem 4.4.

Theorem 4.5 (boundary observation) *Let $u^{(k)}, \partial_t u^{(k)}, \partial_t^2 u^{(k)} \in U (k = 1, 2)$, and $u^{(1)}, u^{(2)}$ satisfy (4.1), (4.58), (4.59) with (4.68). We suppose that $c^{(1)}, c^{(2)} \in C^2(\overline{\Omega})$ satisfy (4.67) with $c^{(1)} = c^{(2)}$ on $\partial\Omega$ and $\nabla c^{(1)} = \nabla c^{(2)}$ on γ, and $R = -u^{(2)}$ satisfies (4.60). Then there exist constants $C > 0$ such that*

$$\|c^{(1)} - c^{(2)}\|_{H^2(\Omega)} \le C \|u^{(1)}(\cdot, t_0) - u^{(2)}(\cdot, t_0)\|_{H^4(\Omega)} + CB, \tag{4.69}$$

where

$$B = \|\nabla \partial_t^3 (u^{(1)} - u^{(2)})\|_{L^2(\Sigma_\delta)} + \|\nabla \partial_t^2 (u^{(1)} - u^{(2)})\|_{L^2(\Sigma_\delta)} + \|\nabla \partial_t (u^{(1)} - u^{(2)})\|_{L^2(\Sigma_\delta)}.$$

Theorem 4.6 (interior observation) *Let $u^{(k)}, \partial_t u^{(k)}, \partial_t^2 u^{(k)} \in U (k = 1, 2)$, and $u^{(1)}, u^{(2)}$ satisfy (4.1), (4.58), (4.59) with (4.68). We suppose that $c^{(1)}, c^{(2)} \in C^2(\overline{\Omega})$ satisfy (4.67) with $c^{(1)} = c^{(2)}$ in $\partial\Omega \cup \omega$ and $R = -u^{(2)}$ satisfies (4.60). Then there exist constants $C > 0$ such that*

$$\|c^{(1)} - c^{(2)}\|_{H^2(\Omega)} \le C \|u^{(1)}(\cdot, t_0) - u^{(2)}(\cdot, t_0)\|_{H^4(\Omega)} + CI, \tag{4.70}$$

where

$$I = \|\partial_t^3 (u^{(1)} - u^{(2)})\|_{L^2(Q_{\omega,\delta})} + \|\partial_t^2 (u^{(1)} - u^{(2)})\|_{L^2(Q_{\omega,\delta})} + \|\partial_t (u^{(1)} - u^{(2)})\|_{L^2(Q_{\omega,\delta})}.$$

Remark 4.5 In the case of diffusion in porous media, the condition $|u^{(2)}(x, t_0)| = |R(x, t_0)| > 0$ for $x \in \overline{\Omega}$ means that the concentration of tracer particles is non-zero.

Remark 4.6 We may determine the first-order coefficients $b_j (j = 1, \ldots, n)$ by n observations via the similar argument used in the proofs of Theorems 4.3 and 4.4.

Determination of the diffusion coefficient. We consider diffusion coefficients $a^{(k)}(k = 1, 2)$ and corresponding solutions $u^{(k)}$. Let us express L as

$$Lu^{(k)}(x,t) = L_2^{(k)} u^{(k)}(x,t), \tag{4.71}$$

where $L_2^{(k)}$ is defined as

$$\begin{aligned} &L_2^{(k)} u^{(k)}(x,t) \\ &= div(a^{(k)}(x)\nabla u^{(k)}(x,t)) - b(x)\cdot \nabla u^{(k)}(x,t) - c(x)u^{(k)}(x,t),\ (x,t) \in Q, \end{aligned}$$

for $k = 1, 2$. We suppose that $a^{(k)} \in C^4(\overline{\Omega})(k = 1, 2), b = (b_1, b_2, \ldots, b_n) \in \{C^3(\overline{\Omega})\}^n$, and $c \in C^3(\overline{\Omega})$. Moreover we assume that there exist constants $m > 0$ and $M_2 > 0$ such that

$$a^{(k)}(x) \ge m,\ x \in \Omega,\ and\ \|a^{(k)}\|_{C^3(\overline{\Omega})} \le M_2 \tag{4.72}$$

for $k = 1, 2$. We investigate the inverse problems of determining the diffusion coefficients $a^{(k)}(k = 1, 2)$ by boundary observations and interior observations.

Set

$$a(x) = a^{(1)}(x) - a^{(2)}(x), u(x,t) = u^{(1)}(x,t) - u^{(2)}(x,t), r(x,t) = u^{(2)}(x,t)$$

for $(x, t) \in Q$. Then by subtracting the equations for $k = 2$ from ones for $k = 1$, we obtain

$$\begin{cases} (\rho_1\partial_t + \rho_2\partial_t^{\frac{1}{2}} - L_2^{(1)})u(x,t) = div(a(x)\nabla r(x,t)), & (x,t) \in Q, \\ u(x,t) = 0, \quad (x,t) \in \partial\Omega \times (0,T), \\ u(x,0) = 0, & x \in \Omega, \end{cases} \tag{4.73}$$

We assume that

$$\begin{cases} r \in C((0,T); C^3(\overline{\Omega})) \cap C((0,T); C^5(\overline{\Omega})) \cap C^2((0,T); \\ C^4(\overline{\Omega})) \cap C^3((0,T); C^2(\overline{\Omega}));\ \partial_t^{\frac{1}{2}} r \in C((0,T); C^3(\overline{\Omega})) \cap C^2((0,T); C^2(\overline{\Omega})) \end{cases} \tag{4.74}$$

Let us introduce weight functions for the Carleman estimates introduced in Sect. 4.1.3. Since we consider two types of observation, we prepare two kinds of distance functions d_1 and d_2. For the symmetric uniformly elliptic operator A, we choose $d_1 \in C^2(\overline{\Omega})$ such that

$$d_1(x) > 0, \quad x \in \Omega, |\nabla d_1(x)| > \sigma_1, \quad x \in \overline{\Omega},$$

$$\sum_{i,j=1}^{n} a_{ij}(x)\partial_i d_1 \nu_j \le 0, \quad x \in \partial\Omega \setminus \gamma,$$

where $\sigma_1 > 0$ is a constant. Let ω_0 be an arbitrarily fixed sub-domain of Ω such that $\omega_0 \subseteq \omega$. We take $d_2 \in C^2(\overline{\Omega})$ such that

$$d_2(x) > 0, \quad x \in \Omega, |\nabla d_2(x)| > \sigma_2, \quad x \in \overline{\Omega \setminus \omega_0}, d_2(x) = 0, \quad x \in \partial\Omega,$$

where $\sigma_2 > 0$ is a constant. The existence of the distance functions d_1 and d_2 is proved in [17, 22, 23]. Then we introduce weight functions $\varphi_k, \psi_k (k = 1, 2)$ as

$$\varphi_k(x,t) = \frac{e^{\lambda d_k(x)}}{t(T-t)}, \ \psi_k(x,t) = \frac{e^{\lambda d_k(x)} - e^{2\lambda \|d_k\|_{C(\overline{\Omega})}}}{t(T-t)}, \quad (x,t) \in Q.$$

For the distance functions d_1 and d_2, we assume that there exists a constant $m_1 > 0$ such that

$$|\nabla r(x, t_0) \cdot \nabla d_1(x)| \geq m_1, \quad x \in \overline{\Omega}, \tag{4.75}$$

or that there exists a constant $m_2 > 0$ such that

$$|\nabla r(x, t_0) \cdot \nabla d_2(x)| \geq m_2, \quad x \in \overline{\Omega \setminus \omega}. \tag{4.76}$$

Let D' be an arbitrary sub-domain such that $\omega \subseteq D' \subseteq \Omega$. Set $D = \Omega \setminus D'$. Henceforth we suppose that $a \equiv 0$ in D.

Now we are ready to state our main results.

Theorem 4.7 (boundary observation) *Let $u^{(k)}, \partial_t u^{(k)}, \partial_t^2 u^{(k)} \in U$, $\nabla u^{(k)} \in U^n (k = 1, 2)$, and $u^{(1)}, u^{(2)}$ satisfy (4.1), (4.58), (4.59) with (4.71). We suppose that $a^{(1)}, a^{(2)} \in C^4(\overline{\Omega})$ satisfy (4.72) with $a^{(1)} = a^{(2)}$ in D and $r = u^{(2)}$ satisfies (4.74) and (4.75). Then there exist constants $C > 0$ such that*

$$\|a^{(1)} - a^{(2)}\|_{H^3(\Omega)} \leq C\|u^{(1)}(\cdot, t_0) - u^{(2)}(\cdot, t_0)\|_{H^5(\Omega)} + CB, \tag{4.77}$$

where

$$B = \|\nabla \partial_t^3 (u^{(1)} - u^{(2)})\|_{L^2(\Sigma_\delta)} + \|\nabla \partial_t^2 (u^{(1)} - u^{(2)})\|_{L^2(\Sigma_\delta)} + \|\nabla \partial_t (u^{(1)} - u^{(2)})\|_{L^2(\Sigma_\delta)}.$$

Theorem 4.8 (interior observation) *Let $u^{(k)}, \partial_t u^{(k)}, \partial_t^2 u^{(k)} \in U$, $\nabla u^{(k)} \in U^n (k = 1, 2)$, and $u^{(1)}, u^{(2)}$ satisfy (4.1), (4.58), (4.59) with (4.71). We suppose that $a^{(1)}, a^{(2)} \in C^4(\overline{\Omega})$ satisfy (4.72) with $a^{(1)} = a^{(2)}$ in $D \cup \omega$ and $r = u^{(2)}$ satisfies (4.74) and (4.76). Then there exist constants $C > 0$ such that*

$$\|a^{(1)} - a^{(2)}\|_{H^3(\Omega)} \leq C\|u^{(1)}(\cdot, t_0) - u^{(2)}(\cdot, t_0)\|_{H^5(\Omega)} + CI, \tag{4.78}$$

where

$$I = \|\partial_t^3 (u^{(1)} - u^{(2)})\|_{L^2(Q_{\omega,\delta})} + \|\partial_t^2 (u^{(1)} - u^{(2)})\|_{L^2(Q_{\omega,\delta})} + \|\partial_t (u^{(1)} - u^{(2)})\|_{L^2(Q_{\omega,\delta})}.$$

Remark 4.7 In the case of one spatial dimension, we may relax some assumptions on $u^{(k)}$, $k = 1, 2$. It depends on the assumptions of the Carleman estimate for the third-order partial differential equations (Lemmas 4.8 and 4.9). See also [24].

Simultaneous determination of the diffusion coefficient and zeroth-order coefficient. Let us consider diffusion coefficients $a^{(k)}(k = 1, 2)$, zeroth-order coefficient $c^{(k)}$, and corresponding solutions $u_l^{(k)}(l = 1, 2)$. We express $L^{(k)}$ as

$$Lu_l^{(k)}(x,t) = L^{(k)}u_l^{(k)}(x,t), \tag{4.79}$$

where $L^{(k)}$ is defined as

$$L^{(k)}u_l^{(k)}(x,t) = div(a^{(k)}(x)\nabla u_l^{(k)}(x,t)) - c^{(k)}(x)u_l^{(k)}(x,t),\ (x,t) \in Q,$$

for $k = 1, 2$. We suppose that $a^{(k)} \in C^4(\overline{\Omega})$ and $c^{(k)} \in C^3(\overline{\Omega})(k = 1, 2)$. We investigate the inverse problems of determining $a^{(k)}$ and $c^{(k)}(k = 1, 2)$ by boundary and interior observations for two inputs, $l = 1, 2$.

We consider

$$\left(\rho_1\partial_t + \rho_2\partial_t^{\frac{1}{2}} - L^{(k)}\right)u_l^{(k)}(x,t) = g_l(x,t), \quad (x,t) \in Q, \tag{4.80}$$

$$u_l^{(k)}(x,t) = h_{1,l}(x,t), \quad (x,t) \in \partial\Omega \times (0,T), \tag{4.81}$$

$$u_l^{(k)}(x,0) = h_{2,l}(x), \quad x \in \Omega, \tag{4.82}$$

for $k = 1, 2, l = 1, 2$.

Set

$$a(x) = a^{(1)}(x) - a^{(2)}(x), c(x) = c^{(1)}(x) - c^{(2)}(x),$$

$$u_l(x,t) = u_l^{(1)}(x,t) - u_l^{(2)}(x,t), r_l(x,t) = u_l^{(2)}(x,t)$$

for $(x,t) \in Q, l = 1, 2$. Then by subtracting the equations for $k = 2$ from ones for $k = 1$, we obtain

$$\begin{cases} \left(\rho_1\partial_t + \rho_2\partial_t^{\frac{1}{2}} - L^{(1)}\right)u_l(x,t) = div(a(x)\nabla r_l(x,t)) - c(x)r_l(x,t), & (x,t) \in Q, \\ u_l(x,t) = 0, & (x,t) \in \partial\Omega \times (0,T), \\ u_l(x,0) = 0, & x \in \Omega, \end{cases} \tag{4.83}$$

for $l = 1, 2$.

We assume that there exists a constant $m_3 > 0$ such that

$$\left|\left(r_2(x,t_0)\nabla r_1(x,t_0) - r_1(x,t_0)\nabla r_2(x,t_0)\right) \cdot \nabla d_1(x)\right| \geq m_3, x \in \overline{\Omega}, \tag{4.84}$$

or that there exists a constant $m_4 > 0$ such that

$$\left|\left(r_2(x,t_0)\nabla r_1(x,t_0) - r_1(x,t_0)\nabla r_2(x,t_0)\right)\cdot\nabla d_2(x)\right| \ge m_4,\ x\in\overline{\Omega\setminus\omega}. \tag{4.85}$$

Let D' be an arbitrary sub-domain such that $\omega\subseteq D'\subseteq\Omega$. Set $D=\Omega\setminus D'$.

Theorem 4.9 (boundary observation) *Let $u^{(k)}, \partial_t u^{(k)}, \partial_t^2 u^{(k)}\in U, \nabla u^{(k)}\in U^n (k=1,2)$, and $u^{(1)}, u^{(2)}$ satisfy (4.80)–(4.82) with (4.79). We suppose that $a^{(1)}, a^{(2)}\in C^4(\overline{\Omega})$ satisfy (4.72) with $a^{(1)}=a^{(2)}$ in D, $c^{(1)}, c^{(2)}\in C^2(\overline{\Omega})$ satisfy (4.67) with $c^{(1)}=c^{(2)}$ on $\partial\Omega$ and $\nabla c^{(1)}=\nabla c^{(2)}$ on γ, and $r=u^{(2)}$ satisfies (4.74), (4.84) and $|r_1(x,t_0)|>0, x\in\overline{\Omega}$. Then there exist constants $C>0$ such that*

$$\|a^{(1)}-a^{(2)}\|_{H^3(\Omega)}+\|c^{(1)}-c^{(2)}\|_{H^2(\Omega)}$$
$$\le C\sum_{l=1}^{2}\|u_l^{(1)}(\cdot,t_0)-u_l^{(2)}(\cdot,t_0)\|_{H5(\Omega)}+CB, \tag{4.86}$$

where

$$B=\sum_{l=1}^{2}\Big(\|\nabla\partial_t^3(u_l^{(1)}-u_l^{(2)})\|_{L^2(\Sigma_\delta)}$$
$$+\|\nabla\partial_t^2(u_l^{(1)}-u_l^{(2)})\|_{L^2(\Sigma_\delta)}+\|\nabla\partial_t(u_l^{(1)}-u_l^{(2)})\|_{L^2(\Sigma_\delta)}\Big).$$

Theorem 4.10 (interior observation) *Let $u^{(k)}, \partial_t u^{(k)}, \partial_t^2 u^{(k)}\in U, \nabla u^{(k)}\in U^n (k=1,2)$, and $u^{(1)}, u^{(2)}$ satisfy (4.80)–(4.82) with (4.79). We suppose that $a^{(1)}, a^{(2)}\in C^4(\overline{\Omega})$ satisfy (4.72) with $a^{(1)}=a^{(2)}$ in D, $c^{(1)}, c^{(2)}\in C^2(\Omega)$ satisfy (4.67) with $c^{(1)}=c^{(2)}$ on $\partial\Omega$, and $\nabla c^{(1)}=\nabla c^{(2)}$, and $r=u^{(2)}$ satisfies (4.74), (4.85) and $|r_1(x,t_0)|>0, x\in\overline{\Omega}$. Then there exist constants $C>0$ such that*

$$\|a^{(1)}-a^{(2)}\|_{H^3(\Omega)}+\|c^{(1)}-c^{(2)}\|_{H^2(\Omega)}$$
$$\le\sum_{l=1}^{2}C\|u_l^{(1)}(\cdot,t_0)-u_l^{(2)}(\cdot,t_0)\|_{H^5(\Omega)}+CI, \tag{4.87}$$

where

$$I=\sum_{l=1}^{2}\Big(\|\partial_t^3(u_l^{(1)}-u_l^{(2)})\|_{L^2(Q_{\omega,\delta})}$$
$$+\|\partial_t^2(u_l^{(1)}-u_l^{(2)})\|_{L^2(Q_{\omega,\delta})}+\|\partial_t(u_l^{(1)}-u_l^{(2)})\|_{L^2(Q_{\omega,\delta})}\Big).$$

4.2.3 Carleman Estimates with Boundary and Interior Data

In this subsection, we establish the Carleman estimates for (4.1) with boundary data. We transform (4.1) into an integer-order partial differential equation. The calculation is similar to Sect. 4.1.

Here we use Lemma 4.1 of Sect. 4.1. Although $\partial_t^{\frac{1}{2}}\partial_t^{\frac{1}{2}} \neq \partial_t$ in general, we may obtain this lemma by applying $\rho_2\partial_t^{\frac{1}{2}} - (\rho_1\partial_t - L)$ to both sides of (4.1) and using $u(x,0) = 0, x \in \Omega$.

Now we are ready to state our Carleman estimates.

Theorem 4.11 (Carleman estimate for (4.1) with boundary data) *Let $p \geq 0$. Suppose that $g(x,t) = 0, (x,t) \in \partial\Omega \times (0,T)$, and $\nabla g(x,t) = 0, (x,t) \in \Sigma$. Then there exists $\lambda_0 > 0$ such that for any $\lambda > \lambda_0$, we can choose $s_0(\lambda) > 0$ for which there exists $C = C(s_0,\lambda) > 0$ such that*

$$\int_Q \Big[(s\varphi_1)^{p-1}\big(|\partial_t^2 u|^2 + \sum_{i,j=1}^n |\partial_t\partial_i\partial_j u|^2\big) + (s\varphi_1)^{p+1}|\nabla\partial_t u|^2 +$$

$$+ (s\varphi_1)^{p+2}|\nabla(\rho_1\partial_t - L)u|^2 + (s\varphi_1)^{p+3}\big(|\partial_t u|^2 + \sum_{i,j=1}^n |\partial_i\partial_j u|^2\big) +$$

$$+ (s\varphi_1)^{p+5}|\nabla u|^2 + (s\varphi_1)^{p+7}|u|^2\Big]e^{2s\psi_1}dxdt \leq$$

$$\leq C\int_Q (s\varphi_1)^{p+1}\big|[\rho_2^2\partial_t - (\rho_1\partial_t - L)^2]u\big|^2 e^{2s\psi_1}dxdt +$$

$$+ C\int_\Sigma \Big[(s\varphi_1)^{p+1}|\nabla\partial_t u|^2 + (s\varphi_1)^{p+2}|\nabla\partial_t^{\frac{1}{2}} u|^2 + (s\varphi_1)^{p+5}|\nabla u|^2\Big]e^{2s\psi_1}dsdt \tag{4.88}$$

for all $s > s_0$ and all $u \in U$ satisfying (4.1) with $u(x,t) = 0, (x,t) \in \partial\Omega \times (0,T)$, and $u(x,0) = 0, x \in \Omega$.

Theorem 4.12 (Carleman estimate for (4.1) with interior data) *Let $p \geq 0$. Suppose that $g(x,t) = 0, (x,t) \in \partial\Omega \times (0,T)$, and $g(x,t) = 0, (x,t) \in Q_\omega$. Then there exists $\lambda_0 > 0$ such that for any $\lambda > \lambda_0$, we can choose $s_0(\lambda) > 0$ for which there exists $C = C(s_0,\lambda) > 0$ such that*

$$\int_Q \Big[(s\varphi_2)^{p-1}\big(|\partial_t^2 u|^2 + \sum_{i,j=1}^n |\partial_t\partial_i\partial_j u|^2\big) + (s\varphi_2)^{p+1}|\nabla\partial_t u|^2 +$$

$$+ (s\varphi_2)^{p+2}|\nabla(\rho_1\partial_t - L)u|^2 + (s\varphi_2)^{p+3}\big(|\partial_t u|^2 + \sum_{i,j=1}^n |\partial_i\partial_j u|^2\big) +$$

$$+ (s\varphi_2)^{p+5}|\nabla u|^2 + (s\varphi_2)^{p+7}|u|^2\big]e^{2s\psi_2}dxdt \le$$

$$\le C\int_Q (s\varphi_2)^{p+1}\big|\big[\rho_2^2\partial_t - (\rho_1\partial_t - L)^2\big]u\big|^2 e^{2s\psi_2}dxdt +$$

$$+ C\int_{Q_\omega} \big[(s\varphi_2)^{p+3}|\partial_t u|^2 + (s\varphi_2)^{p+4}|\partial_t^{\frac{1}{2}} u|^2 + (s\varphi_2)^{p+7}|u|^2\big]e^{2s\psi_2}dxdt \tag{4.89}$$

for all $s > s_0$ *and all* $u \in U$ *satisfying (4.1) with* $u(x,t) = 0$, $(x,t) \in \partial\Omega \times (0,T)$, *and* $u(x,0) = 0$, $x \in \Omega$.

To prove Theorems 4.11 and 4.12, we start with the global Carleman estimates for parabolic equations (see, e.g., [23]) stated in Lemmas 4.4 and 4.5 below.

Lemma 4.4 *Let* $p \ge 0$. *There exists* $\lambda_0 > 0$ *such that for any* $\lambda > \lambda_0$, *we can choose* $s_0(\lambda) > 0$ *for which there exists* $C = C(s_0, \lambda) > 0$ *such that*

$$\int_Q \Big[(s\varphi_1)^{p-1}\Big(|\partial_t v|^2 + \sum_{i,j=1}^n |\partial_i\partial_j v|^2\Big) + (s\varphi_1)^{p+1}|\nabla v|^2 + (s\varphi_1)^{p+3}|v|^2\Big]e^{2s\psi_1}dxdt \le$$

$$\le C\int_Q (s\varphi_1)^p|(\rho_1\partial_t - L)v|^2 e^{2s\psi_1}dxdt + C\int_\Sigma (s\varphi_1)^{p+1}|\nabla v|^2 e^{2s\psi_1}dSdt$$

for all $s > s_0$ *and all* $v \in L^2((0,T); H^2(\Omega)) \cap H^1((0,T); L^2(\Omega))$ *satisfying* $v(x,t) = 0$, $(x,t) \in \partial\Omega \times (0,T)$.

Lemma 4.5 *Let* $p \ge 0$. *There exists* $\lambda_0 > 0$ *such that for any* $\lambda > \lambda_0$, *we can choose* $s_0(\lambda) > 0$ *for which there exists* $C = C(s_0, \lambda) > 0$ *such that*

$$\int_Q \Big[(s\varphi_2)^{p-1}\Big(|\partial_t v|^2 + \sum_{i,j=1}^n |\partial_i\partial_j v|^2\Big) + (s\varphi_2)^{p+1}|\nabla v|^2 + (s\varphi_2)^{p+3}|v|^2\Big]e^{2s\psi_2}dxdt$$

$$\le C\int_Q (s\varphi_2)^p|(\rho_1\partial_t - L)v|^2 e^{2s\psi_2}dxdt + C\int_{Q_\omega} (s\varphi_2)^{p+3}|v|^2 e^{2s\psi_2}dxdt,$$

for all $s > s_0$ *and all* $v \in L^2((0,T); H^2(\Omega)) \cap H^1((0,T); L^2(\Omega))$ *satisfying* $v(x,t) = 0$, $(x,t) \in \partial\Omega \times (0,T)$.

Proof of Theorem 4.11. Throughout the proof, we assume that $s > 1$ is large enough to satisfy $s\varphi > 1$ in Q.

Equation (4.11) yields

$$\rho_1\partial_t w(x,t) - Lw(x,t) = \rho_2^2\partial_t u(x,t) - G(x,t),\ (x,t) \in Q, \tag{4.90}$$

where

$$w(x,t) = \rho_1 \partial_t u(x,t) - Lu(x,t), (x,t) \in Q. \tag{4.91}$$

Since $u(x,t) = 0$, $(x,t) \in \partial\Omega \times (0,T)$, and $g(x,t) = 0$, $(x,t) \in \partial\Omega \times (0,T)$, we have by (4.1),

$$w(x,t) = \rho_1 \partial_t u(x,t) - Lu(x,t) = g(x,t) - \rho_2 \partial_t^{\frac{1}{2}} u(x,t) = 0, (x,t) \in \partial\Omega \times (0,T).$$

Applying Lemma 4.4 to (4.90), we obtain

$$\int_Q [(s\varphi_1)^{p_1-1}|\partial_t w|^2 + (s\varphi_1)^{p_1+1}|\nabla w|^2 + (s\varphi_1)^{p_1+3}|w|^2]e^{2s\psi_1}dxdt \le$$

$$\le C\int_Q (s\varphi_1)^{p_1}|\partial_t u|^2 e^{2s\psi_1}dxdt + C\int_Q (s\varphi_1)^{p_1}|G|^2 e^{2s\psi_1}dxdt +$$

$$+ C\int_\Sigma (s\varphi_1)^{p_1+1}|\nabla w|^2 e^{2s\psi_1}dSdt \tag{4.92}$$

for $p_1 \ge 0$. Next by applying Lemma 4.4 to (4.91), we obtain

$$\int_Q \Big[(s\varphi_1)^{p_2-1}\Big(|\partial_t u|^2 + \sum_{i,j=1}^n |\partial_i\partial_j u|^2\Big) +$$

$$+ (s\varphi_1)^{p_2+1}|\nabla u|^2 + (s\varphi_1)^{p_2+3}|u|^2\Big]e^{2s\psi_1}dxdt \le$$

$$\le C\int_Q (s\varphi_1)^{p_2}|w|^2 e^{2s\psi_1}dxdt + C\int_\Sigma (s\varphi_1)^{p_2+1}|\nabla u|^2 e^{2s\psi_1}dSdt \tag{4.93}$$

for $p_2 \ge 0$.

Putting $p_2 = p_1 + 1$ and substituting the estimate of $|\partial_t u|^2$ in (4.93) into the right-hand side of (4.92), we obtain

$$\int_Q [(s\varphi_1)^{p_1-1}|\partial_t w|^2 + (s\varphi_1)^{p_1+1}|\nabla w|^2 + (s\varphi_1)^{p_1+3}|w|^2]e^{2s\psi_1}dxdt \le$$

$$\le C\int_Q (s\varphi_1)^{p_1+1}|w|^2 e^{2s\psi_1}dxdt + C\int_Q (s\varphi_1)^{p_1}|G|^2 e^{2s\psi_1}dxdt + CB_{1,p_1},$$

where

$$B_{1,p_1} = \int_\Sigma [(s\varphi_1)^{p_1+1}|\nabla w|^2 + (s\varphi_1)^{p_1+2}|\nabla u|^2]e^{2s\psi_1}dSdt.$$

Taking sufficiently large $s > 0$, we can absorb the first term on the right-hand side of the above inequality into the left-hand side and we have

$$\int_Q [(s\varphi_1)^{p_1-1}|\partial_t w|^2 + (s\varphi_1)^{p_1+1}|\nabla w|^2 + (s\varphi_1)^{p_1+3}|w|^2]e^{2s\psi_1}dxdt \leq$$
$$\leq C\int_Q (s\varphi_1)^{p_1}|G|^2 e^{2s\psi_1}dxdt + CB_{1,p_1}. \tag{4.94}$$

By (4.93) with $p_2 = p_1 + 3$ and (4.94), we obtain

$$\int_Q \Big[(s\varphi_1)^{p_1+1}|\nabla(\rho_1\partial_t - L)u|^2 + (s\varphi_1)^{p_1+2}\big(|\partial_t u|^2 + \sum_{i,j=1}^n |\partial_i\partial_j u|^2\big) +$$
$$+ (s\varphi_1)^{p_1+4}|\nabla u|^2 + (s\varphi_1)^{p_1+6}|u|^2\Big]e^{2s\psi_1}dxdt \leq$$
$$\leq C\int_Q (s\varphi_1)^{p_1}|G|^2 e^{2s\psi_1}dxdt + CB_{2,p_1}, \tag{4.95}$$

where

$$B_{2,p_1} = \int_\Sigma [(s\varphi_1)^{p_1+1}|\nabla w|^2 + (s\varphi_1)^{p_1+4}|\nabla u|^2]e^{2s\psi_1}dSdt.$$

Let us choose $p_1 = p + 1$ in (4.94). Then from (4.91) and (4.94), we have

$$\int_Q (s\varphi_1)p|\partial_t(\rho_1\partial_t u - Lu)|^2 e^{2s\psi_1}dxdt \leq C\int_Q (s\varphi_1)^{p+1}|G|^2 e^{2s\psi_1}dxdt + CB_{1,p+1}.$$

Setting $u_0 = \partial_t u$, we obtain

$$\int_Q (s\varphi_1)^p|\rho_1\partial_t u_0 - Lu_0|^2 e^{2s\psi_1}dxdt \leq C\int_Q (s\varphi_1)^{p+1}|G|^2 e^{2s\psi_1}dxdt + CB_{1,p+1}. \tag{4.96}$$

If we use Lemma 4.4 with $v = u_0$ and applying (4.96), we obtain

$$\int_Q \Big[(s\varphi_1)^{p-1}\big(|\partial_t u_0|^2 + \sum_{i,j=1}^n |\partial_i\partial_j u_0|^2\big) +$$
$$+ (s\varphi_1)^{p+1}|\nabla u_0|^2 + (s\varphi_1)^{p+3}|u_0|^2\Big]e^{2s\psi_1}dxdt \leq$$
$$\leq C\int_Q (s\varphi_1)^{p+1}|G|^2 e^{2s\psi_1}dxdt + C\int_\Sigma (s\varphi_1)^{p+1}|\nabla u_0|^2 e^{2s\psi_1}dSdt + CB_{1,p+1}.$$

Recalling $u_0 = \partial_t u$, we have

$$\int_Q \big[(s\varphi_1)^{p-1}\big(|\partial_t^2 u|^2 + \sum_{i,j=1}^n |\partial_t\partial_i\partial_j u|^2\big) + (s\varphi_1)^{p+1}|\nabla\partial_t u|^2$$

$$+ (s\varphi_1)^{p+3}|\partial_t u|^2\big]e^{2s\psi_1}dxdt \le C\int_Q (s\varphi_1)^{p+1}|G|^2 e^{2s\psi_1}dxdt + CB_{3,p},$$

where

$$B_{3,p} = \int_\Sigma [(s\varphi_1)^{p+1}|\nabla\partial_t u|^2 + (s\varphi_1)^{p+2}|\nabla w|^2 + (s\varphi_1)^{p+3}|\nabla u|^2]e^{2s\psi_1}dSdt.$$

Hence using (4.95), we obtain

$$\int_Q \big[(s\varphi_1)^{p-1}\big(|\partial_t^2 u|^2 + \sum_{i,j=1}^n |\partial_t\partial_i\partial_j u|^2\big) + (s\varphi_1)^{p+1}|\nabla\partial_t u|^2 +$$

$$+ (s\varphi_1)^{p+2}|\nabla(\rho_1\partial_t - L)u|^2 + (s\varphi_1)^{p+3}\big(|\partial_t u|^2 + \sum_{i,j=1}^n |\partial_i\partial_j u|^2\big) +$$

$$+ (s\varphi_1)^{p+5}|\nabla u|^2 + (s\varphi_1)^{p+7}|u|^2\big]e^{2s\psi_1}dxdt \le$$

$$\le C\int_Q (s\varphi_1)^{p+1}|G|^2 e^{2s\psi_1}dxdt + CB_{4,p},$$

where

$$B_{4,p} = \int_\Sigma [(s\varphi_1)^{p+1}|\nabla\partial_t u|^2 + (s\varphi_1)^{p+2}|\nabla w|^2 + (s\varphi_1)^{p+5}|\nabla u|^2]e^{2s\psi_1}dSdt.$$

Finally, we consider the boundary term $B_{4,p}$. Since $\nabla g = 0$ on Σ is assumed, $\nabla w = \nabla g - \rho_2\nabla\partial_t^{\frac{1}{2}}u = -\rho_2\nabla\partial_t^{\frac{1}{2}}u$ on Σ. Hence we have

$$\int_Q \big[(s\varphi_1)^{p-1}\big(|\partial_t^2 u|^2 + \sum_{i,j=1}^n |\partial_t\partial_i\partial_j u|^2\big) + (s\varphi_1)^{p+1}|\nabla\partial_t u|^2 +$$

$$+ (s\varphi_1)^{p+2}|\nabla(\rho_1\partial_t - L)u|^2 + (s\varphi_1)^{p+3}\big(|\partial_t u|^2 + \sum_{i,j=1}^n |\partial_i\partial_j u|^2\big) +$$

$$+ (s\varphi_1)^{p+5}|\nabla u|^2 + (s\varphi_1)^{p+7}|u|^2\big]e^{2s\psi_1}dxdt \le$$

$$\le C\int_Q (s\varphi_1)^{p+1}|G|^2 e^{2s\psi_1}dxdt +$$

$$+C\int_{\Sigma}\left[(s\varphi_1)^{p+1}|\nabla\partial_t u|^2+(s\varphi_1)^{p+2}|\nabla\partial_t^{\frac{1}{2}}u|^2+(s\varphi_1)^{p+5}|\nabla u|^2\right]e^{2s\psi_1}dSdt.$$

Thus we obtain (4.88).
Proof of Theorem 4.12. By using Lemma 4.5 instead of Lemma 4.4, we can prove Theorem 4.12 in the same way as Theorem 4.11.

Furthermore we need Carleman estimates for elliptic equations in the proof of the stability estimates in inverse source problems which we will develop in Sect. 4.2.4.

Let us assume that $\widetilde{a_{ij}}\in C^1(\overline{\Omega}), \widetilde{a_{ij}}=\widetilde{a_{ji}}(i,j=1,\dots,n), \widetilde{b_j}\in C(\overline{\Omega})(j=1,\dots,n), \widetilde{c}\in C(\overline{\Omega})$, and that there exists a constant $\widetilde{\mu}>0$ such that

$$\frac{1}{\widetilde{\mu}}|\xi|^2\le\sum_{i,j=1}^{n}\widetilde{a}_{ij}(x)\xi_i\xi_j\le\widetilde{\mu}|\xi|2, \xi=(\xi_1,\dots,\xi_n)\in R^n, x\in\overline{\Omega}.$$

We consider the following symmetric uniformly elliptic operator:

$$\widetilde{L}\widetilde{v}(x):=\sum_{i,j=1}^{n}\partial_i(\widetilde{a}_{ij}(x)\partial_j\widetilde{v}(x))-\sum_{j=1}^{n}\widetilde{b}_j(x)\partial_j\widetilde{v}(x)-\widetilde{c}(x)\widetilde{v}(x), x\in\Omega.$$

Set $\widetilde{\varphi}_k(x):=\varphi_k(x,t_0), x\in\Omega$, and $\widetilde{\psi}_k(x):=\psi_k(x,t_0), x\in\Omega$, for $k=1,2$. Then we have the following lemmas.

Lemma 4.6 *Let $p\ge 0$. There exists $\lambda_0>0$ such that for any $\lambda>\lambda_0$, we can choose $s_0(\lambda)>0$ for which there exists $C=C(s_0,\lambda)>0$ such that*

$$\int_{\Omega}\left[(s\widetilde{\varphi}_1)^{p-1}\sum_{i,j=1}^{n}|\partial_i\partial_j\widetilde{v}|^2+(s\widetilde{\varphi}-1)^{p+1}|\nabla\widetilde{v}|^2+(s\widetilde{\varphi}_1)^{p+3}|\widetilde{v}|^2\right]e^{2s\widetilde{\psi}_1}dx\le$$

$$\le C\int_{\Omega}(s\widetilde{\varphi}_1)^p|\widetilde{L}\widetilde{v}|^2e^{2s\widetilde{\psi}_1}dx+C\int_{\gamma}(s\widetilde{\varphi}_1)p+1|\nabla\widetilde{v}|^2e^{2s\widetilde{\psi}_1}dS,$$

for all $s>s_0$ and all $\widetilde{v}\in H^2(\Omega)$ satisfying $\widetilde{v}(x)=0, x\in\partial\Omega$.

Lemma 4.7 *Let $p\ge 0$. There exists $\lambda_0>0$ such that for any $\lambda>\lambda_0$, we can choose $s_0(\lambda)>0$ for which there exists $C=C(s_0,\lambda)>0$ such that*

$$\int_{\Omega}\left[(s\widetilde{\varphi}_2)^{p-1}\sum_{i,j=1}^{n}|\partial_i\partial_j\widetilde{v}|^2+(s\widetilde{\varphi}_2)^{p+1}|\nabla\widetilde{v}|^2+(s\widetilde{\varphi}_2)^{p+3}|\widetilde{v}|^2\right]e^{2s\widetilde{\psi}_2}dx\le$$

$$\le C\int_{\Omega}(s\widetilde{\varphi}_2)^p|\widetilde{L}\widetilde{v}|^2e^{2s\widetilde{\psi}_2}dx+C\int_{\omega}(s\widetilde{\varphi}_2)^{p+3}|\widetilde{v}|^2e^{2s\widetilde{\psi}_2}dx$$

for all $s>s_0$ and all $\widetilde{v}\in H^2(\Omega)$ satisfying $\widetilde{v}(x)=0, x\in\partial\Omega$.

These lemmas can be shown in the same manner as the parabolic case by means of integration by parts. Hence we omit the proofs of these lemmas here.

We conclude this subsection by introducing Carleman estimates for the third-order partial differential equations which we use in the proof of the stability estimates in inverse problems of determining the diffusion coefficients.

Let $p = (p_1, \dots, p_n) \in \{C^1(\overline{\Omega})\}^n$.

Lemma 4.8 *We assume that there exists $m_1 > 0$ such that $|p(x) \cdot \nabla d_1(x)| \geq m_1, x \in \overline{\Omega}$. Then there exists $\lambda_0 > 0$ such that for any $\lambda > \lambda_0$, we can choose $s_0(\lambda) > 0$ for which there exists $C = C(s_0, \lambda) > 0$ such that*

$$\int_\Omega \Big[s\widetilde{\varphi}_1 \sum_{i,j,k=1}^n |\partial_i \partial_j \partial_k \widetilde{v}|^2 + (s\widetilde{\varphi}_1)^2 |\nabla \Delta \widetilde{v}|^2 +$$

$$+ (s\widetilde{\varphi}_1)^3 \sum_{i,j=1}^n |\partial_i \partial_j \widetilde{v}|^2 + (s\widetilde{\varphi}_1)^5 (|\nabla \widetilde{v}|^2 + |\widetilde{v}|^2) \Big] e^{2s\widetilde{\psi}_1} dx \leq$$

$$\leq C \int_\Omega (|\nabla(p \cdot \nabla \Delta \widetilde{v})|^2 + |p \cdot \nabla \Delta \widetilde{v}|^2) e^{2s\widetilde{\psi}_1} dx$$

for all $s > s_0$ and all $\widetilde{v} \in H^4(\Omega)$ satisfying $|\widetilde{v}(x)| = |\nabla \widetilde{v}(x)| = |\Delta \widetilde{v}(x)| = |\nabla \Delta \widetilde{v}(x)| = 0, x \in \partial\Omega$, and $|\nabla \partial_k \widetilde{v}(x)| = 0, x \in \gamma (k = 1, 2, \dots, n)$.

Lemma 4.9 *We assume that there exists $m_2 > 0$ such that $|p(x) \cdot \nabla d_2(x)| \geq m_2, x \in \Omega \setminus \omega$. Then there exists $\lambda_0 > 0$ such that for any $\lambda > \lambda_0$, we can choose $s_0(\lambda) > 0$ for which there exists $C = C(s_0, \lambda) > 0$ such that*

$$\int_\Omega \Big[s\widetilde{\varphi}_2 \sum_{i,j,k=1}^n |\partial_i \partial_j \partial_k \widetilde{v}|^2 + (s\widetilde{\varphi}_2)^2 |\nabla \Delta \widetilde{v}|^2 +$$

$$+ (s\widetilde{\varphi}_2)^3 \sum_{i,j=1}^n |\partial_i \partial_j \widetilde{v}|^2 + (s\widetilde{\varphi}_2)^5 (|\nabla \widetilde{v}|^2 + |\widetilde{v}|^2) \Big] e^{2s\widetilde{\psi}_2} dx \leq$$

$$\leq C \int_\Omega (|\nabla(p \cdot \nabla \Delta \widetilde{v})|^2 + |p \cdot \nabla \widetilde{v}|^2) e^{2s\widetilde{\psi}_2} dx,$$

for all $s > s_0$ and all $\widetilde{v} \in H^4(\Omega)$ satisfying $|\widetilde{v}(x)| = |\nabla \widetilde{v}(x)| = |\Delta \widetilde{v}(x)| = |\nabla \Delta \widetilde{v}(x)| = 0, x \in \partial\Omega$, and $\widetilde{v}(x) = 0, x \in \omega$.

To establish the Carleman estimate for $p \cdot \nabla \Delta \widetilde{v}$, we start by proving the first-order partial differential equations $p \cdot \nabla \widetilde{v}$.

Lemma 4.10 *We assume that there exists* $m_1 > 0$ *such that* $|p(x) \cdot \nabla d_1(x)| \geq m_1, x \in \overline{\Omega}$. *Then there exists* $\lambda_0 > 0$ *such that for any* $\lambda > \lambda_0$, *we can choose* $s_0(\lambda) > 0$ *for which there exists* $C = C(s_0, \lambda) > 0$ *such that*

$$\int_\Omega (s\widetilde{\varphi}_1)^2(|\nabla \widetilde{v}|^2 + |\widetilde{v}|^2)e^{2s\widetilde{\psi}_1}dx \leq C \int_\Omega (|\nabla(p \cdot \nabla \widetilde{v})|^2 + |p \cdot \nabla \widetilde{v}|^2)e^{2s\widetilde{\psi}_1}dx$$

for all $s > s_0$ *and all* $\widetilde{v} \in H^2(\Omega)$ *satisfying* $|\widetilde{v}(x)| = |\nabla \widetilde{v}(x)| = 0, x \in \partial\Omega$.

Lemma 4.11 *We assume that there exists* $m_2 > 0$ *such that* $|p(x) \cdot \nabla d_2(x)| \geq m_2, x \in \overline{\Omega \setminus \omega}$. *Then there exists* $\lambda_0 > 0$ *such that for any* $\lambda > \lambda_0$, *we can choose* $s_0(\lambda) > 0$ *for which there exists* $C = C(s_0, \lambda) > 0$ *such that*

$$\int_\Omega (s\widetilde{\varphi}_2)^2(|\nabla \widetilde{v}|^2 + |\widetilde{v}|^2)e^{2s\widetilde{\psi}_2}dx \leq C \int_\Omega (|\nabla(p \cdot \nabla \widetilde{v})|^2 + |p \cdot \nabla \widetilde{v}|^2)e^{2s\widetilde{\psi}_2}dx$$

for all $s > s_0$ *and all* $\widetilde{v} \in H^2(\Omega)$ *satisfying* $|\widetilde{v}(x)| = |\nabla \widetilde{v}(x)| = 0, x \in \partial\Omega$, *and* $\widetilde{v}(x) = 0, x \in \omega$.

Proof of Lemma 4.10. Setting $\widetilde{w} = \widetilde{v}e^{s\widetilde{\psi}_1}$ in Ω, we have

$$e^{s\widetilde{\psi}_1}(p \cdot \nabla \widetilde{v}) = p \cdot \nabla \widetilde{w} - s\lambda\widetilde{\varphi}_1(p \cdot \nabla d_1)\widetilde{w} \text{ in } \Omega.$$

Taking the weighted L^2 norm, by integrating by parts we obtain

$$\int_\Omega |p \cdot \nabla \widetilde{v}|^2 e^{2s\widetilde{\psi}_1}dx = \int_\Omega |p \cdot \nabla \widetilde{w}|^2 dx + \int_\Omega s^2\lambda^2\widetilde{\varphi}_1^2(p \cdot \nabla d_1)^2|\widetilde{w}|^2 dx -$$

$$-2\int_\Omega s\lambda\widetilde{\varphi}_1(p \cdot \nabla d_1)\sum_{j=1}^n p_j\widetilde{w}\partial_j\widetilde{w}dx \geq$$

$$\geq \int_\Omega s^2\lambda^2\widetilde{\varphi}_1^2(p \cdot \nabla d_1)^2|\widetilde{w}|^2dx - \int_\Omega s\lambda\widetilde{\varphi}_1(p \cdot \nabla d_1)\sum_{j=1}^n p_j\partial_j(\widetilde{w})^2dx =$$

$$= \int_\Omega s^2\lambda^2\widetilde{\varphi}_1^2(p \cdot \nabla d_1)^2|\widetilde{w}|^2dx + \int_\Omega s\lambda^2\widetilde{\varphi}_1(p \cdot \nabla d_1)^2|\widetilde{w}|^2dx +$$

$$+ \int_\Omega s\lambda\widetilde{\varphi}_1[(p \cdot \nabla d_1)(div p) + p \cdot \nabla(p \cdot \nabla d_1)]|\widetilde{w}|^2dx.$$

Hence we have

$$\int_\Omega s^2\lambda^2\widetilde{\varphi}_1^2|\widetilde{w}|^2dx \leq C\int_\Omega |p \cdot \nabla \widetilde{v}|^2e^{2s\widetilde{\psi}_1}dx + C\int_\Omega (s\lambda^2\widetilde{\varphi}_1 + s\lambda\widetilde{\varphi}_1)|\widetilde{w}|^2dx.$$

Taking sufficiently large $s > 0$, we may absorb the second term on the right-hand side of the above inequality into the left-hand side and we see that

$$\int_\Omega s^2\lambda^2\widetilde{\varphi}_1^2|\widetilde{w}|^2 dx \le C\int_\Omega |p\cdot\nabla\widetilde{v}|^2 e^{2s\widetilde{\psi}_1} dx,$$

that is,

$$\int_\Omega s^2\lambda^2\widetilde{\varphi}_1^2|\widetilde{v}|^2 e^{2s\widetilde{\psi}_1} dx \le C\int_\Omega |p\cdot\nabla\widetilde{v}|^2 e^{2s\widetilde{\psi}_1} dx. \tag{4.97}$$

Set $\widetilde{v}_k = \partial_k\widetilde{v}$ in Ω for $k = 1, 2, \ldots, n$. We consider

$$p\cdot\nabla\widetilde{v}_k = \partial_k(p\cdot\nabla\widetilde{v}) - (\partial_k p)\cdot\nabla\widetilde{v}.$$

Applying the estimate (4.97) to the above equation, we may obtain

$$\int_\Omega s^2\lambda^2\widetilde{\varphi}_1^2|\widetilde{v}_k|^2 e^{2s\widetilde{\psi}_1} dx \le C\int_\Omega |p\cdot\nabla\widetilde{v}_k|^2 e^{2s\widetilde{\psi}_1} dx \le$$

$$\le C\int_\Omega |\partial_k(p\cdot\nabla\widetilde{v})|^2 e^{2s\widetilde{\psi}_1} dx + C\int_\Omega |(\partial_k p)\cdot\nabla\widetilde{v}|^2 e^{2s\widetilde{\psi}_1} dx.$$

Hence we have

$$\int_\Omega s^2\lambda^2\widetilde{\varphi}_1^2|\nabla\widetilde{v}|^2 e^{2s\widetilde{\psi}_1} dx \le C\int_\Omega |\nabla(p\cdot\nabla\widetilde{v})|^2 e^{2s\widetilde{\psi}_1} dx + C\int_\Omega |\nabla\widetilde{v}|^2 e^{2s\widetilde{\psi}_1} dx.$$

Choosing sufficiently large $s > 0$, we can absorb the second term on the right-hand side of the above inequality into the left-hand side and we may get

$$\int_\Omega s^2\lambda^2\widetilde{\varphi}_1^2|\nabla\widetilde{v}|^2 e^{2s\widetilde{\psi}_1} dx \le C\int_\Omega |\nabla(p\cdot\nabla\widetilde{v})|^2 e^{2s\widetilde{\psi}_1} dx.$$

Combining this with (4.97), we obtain the Carleman estimate of Lemma 4.10. Thus we conclude Lemma 4.10.

Proof of Lemma 4.11. By an argument similar to that used in the proof of Lemma 4.10, we may obtain Lemma 4.11.

Remark 4.8 In one spatial dimension, the assumption $|\nabla\widetilde{v}| = 0$ on $\partial\Omega$ is not necessary in Lemmas 4.10 and 4.11. In this case, we have the following Carleman estimate by integration by parts:

$$\int_\Omega (s\widetilde{\varphi}_k)^2(|\partial_1\widetilde{v}|^2 + |\widetilde{v}|^2)e^{2s\widetilde{\psi}_k} dx \le C\int_\Omega |p_1\partial_1\widetilde{v}|^2 e^{2s\widetilde{\psi}_k} dx$$

for $k = 1, 2$.

Now we are ready to prove Lemmas 4.8 and 4.9.

Proof of Lemma 4.8. Set $\widetilde{y} = \Delta\widetilde{v}$ in Ω. By $|\Delta\widetilde{v}(x)| = |\nabla\Delta\widetilde{v}(x)| = 0$, $x \in \partial\Omega$, we see that $|\widetilde{y}(x)| = |\nabla\widetilde{y}(x)| = 0$, $x \in \partial\Omega$. By Lemma 4.10, we obtain

$$\int_\Omega (s\widetilde{\varphi}_1)^2(|\nabla\widetilde{y}|^2 + |\widetilde{y}|2)e^{2s\widetilde{\psi}_1}dx \le C\int_\Omega (|\nabla(p\cdot\nabla\widetilde{y})|^2 + |p\cdot\nabla\widetilde{y}|^2)e^{2s\widetilde{\psi}_1}dx,$$

that is,

$$\int_\Omega (s\widetilde{\varphi}_1)^2(|\nabla\widetilde{v}|^2 + |\widetilde{v}|^2)e^{2s\widetilde{\psi}_1}dx \le$$

$$\le C\int_\Omega (|\nabla(p\cdot\nabla\widetilde{v})|^2 + |p\cdot\nabla\widetilde{v}|^2)e^{2s\widetilde{\psi}_1}dx. \tag{4.98}$$

Next we use the Carleman estimate for elliptic equations to estimate the left-hand side of the above inequality. By Lemma 4.6 with $p = 2$, we have

$$\int_\Omega \Big[s\widetilde{\varphi}_1 \sum_{i,j=1}^n |\partial_i\partial_j\widetilde{v}|^2 + (s\widetilde{\varphi}_1)^3|\nabla\widetilde{v}|^2 + (s\widetilde{\varphi}_1)^5|\widetilde{v}|^2\Big]e^{2s\widetilde{\psi}_1}dx \le$$

$$\le C\int_\Omega (s\widetilde{\varphi}_1)^2|\widetilde{v}|^2e^{2s\widetilde{\psi}_1}dx. \tag{4.99}$$

Setting $\widetilde{v}_k = \partial_k\widetilde{v}$ in Ω for $k = 1, 2, \ldots, n$ and using Lemma 4.6 again, we see that

$$\int_\Omega \Big[s\widetilde{\varphi}_1 \sum_{i,j=1}^n |\partial_i\partial_j\widetilde{v}_k|^2 + (s\widetilde{\varphi}_1)^3|\nabla\widetilde{v}_k|^2 + (s\widetilde{\varphi}_1)^5|\widetilde{v}_k|^2\Big]e^{2s\widetilde{\psi}_1}dx \le$$

$$\le C\int_\Omega (s\widetilde{\varphi}_1)^2|\widetilde{v}_k|^2e^{2s\widetilde{\psi}_1}dx,$$

that is,

$$\int_\Omega \Big[s\widetilde{\varphi}_1 \sum_{i,j,k=1}^n |\partial_i\partial_j\partial_k\widetilde{v}|^2 + (s\widetilde{\varphi}_1)^3 \sum_{i,j=1}^n |\partial_i\partial_j\widetilde{v}|^2 + (s\widetilde{\varphi}_1)^5|\nabla\widetilde{v}|^2\Big]e^{2s\widetilde{\psi}_1}dx \le$$

$$\le C\int_\Omega (s\widetilde{\varphi}_1)^2|\nabla\Delta\widetilde{v}|^2e^{2s\widetilde{\psi}_1}dx. \tag{4.100}$$

Summing up the inequalities (4.98)–(4.100), we may obtain the Carleman estimate of Lemma 4.8.

Proof of Lemma 4.9. Using Lemmas 4.11 and 4.7 instead of Lemmas 4.10 and 4.6, we may prove Lemma 4.9 in the same way as Lemma 4.8.

4.2.4 Proof of Stability Estimates in Inverse Source and Coefficient Problems

Hereafter we let C denote a generic constant which is independent of s, x, t and let $C(s)$ denote a generic constant which is independent of x, t but depends on s.

Stability in inverse source problems.

Proof of Theorem 4.3. By using Lemma 4.1 (in Sect. 4.1) for (4.62), we obtain

$$\rho_2^2 \partial_t u(x,t) - (\rho_1 \partial_t - L)^2 u(x,t) = F(x,t), (x,t) \in Q, \tag{4.101}$$

where we introduced $F(x,t)$ as

$$F(x,t) = \left[\rho_2 \partial_t^{\frac{1}{2}} - (\rho_1 \partial_t - L)\right](f(x)R(x,t)) + \rho_2 f(x) \frac{R(x,0)}{\sqrt{\pi t}} =$$

$$= R(x,t) \sum_{i,j=1}^{n} \partial_i (a_{ij}(x) \partial_j f(x)) +$$

$$+ \sum_{j=1}^{n} \big(2 \sum_{i=1}^{n} a_{ij}(x) \partial_i R(x,t) - b_j(x) R(x,t)\big) \partial_j f(x) +$$

$$+ \left[\rho_2 \partial_t^{\frac{1}{2}} R(x,t) - \rho_1 \partial_t R(x,t) + \sum_{i,j=1}^{n} \partial_i (a_{ij}(x) \partial_j R(x,t)) -\right.$$

$$\left. - \sum_{j=1}^{n} b_j(x) \partial_j R(x,t) - c(x) R(x,t) + \frac{\rho_2 R(x,0)}{\sqrt{\pi t}} \right] f(x), (x,t) \in Q. \tag{4.102}$$

Let us set $y = \partial_t u$, $z = \partial_t^2 u$ in Q. By differentiating (4.101) with respect to t, we have

$$\rho_2^2 \partial_t y(x,t) - (\rho_1 \partial_t - L)^2 y(x,t) = \partial_t F(x,t), (x,t) \in Q, \tag{4.103}$$

$$\rho_2^2 \partial_t z(x,t) - (\rho_1 \partial_t - L)^2 z(x,t) = \partial_t^2 F(x,t), (x,t) \in Q. \tag{4.104}$$

Since $u(x,t) = 0$, $(x,t) \in \partial\Omega \times (0,T)$, we see that

$$y(x,t) = z(x,t) = 0, (x,t) \in \partial\Omega \times (0,T).$$

To use the Carleman estimate in Q_δ, we introduce the weight functions. Set

$$\varphi_{\delta,1}(x,t) = \frac{e^{\lambda d_1(x)}}{(t - t_0 + \delta)(t_0 + \delta - t)},$$

$$\psi_{\delta,1}(x,t) = \frac{e^{\lambda d_1(x)} - e^{2\lambda \|d_1\|_{C(\overline{\Omega})}}}{(t-t_0+\delta)(t_0+\delta-t)}, \ (x,t) \in Q_\delta.$$

Fixing $\lambda > 0$ and applying Theorem 4.11 ($p = 0$) to (4.103) and (4.104) in Q_δ, we have

$$\int_{Q_\delta} \Big[(s\varphi_{\delta,1})^3\big(|\partial_t y|^2 + |\partial_t z|^2 + \sum_{i,j=1}^n |\partial_i\partial_j y|^2 + \sum_{i,j=1}^n |\partial_i\partial_j z|^2\big) +$$

$$+ (s\varphi_{\delta,1})^5(|\nabla y|^2 + |\nabla z|^2) + (s\varphi_{\delta,1})^7(|y|^2 + |z|^2)\Big]e^{2s\psi_{\delta,1}}dxdt \le$$

$$\le C\int_{Q_\delta} s\varphi_{\delta,1}(|\partial_t F|^2 + |\partial_t^2 F|^2)e^{2s\psi_{\delta,1}}dxdt + C\widetilde{B}, \tag{4.105}$$

where

$$\widetilde{B} = \int_{\Sigma_\delta} \Big[s\varphi_{\delta,1}(|\nabla\partial_t y|^2 + |\nabla\partial_t z|^2) +$$

$$+ (s\varphi_{\delta,1})^2\big(|\nabla\partial_t^{\frac{1}{2}} y|^2 + |\nabla\partial_t^{\frac{1}{2}} z|^2\big) + (s\varphi_{\delta,1})^5(|\nabla y|^2 + |\nabla z|^2)\Big]e^{2s\psi_{\delta,1}}dSdt.$$

We can estimate $\widetilde{B}$ by B^2. We note that $\partial_t^{\frac{1}{2}}\partial_t^m = \partial_t^{m+\frac{1}{2}}$, $m \in N$, and Lemma A.1. Since there exist constants $C_k(s) > 0$ such that $\varphi_{\delta,1}^k e^{2s\psi_{\delta,1}} \le C_k(s)$ on Σ_δ for $k = 0, 1, 2, \ldots$, we have

$$\widetilde{B} \le C\int_{\Sigma_\delta} \Big[s\varphi_{\delta,1}|\nabla\partial_t^3 u|^2 + (s\varphi_{\delta,1})^2\big(|\nabla\partial_t^{\frac{1}{2}}\partial_t u|^2 + |\nabla\partial_t^{\frac{1}{2}}\partial_t^2 u|^2\big) +$$

$$+ (s\varphi_{\delta,1})^5(|\nabla\partial_t u|^2 + |\nabla\partial_t^2 u|^2)\Big]e^{2s\psi_{\delta,1}}dSdt \le C(s)B^2.$$

Note that

$$\int_{Q_\delta} s\varphi_{\delta,1}(|\partial_t F|^2 + |\partial_t^2 F|^2)e^{2s\psi_{\delta,1}}dxdt \le C\int_{Q_\delta} s\varphi_{\delta,1}\sum_{|\alpha|\le 2}|\partial_x^\alpha f|^2 e^{2s\psi_{\delta,1}}dxdt.$$

This together with (4.105) gives

$$\int_{Q_\delta} \Big[(s\varphi_{\delta,1})^3\big(|\partial_t y|^2 + |\partial_t z|^2 + \sum_{i,j=1}^n |\partial_i\partial_j y|^2 + \sum_{i,j=1}^n |\partial_i\partial_j z|^2 +$$

$$+ (s\varphi_{\delta,1})^5(|\nabla y|^2 + |\nabla z|^2) + (s\varphi_{\delta,1})^7(|y|^2 + |z|^2)\Big]e^{2s\psi_{\delta,1}}dxdt \le$$

$$\leq C \int_{Q_\delta} s\varphi_{\delta,1} \sum_{|\alpha|\leq 2|} \partial_x^\alpha f|^2 e^{2s\psi_{\delta,1}} dxdt + C(s)B^2. \tag{4.106}$$

Let us expand the left-hand side of (4.101). We have

$$\rho_2^2\partial_t u(x,t) - \rho_1^2\partial_t^2 u(x,t) + 2\rho_1\partial_t Lu(x,t) - L^2u(x,t) = F(x,t), (x,t) \in Q.$$

In particular at $t = t_0$, we have

$$\rho_2^2\partial_t u(x,t_0) - \rho_1^2\partial_t^2 u(x,t_0) + 2\rho_1\partial_t Lu(x,t_0) - L^2u(x,t_0) = F(x,t_0), x \in \Omega. \tag{4.107}$$

Taking the weighted L^2 norm of (4.107) in Ω, we obtain

$$\int_\Omega \varphi_{\delta,1}(x,t_0)|F(x,t_0)|^2 e^{2s\psi_{\delta,1}(x,t_0)}dx \leq$$

$$\leq C\sum_{k=1}^{3} J_k + C\int_\Omega \varphi_{\delta,1}(x,t_0) \sum_{|\alpha|\leq 4} |\partial_x^\alpha u(x,t_0)|^2 e^{2s\psi_{\delta,1}(x,t_0)}dx, \tag{4.108}$$

where

$$J_1 = \int_\Omega \varphi_{\delta,1}(x,t_0)|\partial_t u(x,t_0)|^2 e^{2s\psi_{\delta,1}(x,t_0)}dx,$$

$$J_2 = \int_\Omega \varphi_{\delta,1}(x,t_0)|\partial_t^2 u(x,t_0)|^2 e^{2s\psi_{\delta,1}(x,t_0)}dx,$$

$$J_3 = \int_\Omega \varphi_{\delta,1}(x,t_0)|\partial_t Lu(x,t_0)|^2 e^{2s\psi_{\delta,1}(x,t_0)}dx.$$

Let us estimate J_1, J_2, J_3. We assume that $s > 1$ is large enough to satisfy $s\varphi_{\delta,1} > 1$ in Q. We note that $\partial_t\psi_{\delta,1}(x,t) \leq C\varphi_{\delta,1}^2(x,t)$ for $(x,t) \in Q$ and

$$J_1 = \int_{t_0-\delta}^{t_0}\int_\Omega \partial_t(\varphi_{\delta,1}|y|^2 e^{2s\psi_{\delta,1}})dxdt \leq$$

$$\leq C\int_{t_0-\delta}^{t_0}\int \Omega[\varphi_{\delta,1}^2|y|^2 + \varphi_{\delta,1}|\partial_t y||y| + s\varphi_{\delta,1}^3|y|^2]e^{2s\psi_{\delta,1}}dxdt \leq$$

$$\leq C\int_{Q_\delta} s\varphi_{\delta,1}^3(|y|^2 + |z|^2)e^{2s\psi_{\delta,1}}dxdt.$$

Combining this with (4.106), we may estimate the right-hand side of the above inequality and we obtain

$$J_1 \le \frac{C}{s^5} \int_{Q_\delta} \varphi_{\delta,1} \sum_{|\alpha| \le 2} |\partial_x^\alpha f|^2 e^{2s\psi_{\delta,1}} dxdt + C(s)B^2. \tag{4.109}$$

Similarly, we obtain

$$J_2 = \int_{t_0-\delta}^{t_0} \int_\Omega \partial_t (\varphi_{\delta,1} |\partial_t y|^2 e^{2s\psi_{\delta,1}}) dxdt$$

$$\le C \int_{t_0-\delta}^{t_0} \int_\Omega [\varphi_{\delta,1}^2 |\partial_t y|^2 + \varphi_{\delta,1} |\partial_t^2 y||\partial_t y| + s\varphi_{\delta,1}^3 |\partial_t y|^2] e^{2s\psi_{\delta,1}} dxdt \le$$

$$\le C \int_{Q_\delta} s\varphi_{\delta,1}^3 (|\partial_t y|^2 + |\partial_t z|^2) e^{2s\psi_{\delta,1}} dxdt.$$

Putting this together with (4.106), we see that

$$J_2 \le \frac{C}{s} \int_{Q_\delta} \varphi_{\delta,1} \sum_{|\alpha| \le 2} |\partial_x^\alpha f|^2 e^{2s\psi_{\delta,1}} dxdt + C(s)B^2. \tag{4.110}$$

Moreover we have

$$J_3 = \int_{t_0-\delta}^{t_0} \int_\Omega \partial_t (\varphi_{\delta,1} |Ly|^2 e^{2s\psi_{\delta,1}}) dxdt$$

$$\le C \int_{t_0-\delta}^{t_0} \int \Omega [\varphi_{\delta,1}^2 |Ly|^2 + \varphi_{\delta,1} |\partial_t Ly||Ly| + s\varphi_{\delta,1}^3 |Ly|^2] e^{2s\psi_{\delta,1}} dxdt$$

$$\le C \int_{Q_\delta} s\varphi_{\delta,1}^3 (|Ly|^2 + |Lz|^2) e^{2s\psi_{\delta,1}} dxdt$$

$$\le C \int_{Q_\delta} s\varphi_{\delta,1}^3 \Big(\sum_{|\alpha| \le 2} |\partial_x^\alpha y|^2 + \sum_{|\alpha| \le 2} |\partial_x^\alpha z|^2 \Big) e^{2s\psi_{\delta,1}} dxdt.$$

This together with (4.106) gives

$$J_3 \le \frac{C}{s} \int_{Q_\delta} \varphi_{\delta,1} \sum_{|\alpha| \le 2} |\partial_x^\alpha f|^2 e^{2s\psi_{\delta,1}} dxdt + C(s)B^2. \tag{4.111}$$

By (4.108) through (4.111), we have

$$\int_\Omega \varphi_{\delta,1}(x, t_0) |F(x, t_0)|^2 e^{2s\psi_{\delta,1}(x,t_0)} dx$$

$$\le \frac{C}{s}\int_{Q_\delta} \varphi_{\delta,1} \sum_{|\alpha|\le 2} |\partial_x^\alpha f|^2 e^{2s\psi_{\delta,1}} dxdt$$

$$+ C \int \Omega\varphi_{\delta,1}(x, t_0) \sum_{|\alpha|\le 4} |\partial_x^\alpha u(x, t_0)|^2 e^{2s\psi_{\delta,1}(x,t_0)} dx + C(s)B^2. \tag{4.112}$$

We will estimate the left-hand side of the inequality (4.112) from below using the Carleman estimate for the elliptic equation stated in Lemma 4.6 ($p = 1$). By (4.102) at $t = t_0$, we have

$$\sum_{i,j=1}^{n} \partial_i (a_{ij}(x)\partial_j \widetilde{f}(x))$$

$$+ \frac{1}{R(x, t_0)} \sum_{j=1}^{n} \big(2 \sum_{i=1}^{n} a_{ij}(x)\partial_i R(x, t_0) - b_j(x)R(x, t_0)\big)\partial_j \widetilde{f}(x)$$

$$+ \frac{1}{R(x, t_0)} \big[\rho_2 \partial_t^{\frac{1}{2}} R(x, t_0) - \rho_1 \partial_t R(x, t_0) + \sum_{i,j=1}^{n} \partial_i (a_{ij}(x)\partial_j R(x, t_0))$$

$$- \sum_{j=1}^{n} b_j(x)\partial_j R(x, t_0) - c(x)R(x, t_0) + \frac{\rho_2 R(x, 0)}{\sqrt{\pi t_0}}\big]\widetilde{f}(x) = \frac{F(x, t_0)}{R(x, t_0)}, x \in \Omega. \tag{4.113}$$

We note that $f(x) = 0, x \in \partial\Omega$, and $\nabla f(x) = 0, x \in \gamma$, are assumed. Applying Lemma 4.6 to (4.113) in Ω, we obtain

$$\frac{1}{s}\int_\Omega \varphi_{\delta,1}(x, t_0) \sum_{|\alpha|\le 2} |\partial_x^\alpha f(x)|^2 e^{2s\psi_{\delta,1}(x,t_0)} dx \le$$

$$\le \frac{C}{s}\int_\Omega \sum_{|\alpha|\le 2} |\partial_x^\alpha f(x)|^2 e^{2s\psi_{\delta,1}(x,t_0)} dx \le \frac{C}{s}\int_\Omega \Big(\sum_{i,j=1}^{n} |\partial_i\partial_j f(x)|^2$$

$$+ (s\varphi_{\delta,1}(x, t_0))2|\nabla f(x)|^2 + (s\varphi_{\delta,1}(x, t_0))^4 |f(x)|^2 \Big) e^{2s\psi_{\delta,1}(x,t0)} dx$$

$$\le C\int_\Omega \varphi_{\delta,1}(x, t_0) \Big|\frac{F(x, t_0)}{R(x, t_0)}\Big|^2 e^{2s\psi(x,t_0)} dx$$

$$\le C\int_\Omega \varphi_{\delta,1}(x, t_0) |F(x, t_0)|^2 e^{2s\psi(x,t_0)} dx. \tag{4.114}$$

By (4.112) and (4.114), we obtain

$$\frac{1}{s}\int_{\Omega}\varphi_{\delta,1}(x,t_0)\sum_{|\alpha|\le 2}|\partial_x^\alpha f(x)|^2 e^{2s\psi_{\delta,1}(x,t_0)}dx$$

$$\le \frac{C}{s}\int_{Q_\delta}\varphi_{\delta,1}\sum_{|\alpha|\le 2}|\partial_x^\alpha f|^2 e^{2s\psi_{\delta,1}}dxdt$$

$$+C\int_{\Omega}\varphi_{\delta,1}(x,t_0)\sum_{|\alpha|\le 4}|\partial_x^\alpha u(x,t_0)|^2 e^{2s\psi_{\delta,1}(x,t0)}dx + C(s)B^2. \tag{4.115}$$

Let us estimate the first integral term on the right-hand side of (4.115).

$$\int_{Q_\delta}\varphi_{\delta,1}\sum_{|\alpha|\le 2}|\partial_x^\alpha f|^2 e^{2s\psi_{\delta,1}}dxdt$$

$$=\int_{\Omega}\varphi_{\delta,1}(x,t_0)\sum_{|\alpha|\le 2}|\partial_x^\alpha f|^2 e^{2s\psi_{\delta,1}(x,t0)}h_s(x)dx,$$

where

$$h_s(x)=\frac{1}{\varphi_{\delta,1}(x,t_0)}\int_{t_0-\delta}^{t_0+\delta}\varphi_{\delta,1}e^{-2s(\psi_{\delta,1}(x,t_0)-\psi_{\delta,1}(x,t))}dt,\ x\in\Omega.$$

Since $\psi_{\delta,1}(x,t_0)-\psi_{\delta,1}(x,t)\ge 0$, $(x,t)\in Q_\delta$, h_s converges pointwise to 0 in Ω as $s\to\infty$ by Lebesgue's dominated convergence theorem. Moreover by Dini's theorem, we see that h_s converges uniformly to 0 in Ω as $s\to\infty$. Hence, taking sufficiently large $s>0$, we can absorb the first term on the right-hand side of (4.115) into the left-hand side and obtain

$$\frac{1}{s}\int_{\Omega}\varphi_{\delta,1}(x,t_0)\sum_{|\alpha|\le 2}|\partial_x^\alpha f(x)|^2 e^{2s\psi_{\delta,1}(x,t_0)}dx\le$$

$$\le C\int_{\Omega}\varphi_{\delta,1}(x,t_0)\sum_{|\alpha|\le 4}|\partial_x^\alpha u(x,t_0)|^2 e^{2s\psi_{\delta,1}(x,t_0)}dx + C(s)B^2. \tag{4.116}$$

Fix $s>0$. Noting that $\varphi_{\delta,1}(x,t_0)e^{2s\psi_{\delta,1}(x,t_0)}$ has its upper and lower bounds in Ω, we see that

$$\|f\|_{H^2(\Omega)}\le C\|u(\cdot,t_0)\|_{H^4(\Omega)}+CB.$$

Thus we obtain the stability estimate.

Proof of Theorem 4.4. We may prove Theorem 4.4 by an argument similar to that used in the proof of Theorem 4.3. In the proof, Theorem 4.12 and Lemma 4.7 are used instead of Theorem 4.11 and Lemma 4.6.

Stability for the diffusion coefficient. Next we prove Theorems 4.7 and 4.8. The proofs are very similar to the proofs of Theorems 4.3 and 4.4.

Proof of Theorem 4.7. Applying Lemma 4.1 to (4.73), we obtain

$$\rho_2^2\partial_t u(x,t) - (\rho_1\partial_t - L_2^{(1)})^2 u(x,t) = \widetilde{F}(x,t),\ (x,t) \in Q, \tag{4.117}$$

where

$$\widetilde{F}(x,t) = \left[\rho_2\partial_t^{\frac{1}{2}} - (\rho_1\partial_t - L_2^{(1)})\right](div(a(x)\nabla r(x,t))) + \frac{\rho_2 div(a(x)\nabla r(x,0))}{\sqrt{\pi t}} =$$

$$= a^{(1)}(x)\nabla r(x,t)\cdot\nabla\Delta a(x) + 2a^{(1)}(x)\sum_{i,j=1}^{n}(\partial_i\partial_j r(x,t))(\partial_i\partial_j a(x)) +$$

$$+ a^{(1)}(x)\Delta r(x,t)\Delta a(x) + (\nabla a^{(1)}(x) - b(x))\cdot(\nabla r(x,t)\cdot\nabla)\nabla a(x) +$$

$$+ \left[(\rho_2\partial_t^{\frac{1}{2}} - \rho_1\partial_t)\nabla r(x,t) + 3a^{(1)}(x)\nabla\Delta r(x,t) + (\Delta r(x,t))\nabla a^{(1)}(x) -\right.$$

$$\left. - (\Delta r(x,t))b(x) - c(x)\nabla r(x,t) + \frac{\rho_2\nabla r(x,0)}{\sqrt{\pi t}}\right]\cdot\nabla a(x) +$$

$$+ (\nabla a^{(1)}(x) - b(x))\cdot(\nabla a(x)\cdot\nabla)\nabla r(x,t) +$$

$$+ \left[(\rho_2\partial_t^{\frac{1}{2}} - \rho_1\partial_t)r(x,t) + \left(\nabla a^{(1)}(x)\cdot\nabla r(x,t)\right) + a^{(1)}(x)\Delta^2 r(x,t) -\right.$$

$$\left. - (b(x)\cdot\nabla\Delta r(x,t)) - c(x)\Delta r(x,t) + \frac{\rho_2\Delta r(x,0)}{\sqrt{\pi t}}\right]a(x),\ (x,t) \in Q. \tag{4.118}$$

Setting $y = \partial_t u$, $z = \partial_t^2 u$ in Q and differentiating (4.117) with respect to t, we have

$$\rho_2^2\partial_t y(x,t) - (\rho_1\partial_t - L_2^{(1)})^2 y(x,t) = \partial_t\widetilde{F}(x,t),\ (x,t) \in Q, \tag{4.119}$$

$$\rho_2^2\partial_t z(x,t) - (\rho_1\partial_t - L_2^{(1)})^2 z(x,t) = \partial_t^2\widetilde{F}(x,t),\ (x,t) \in Q. \tag{4.120}$$

Since $u(x,t) = 0$, $(x,t) \in \partial\Omega \times (0,T)$, we see that

$$y(x,t) = z(x,t) = 0,\ (x,t) \in \partial\Omega \times (0,T).$$

Fixing $\lambda > 0$ and applying Theorem 4.11 ($p = 1$) to (4.119) and (4.120) in Q_δ, we have

$$\int_{Q_\delta} \big[(s\varphi_{\delta,1})^2(|\nabla \partial_t y|^2 + |\nabla \partial_t z|^2) +$$

$$+ (s\varphi_{\delta,1})^3\big(|\nabla(\rho_1\partial_t - L_2^{(1)})y|^2 + |\nabla(\rho_1\partial_t - L_2^{(1)})z|^2\big) +$$

$$+ (s\varphi_{\delta,1})^4\left(|\partial_t y|^2 + |\partial_t z|^2 + \sum_{i,j=1}^{n} |\partial_i\partial_j y|^2 + \sum_{i,j=1}^{n} |\partial_i\partial_j z|^2\right) +$$

$$+ (s\varphi_{\delta,1})^6(|\nabla y|^2 + |\nabla z|^2) + (s\varphi_{\delta,1})^8(|y|^2 + |z|^2)\big]e^{2s\psi_{\delta,1}}dxdt \le$$

$$\le C\int_{Q_\delta} (s\varphi_{\delta,1})^2\big(|\partial_t \widetilde{F}|^2 + |\partial_t^2 \widetilde{F}|^2\big)e^{2s\psi_{\delta,1}}dxdt + C\widehat{B}, \tag{4.121}$$

where

$$\hat{B} = \int_{\Sigma_\delta} \big[(s\varphi_{\delta,1})^2(|\nabla \partial_t y|^2 + ||\nabla \partial_t z|^2) +$$

$$+ (s\varphi_{\delta,1})^3\big(|\nabla \partial_t^{\frac{1}{2}} y|^2 + |\nabla \partial_t^{\frac{1}{2}} z|2\big) + (s\varphi_{\delta,1})^6(|\nabla y|2 + |\nabla z|2)\big]e^{2s\psi_{\delta,1}}dSdt.$$

As we have seen in the proof of Theorem 4.3, we may obtain $\widehat{B} \le C(s)B^2$.

Note that

$$\int_{Q_\delta} (s\varphi_{\delta,1})^2\big(|\partial_t \widetilde{F}|^2 + |\partial_t^2 \widetilde{F}|^2\big)e^{2s\psi_{\delta,1}}dxdt \le$$

$$\le C\int_{Q_\delta} (s\varphi_{\delta,1})^2\left(|\nabla \Delta a|^2 + \sum_{|\alpha|\le 2} |\partial_x^\alpha a|^2\right)e^{2s\psi_{\delta,1}}dxdt.$$

This together with (4.121) gives

$$\int_{Q_\delta} \big[(s\varphi_{\delta,1})^2(|\nabla \partial_t y|^2 + |\nabla \partial_t z|^2) +$$

$$+ (s\varphi_{\delta,1})^3\big(|\nabla(\rho_1\partial_t - L_2^{(1)})y|^2 + |\nabla(\rho_1\partial_t - L_2^{(1)})z|^2\big) +$$

$$+ (s\varphi_{\delta,1})^4\left(|\partial_t y|^2 + |\partial_t z|^2 + \sum_{i,j=1}^{n} |\partial_i\partial_j y|^2 + \sum_{i,j=1}^{n} |\partial_i\partial_j z|^2\right) +$$

$$+ (s\varphi_{\delta,1})^6(|\nabla y|^2 + |\nabla z|^2) + (s\varphi_{\delta,1})^8(|y|^2 + |z|^2)\big]e^{2s\psi_{\delta,1}}dxdt \le$$

$$\le C\int_{Q_\delta} (s\varphi_{\delta,1})^2\big(|\nabla a|^2 + \sum_{|\alpha|\le 2} |\partial_x^\alpha a|^2\big)e^{2s\psi_{\delta,1}}dxdt + C(s)B^2. \tag{4.122}$$

Let us expand the left-hand side of (4.117). We have

$$\rho_2^2\partial_t u(x,t) - \rho_1^2\partial_t^2 u(x,t) + 2\rho_1\partial_t L_2^{(1)}u(x,t) - (L_2^{(1)})^2 u(x,t) = \widetilde{F}(x,t),\ (x,t)\in Q.$$

Moreover we have

$$\rho_2^2\nabla\partial_t u(x,t) - \rho_1^2\nabla\partial_t^2 u(x,t) + 2\rho_1\nabla\partial_t L_2^{(1)}u(x,t) - \nabla(L_2^{(1)})^2 u(x,t) = \nabla\widetilde{F}(x,t),\ (x,t)\in Q.$$

In particular at $t = t_0$, we have

$$\rho_2^2\partial_t u(x,t_0) - \rho_1^2\partial_t^2 u(x,t_0) + 2\rho_1\partial_t L_2^{(1)}u(x,t_0) - (L_2^{(1)})^2 u(x,t_0) = \widetilde{F}(x,t_0),\ x\in\Omega. \quad (4.123)$$

and

$$\rho_2^2\nabla\partial_t u(x,t_0) - \rho_1^2\nabla\partial_t^2 u(x,t_0) + 2\rho_1\nabla\partial_t L_2^{(1)}u(x,t_0) - \nabla(L_2^{(1)})^2 u(x,t_0) = \nabla\widetilde{F}(x,t_0),\ x\in\Omega. \quad (4.124)$$

Taking the weighted L^2 norm of (4.123) and (4.124) in Ω, we obtain

$$\int_\Omega \left(|\widetilde{F}(x,t_0)|^2 + |\nabla\widetilde{F}(x,t_0)|^2\right)e^{2s\psi_{\delta,1}(x,t_0)}dx \le \le C\sum_{k=1}^{6}\widetilde{J}_k + C\int_\Omega \sum_{|\alpha|\le 5} |\partial_x^\alpha u(x,t_0)|2e^{2s\psi_{\delta,1}(x,t_0)}dx, \quad (4.125)$$

where

$$\widetilde{J}_1 = \int_\Omega |\partial_t u(x,t_0)|^2 e^{2s\psi_{\delta,1}(x,t_0)}dx,\ \widetilde{J}_2 = \int_\Omega |\partial_t^2 u(x,t_0)|^2 e^{2s\psi_{\delta,1}(x,t_0)}dx,$$

$$\widetilde{J}_3 = \int_\Omega |\partial_t L_2^{(1)}u(x,t_0)|^2 e^{2s\psi_{\delta,1}(x,t_0)}dx,\ \widetilde{J}_4 = \int_\Omega |\nabla\partial_t u(x,t_0)|^2 e^{2s\psi_{\delta,1}(x,t_0)}dx,$$

$$\widetilde{J}_5 = \int_\Omega |\nabla\partial_t^2 u(x,t_0)|^2 e^{2s\psi_{\delta,1}(x,t_0)}dx,\ \widetilde{J}_6 = \int_\Omega |\nabla\partial_t L_2^{(1)}u(x,t_0)|^2 e^{2s\psi_{\delta,1}(x,t_0)}dx.$$

Henceforth we estimate $\widetilde{J}_1$ through $\widetilde{J}_6$ by using the Carleman estimate. We assume that $s > 1$ is large enough to satisfy $s\varphi_{\delta,1} > 1$ in Q. We note that $\partial_t \psi_{\delta,1}(x,t) \le C\varphi_{\delta,1}^2(x,t)$ for $(x,t) \in Q$.

$$\widetilde{J}_1 = \int_{t_0-\delta}^{t_0}\int_\Omega \partial_t(|y|^2 e^{2s\psi_{\delta,1}})dxdt \le$$

$$\le C\int_{t_0-\delta}^{t_0}\int_\Omega (|\partial_t y||y| + s\varphi_{\delta,1}^2|y|^2)e^{2s\psi_{\delta,1}}dxdt \le$$

$$\le C\int_{Q_\delta} s\varphi_{\delta,1}^2(|y|^2 + |z|^2)e^{2s\psi_{\delta,1}}dxdt.$$

Combining this with (4.122), we may estimate the right-hand side of the above inequality and we obtain

$$\widetilde{J}_1 \le \frac{C}{s^5}\int_{Q_\delta} \varphi_{\delta,1}^2\big(|\nabla\Delta a|^2 + \sum_{|\alpha|\le 2}|\partial_x^\alpha a|^2\big)e^{2s\psi_{\delta,1}}dxdt + C(s)B^2. \tag{4.126}$$

Similarly, we obtain

$$\widetilde{J}_2 = \int_{t_0-\delta}^{t_0}\int_\Omega \partial_t(|\partial_t y|^2 e^{2s\psi_{\delta,1}})dxdt \le$$

$$\le C\int_{t_0-\delta}^{t_0}\int_\Omega (|\partial_t^2 y||\partial_t y| + s\varphi_{\delta,1}^2|\partial_t y|^2)e^{2s\psi_{\delta,1}}dxdt \le$$

$$\le C\int_{Q_\delta} s\varphi_{\delta,1}^2(|\partial_t y|^2 + |\partial_t z|^2)e^{2s\psi_{\delta,1}}dxdt.$$

Putting this together with (4.122), we see that

$$\widetilde{J}_2 \le \frac{C}{s}\int_{Q_\delta} \varphi_{\delta,1}^2\big(|\nabla\Delta a|^2 + \sum_{|\alpha|\le 2}|\partial_x^\alpha a|^2\big)e^{2s\psi_{\delta,1}}dxdt + C(s)B^2. \tag{4.127}$$

Moreover we have

$$\widetilde{J}_3 = \int_{t_0-\delta}^{t_0}\int_\Omega \partial_t\big(|L_2^{(1)}y|^2 e^{2s\psi_{\delta,1}}\big)dxdt \le$$

$$\le C\int_{t_0-\delta}^{t_0}\int_\Omega \big(|L_2^{(1)}\partial_t y||L_2^{(1)}y| + s\varphi_{\delta,1}^2|L_2^{(1)}y|^2\big)e^{2s\psi_{\delta,1}}dxdt \le$$

$$\le C\int_{Q_\delta} s\varphi_{\delta,1}^2\left(|L_2^{(1)}y|^2+|L_2^{(1)}z|^2\right)e^{2s\psi_{\delta,1}}dxdt \le$$

$$\le C\int_{Q_\delta} s\varphi_{\delta,1}^2\left(\sum_{|\alpha|\le 2}|\partial_x^\alpha y|^2+\sum_{|\alpha|\le 2}|\partial_x^\alpha z|^2\right)e^{2s\psi_{\delta,1}}dxdt.$$

This together with (4.122) gives

$$\widetilde{J}_3 \le \frac{C}{s}\int_{Q_\delta}\varphi_{\delta,1}^2\left(|\nabla\Delta a|^2+\sum_{|\alpha|\le 2}|\partial_x^\alpha a|^2\right)e^{2s\psi_{\delta,1}}dxdt + C(s)B^2. \tag{4.128}$$

We have

$$\widetilde{J}_4 = \int_{t_0-\delta}^{t_0}\int_\Omega \partial_t(|\nabla y|^2 e^{2s\psi_{\delta,1}})dxdt \le$$

$$\le C\int_{t_0-\delta}^{t_0}\int_\Omega (|\nabla\partial_t y||\nabla y| + s\varphi_{\delta,1}^2|\nabla y|^2)e^{2s\psi_{\delta,1}}dxdt \le$$

$$\le C\int_{Q_\delta} s\varphi_{\delta,1}^2(|\nabla y|^2+|\nabla z|^2)e^{2s\psi_{\delta,1}}dxdt.$$

Combining this with (4.122), we may estimate the right-hand side of the above inequality and we obtain

$$\widetilde{J}_4 \le \frac{C}{s^3}\int_{Q_\delta}\varphi_{\delta,1}^2\left(|\nabla\Delta a|^2+\sum_{|\alpha|\le 2}|\partial_x^\alpha a|^2\right)e^{2s\psi_{\delta,1}}dxdt + C(s)B^2. \tag{4.129}$$

Similarly, we obtain

$$\widetilde{J}_5 = \int_{t_0-\delta}^{t_0}\int_\Omega \partial_t(|\nabla\partial_t y|^2 e^{2s\psi_{\delta,1}})dxdt \le$$

$$\le C\int_{t_0-\delta}^{t_0}\int_\Omega (|\nabla\partial_t^2 y||\nabla\partial_t y| + s\varphi_{\delta,1}^2|\nabla\partial_t y|^2)e^{2s\psi_{\delta,1}}dxdt \le$$

$$\le C\int_{Q_\delta} s\varphi_{\delta,1}^2(|\nabla\partial_t y|^2+|\nabla\partial_t z|^2)e^{2s\psi_{\delta,1}}dxdt.$$

Putting this together with (4.122), we see that

$$\widetilde{J}_5 \le C\int_{Q_\delta} s\varphi_{\delta,1}^2\left(|\nabla\Delta a|^2+\sum_{|\alpha|\le 2}|\partial_x^\alpha a|^2\right)e^{2s\psi_{\delta,1}}dxdt + C(s)B^2. \tag{4.130}$$

Moreover we have

$$\widetilde{J}_6 = \int_{t_0-\delta}^{t_0} \int_\Omega \partial_t (|\nabla L_2^{(1)} y|^2 e^{2s\psi_{\delta,1}}) dxdt \le$$

$$\le C \int_{t_0-\delta}^{t_0} \int_\Omega (|\nabla L_2^{(1)} \partial_t y||\nabla L_1 y| + s\varphi_{\delta,1}^2 |\nabla L_2^{(1)} y|^2) e^{2s\psi_{\delta,1}} dxdt \le$$

$$\le C \int_{Q_\delta} s\varphi_{\delta,1}^2 (|\nabla L_2^{(1)} y|^2 + |\nabla L_2^{(1)} z|^2) e^{2s\psi_{\delta,1}} dxdt \le$$

$$\le C \int_{Q_\delta} s\varphi_{\delta,1}^2 (|\nabla \partial_t y|^2 +$$

$$+ |\nabla(\rho_1 \partial_t - L_2^{(1)}) y|^2 + |\nabla \partial_t z|^2 + |\nabla(\rho_1 \partial_t - L_2^{(1)}) z|^2) e^{2s\psi_{\delta,1}} dxdt.$$

This together with (4.122) gives

$$\widetilde{J}_6 \le C \int_{Q_\delta} s\varphi_{\delta,1}^2 \left(|\nabla \Delta a|^2 + \sum_{|\alpha| \le 2} |\partial_x^\alpha a|^2 \right) e^{2s\psi_{\delta,1}} dxdt + C(s) B^2. \tag{4.131}$$

Summing up the estimate of (4.125) through (4.131), we have

$$\int_\Omega (|\widetilde{F}(x, t_0)|^2 + |\nabla \widetilde{F}(x, t_0)|^2) e^{2s\psi_{\delta,1}(x,t_0)} dx \le$$

$$\le C \int_{Q_\delta} s\varphi_{\delta,1}^2 \left(|\nabla \Delta a|^2 + \sum_{|\alpha| \le 2} |\partial_x^\alpha a|^2 \right) e^{2s\psi_{\delta,1}} dxdt +$$

$$+ C \int_\Omega \sum_{|\alpha| \le 5} |\partial_x^\alpha u(x, t_0)|^2 e^{2s\psi_{\delta,1}(x,t_0)} dx + C(s) B^2. \tag{4.132}$$

Let us estimate the left-hand side of the inequality (4.132) from below. By (4.118) at $t = t_0$, we have

$$a^{(1)}(x) \nabla r(x, t_0) \cdot \nabla a(x) =$$

$$= \widetilde{F}(x, t_0) - 2a^{(1)}(x) \sum_{i,j=1}^n (\partial_i \partial_j r(x, t_0))(\partial_i \partial_j a(x)) -$$

$$- a^{(1)}(x) \Delta r(x, t_0) \Delta a(x) - (\nabla a^{(1)}(x) - b(x)) \cdot (\nabla r(x, t_0) \cdot \nabla) \nabla a(x) -$$

$$- \big[(\rho_2\partial_t^{\frac{1}{2}} - \rho_1\partial_t)\nabla r(x,t_0) + 3a^{(1)}(x)\nabla\Delta r(x,t_0) + (\Delta r(x,t_0))\nabla a^{(1)}(x) -$$

$$- (\Delta r(x,t_0))b(x) - c(x)\nabla r(x,t_0) + \frac{\rho_2\nabla r(x,0)}{\sqrt{\pi t_0}}\big] \cdot \nabla a(x) -$$

$$- (\nabla a^{(1)}(x) - b(x)) \cdot (\nabla a(x) \cdot \nabla)\nabla r(x,t_0) -$$

$$- \big[(\rho_2\partial_t^{\frac{1}{2}} - \rho_1\partial_t)\Delta r(x,t_0) + \big(\nabla a^{(1)}(x) \cdot \nabla\Delta r(x,t_0)\big) + a^{(1)}(x)\Delta^2 r(x,t_0) -$$

$$- (b(x) \cdot \nabla\Delta r(x,t_0)) - c(x)\Delta r(x,t_0) + \frac{\rho_2\Delta r(x,0)}{\sqrt{\pi t_0}}\big]a(x), \ x \in \Omega. \qquad (4.133)$$

Note that

$$|\nabla r(x,t_0) \cdot \nabla d_1(x)| \ge m_1 > 0, \ x \in \Omega,$$

and $a \in H^4(\Omega)$ satisfies $a(x) = 0, \ x \in D$. Let us apply Lemma 4.8 to (4.133) in Ω. Then we obtain

$$\int_\Omega \big[s\varphi_{\delta,1}(x,t_0) \sum_{i,j,k=1}^n |\partial_i\partial_j\partial_k a(x)|^2 + (s\varphi_{\delta,1}(x,t_0))^2|\nabla\Delta a(x)|^2 +$$

$$+ (s\varphi_{\delta,1}(x,t_0))^3 \sum_{i,j=1}^n |\partial_i\partial_j a(x)|^2 + (s\varphi_{\delta,1}(x,t_0))^5(|\nabla a(x)|^2$$

$$+ |a(x)|^2)\big]e^{2s\psi_{\delta,1}(x,t_0)}dx \le C\int_\Omega (|\widetilde{F}(x,t_0)|^2 + |\nabla\widetilde{F}(x,t_0)|^2)e^{2s\psi_{\delta,1}(x,t_0)}dx$$

$$+ C\int_\Omega \sum_{|\alpha|\le 3} |\partial_x^\alpha a(x)|^2 e^{2s\psi_{\delta,1}(x,t_0)}dx.$$

Combining this with (4.132), we obtain

$$\int_\Omega \big[s\varphi_{\delta,1}(x,t_0) \sum_{i,j,k=1}^n |\partial_i\partial_j\partial_k a(x)|^2 + (s\varphi_{\delta,1}(x,t_0))^2|\nabla\Delta a(x)|^2 +$$

$$+ (s\varphi_{\delta,1}(x,t_0))^3 \sum_{i,j=1}^n |\partial_i\partial_j a(x)|^2 +$$

$$+ (s\varphi_{\delta,1}(x,t_0))^5(|\nabla a(x)|^2 + |a(x)|^2)\big]e^{2s\varphi_{\delta,1}(x,t_0)}dx \le$$

$$
\begin{aligned}
&\le C\int_{Q_\delta} s\varphi_{\delta,1}^2(x,t_0)\big(|\nabla\Delta a|^2+\sum_{|\alpha|\le 2}|\partial_x^\alpha a|^2\big)e^{2s\varphi_{\delta,1}}dxdt+\\
&+C\int_\Omega \sum_{|\alpha|\le 3}|\partial_x^\alpha a(x)|^2 e^{2s\varphi_{\delta,1}(x,t_0)}dx+\\
&+C\int_\Omega \sum_{|\alpha|\le 5}|\partial_x^\alpha u(x,t_0)|^2 e^{2s\varphi_{\delta,1}(x,t_0)}dx+C(s)B^2\le\\
&\le C\int_\Omega\int_{Q_\delta}\Big[\sum_{i,j,k=1}^n|\partial_i\partial_j\partial_k a(x)|^2+\\
&+s\varphi_{\delta,1}^2(x,t_0)\big(|\nabla\Delta a(x)|^2+\sum_{|\alpha|\le 2}|\partial_x^\alpha a(x)|^2\big)\Big]e^{2s\varphi_{\delta,1}(x,t_0)}dx+\\
&+C\int_\Omega \sum_{|\alpha|\le 5}|\partial_x^\alpha u(x,t_0)|^2 e^{2s\varphi_{\delta,1}(x,t_0)}dx+C(s)B^2. \qquad (4.134)
\end{aligned}
$$

In the last inequality, we used the fact that

$$\varphi_{\delta,1}^2(x,t)e^{2s\psi_{\delta,1}(x,t)}\le\varphi_{\delta,1}^2(x,t_0)e^{2s\psi_{\delta,1}(x,t_0)},\ (x,t)\in Q_\delta,$$

for large $s>0$.

Choose sufficiently large $s>0$ and absorb the first term on the right-hand side of (4.134) into the left-hand side. Then we obtain

$$
\begin{aligned}
&\int_\Omega\Big[s\varphi_{\delta,1}(x,t_0)\sum_{i,j,k=1}^n|\partial_i\partial_j\partial_k a(x)|^2+(s\varphi_{\delta,1}(x,t_0))^2|\nabla\Delta a(x)|^2+\\
&+(s\varphi_{\delta,1}(x,t_0))^3\sum_{i,j=1}^n|\partial_i\partial_j a(x)|^2+\\
&+(s\varphi_{\delta,1}(x,t_0))^5(|\nabla a(x)|^2+|a(x)|^2)\Big]e^{2s\psi_{\delta,1}(x,t_0)}dx\le\\
&\le C\int_\Omega\sum_{|\alpha|\le 5}|\partial_x^\alpha u(x,t_0)|^2 e^{2s\psi_{\delta,1}(x,t_0)}dx+C(s)B^2. \qquad (4.135)
\end{aligned}
$$

Fix $s>0$. Noting that $\varphi_{\delta,1}(\cdot,t_0)e^{2s\psi_{\delta,1}(\cdot,t_0)}$ has its upper and lower bounds in Ω, we see that

$$\|a\|_{H^3(\Omega)}\le C\|u(\cdot,t_0)\|_{H^5(\Omega)}+CB.$$

Thus we obtain the stability estimate (4.77).
Proof of Theorem 4.8. *We may prove Theorem* 4.8 in the same way as Theorem 4.7.

Stability in simultaneous determination. Next we prove Theorems 4.9 and 4.10 by the combination of the proofs of Theorems 4.7 and 4.8 and the proofs of Theorems 4.3 and 4.4.

Proof of Theorem 4.9. Applying Lemma 4.1 (of Sect. 4.1) to (4.83), we obtain

$$\rho_2^2 \partial_t u_l(x,t) - (\rho_1 \partial_t - L^{(1)})^2 u_l(x,t) = F_l(x,t),\ (x,t) \in Q, \tag{4.136}$$

where

$$\begin{aligned}
F_l(x,t) = &\left[\rho_2 \partial_t^{\frac{1}{2}} - (\rho_1 \partial_t - L^{(1)})\right](div(a(x)\nabla r_l(x,t)) - c(x) r_l(x,t)) + \\
&+ \frac{\rho_2 div(a(x)\nabla r_l(x,0))}{\sqrt{\pi t}} - \sqrt{\rho_2 c(x) r_l(x,0)}\sqrt{\pi t} = \\
&= a^{(1)}(x)\nabla r_l(x,t)\cdot\nabla\Delta a(x) + 2a^{(1)}(x)\sum_{i,j=1}^{n}(\partial_i\partial_j r_l(x,t))(\partial_i\partial_j a(x)) + \\
&+ a^{(1)}(x)\Delta r_l(x,t)\Delta a(x) + \nabla a^{(1)}(x)\cdot(\nabla r_l(x,t)\cdot\nabla)\nabla a(x) + \\
&+ \left[(\rho_2\partial_t^{\frac{1}{2}} - \rho_1\partial_t)\nabla r_l(x,t) + 3a^{(1)}(x)\nabla\Delta r_l(x,t) + (\Delta r_l(x,t))\nabla a^{(1)}(x) - \right.\\
&\left. - c^{(1)}(x)\nabla r_l(x,t) + \frac{\rho_2\nabla r_l(x,0)}{\sqrt{\pi t}}\right]\cdot\nabla a(x) + \\
&+ \nabla a^{(1)}(x)\cdot(\nabla a(x)\cdot\nabla)\nabla r_l(x,t) + \\
&+ \left[(\rho_2\partial_t^{\frac{1}{2}} - \rho_1\partial_t)\Delta r_l(x,t) + \left(\nabla a^{(1)}(x)\cdot\nabla\Delta r_l(x,t)\right) + a^{(1)}(x)\Delta^2 r_l(x,t) - \right.\\
&\left. - c^{(1)}(x)\Delta r_l(x,t) + \frac{\rho_2\Delta r_l(x,0)}{\sqrt{\pi t}}\right] a(x) - \\
&- r_l(x,t) div(a^{(1)}(x))\nabla c(x) - 2a^{(1)}(x)\nabla r_l(x,t)\cdot\nabla c(x) - \\
&- \left[(\rho_2\partial_t^{\frac{1}{2}} - \rho_1\partial_t) r_l(x,t) + div(a^{(1)}(x)\nabla r_l(x,t)) - \right.\\
&\left. - c^{(1)}(x) r_l(x,t) + \frac{\rho_2 r_l(x,0)}{\sqrt{\pi t}}\right] c(x),\ (x,t)\in Q,
\end{aligned} \tag{4.137}$$

for $l = 1, 2$.

Setting $y_l = \partial_t u_l$, $z_l = \partial_t^2 u_l$ in Q for $l = 1, 2$ and differentiating (4.136) with respect to t, we have

$$\rho_2^2 \partial_t y_l(x,t) - (\rho_1 \partial_t - L^{(1)})^2 y_l(x,t) = \partial_t F_l(x,t), \ (x,t) \in Q, \tag{4.138}$$

$$\rho_2^2 \partial_t z_l(x,t) - (\rho_1 \partial_t - L^{(1)})^2 z_l(x,t) = \partial_t^2 F_l(x,t), \ (x,t) \in Q, \tag{4.139}$$

for $l = 1, 2$. Since $u_l(x,t) = 0$, $(x,t) \in \partial\Omega \times (0,T)$, we see that

$$y_l(x,t) = z_l(x,t) = 0, \ (x,t) \in \partial\Omega \times (0,T),$$

for $l = 1, 2$.

Fixing $\lambda > 0$ and applying Theorem 4.11 ($p = 1$) to (4.138) and (4.139) in Q_δ, we have

$$\begin{aligned}
&\int_{Q_\delta} \Big[(s\varphi_{\delta,1})^2(|\nabla \partial_t y_l|^2 + |\nabla \partial_t z_l|^2) + \\
&+ (s\varphi_{\delta,1})^3\big(|\nabla(\rho_1\partial_t - L^{(1)})y_l|^2 + |\nabla(\rho_1\partial_t - L^{(1)})z_l|^2\big) + \\
&+ (s\varphi_{\delta,1})^4\big(|\partial_t y_l|^2 + |\partial_t z_l|^2 + \sum_{i,j=1}^n |\partial_i\partial_j y_l|^2 + \sum_{i,j=1}^n |\partial_i\partial_j z_l|^2\big) + \\
&+ (s\varphi_{\delta,1})^6(|\nabla y_l|^2 + |\nabla z_l|^2) + (s\varphi_{\delta,1})^8(|y_l|^2 + |z_l|^2)\Big] e^{2s\psi_{\delta,1}} dxdt \le \\
&\le C \int_{Q_\delta} (s\varphi_{\delta,1})^2(|\partial_t F_l|^2 + |\partial_t^2 F_l|^2) e^{2s\psi_{\delta,1}} dxdt + C\widehat{B}_l,
\end{aligned} \tag{4.140}$$

where

$$\begin{aligned}
\widehat{B}_l = &\int_{\Sigma_\delta} \Big[(s\varphi_{\delta,1})^2(|\nabla\partial_t y_l|^2 + |\nabla\partial_t z_l|^2) + (s\varphi_{\delta,1})^3\big(|\nabla\partial_t^{\frac{1}{2}} y_l|^2 + |\nabla\partial_t^{\frac{1}{2}} z_l|^2\big) + \\
&+ (s\varphi_{\delta,1})^6(|\nabla y_l|^2 + |\nabla z_l|^2)\Big] e^{2s\psi_{\delta,1}} dSdt,
\end{aligned}$$

for $l = 1, 2$. As we have seen in the proof of Theorem 4.3, we may obtain $\widehat{B}_1 + \widehat{B}_2 \le C(s)B^2$.

Note that

$$\begin{aligned}
&\int_{Q_\delta} (s\varphi_{\delta,1})^2(|\partial_t F_l|^2 + |\partial_t^2 F_l|^2) e^{2s\psi_{\delta,1}} dxdt \le \\
&\le C \int_{Q_\delta} (s\varphi_{\delta,1})^2\Big(|\nabla\Delta a|^2 + \sum_{|\alpha|\le 2} |\partial_x^\alpha a|^2 + \sum_{|\alpha|\le 2} |\partial_x^\alpha c|^2\Big) e^{2s\psi_{\delta,1}} dxdt
\end{aligned}$$

for $l = 1, 2$. This together with (4.140) gives

$$\int_{Q_\delta} \big[(s\varphi_{\delta,1})^2(|\nabla\partial_t y_l|^2 + |\nabla\partial_t z_l|^2) +$$

$$+ (s\varphi_{\delta,1})^3\big(|\nabla(\rho_1\partial_t - L^{(1)})y_l|^2 + |\nabla(\rho_1\partial_t - L^{(1)})z_l|^2\big) +$$

$$+ (s\varphi_{\delta,1})^4\left(|\partial_t y_l|^2 + |\partial_t z_l|^2 + \sum_{i,j=1}^{n} |\partial_i\partial_j y_l|^2 + \sum_{i,j=1}^{n} |\partial_i\partial_j z_l|^2\right) +$$

$$+ (s\varphi_{\delta,1})^6(|\nabla y_l|^2 + |\nabla z_l|^2) + (s\varphi_{\delta,1})^8(|y_l|^2 + |z_l|^2)\big]e^{2s\psi_{\delta,1}}dxdt \le$$

$$\le C\int_{Q_\delta} (s\varphi_{\delta,1})^2\left(|\nabla\Delta a|^2 + \sum_{|\alpha|\le 2} |\partial_x^\alpha a|^2 + \sum_{|\alpha|\le 2} |\partial_x^\alpha c|^2\right)e^{2s\psi_{\delta,1}}dxdt + C(s)B^2 \tag{4.141}$$

for $l = 1, 2$.

Let us expand the left-hand side of (4.136). We have

$$\rho_2^2\partial_t u_l(x,t) - \rho_1^2\partial_t^2 u_l(x,t)$$
$$+ 2\rho_1\partial_t L^{(1)}u_l(x,t) - (L^{(1)})^2 u_l(x,t) = F_l(x,t),\ (x,t) \in Q,$$

for $l = 1, 2$. Moreover we have

$$\rho_2^2\nabla\partial_t u_l(x,t) - \rho_1^2\nabla\partial_t^2 u_l(x,t)$$
$$+ 2\rho_1\nabla\partial_t L^{(1)}u_l(x,t) - \nabla(L^{(1)})^2 u_l(x,t) = \nabla F_l(x,t),\ (x,t) \in Q,$$

for $l = 1, 2$.

In particular at $t = t_0$, we have

$$\rho_2^2\partial_t u_l(x,t_0) - \rho_1^2\partial_t^2 u_l(x,t_0)$$
$$+ 2\rho_1\partial_t L^{(1)}u_l(x,t_0) - (L^{(1)})^2 u_l(x,t_0) = F_l(x,t_0),\ x \in \Omega, \tag{4.142}$$

and

$$\rho_2^2\nabla\partial_t u_l(x,t_0) - \rho_1^2\nabla\partial_t^2 u_l(x,t_0)$$
$$+ 2\rho_1\nabla\partial_t L^{(1)}u_l(x,t_0) - \nabla(L^{(1)})^2 u_l(x,t_0) = \nabla F_l(x,t_0),\ x \in \Omega, \tag{4.143}$$

for $l = 1, 2$. Taking the weighted L^2 norm of (4.142) and (4.143) in Ω, we obtain

$$\int_\Omega ((s\varphi_{\delta,1})^2|F_l(x,t_0)|^2 + |\nabla F_l(x,t_0)|^2)e^{2s\psi_{\delta,1}(x,t_0)}dx \le$$

$$\le C\sum_{k=1}^{6} J_{k,l} + C\int_\Omega \sum_{|\alpha|\le 5}(s\varphi_{\delta,1}(x,t_0))^2|\partial_x^\alpha u_l(x,t_0)|^2 e^{2s\psi_{\delta,1}(x,t_0)}dx, \tag{4.144}$$

where

$$J_{1,l} = \int_\Omega (s\varphi_{\delta,1}(x,t_0))^2|\partial_t u_l(x,t_0)|^2 e^{2s\psi_{\delta,1}(x,t_0)}dx,$$

$$J_{2,l} = \int_\Omega (s\varphi_{\delta,1}(x,t_0))^2|\partial_t^2 u_l(x,t_0)|^2 e^{2s\psi_{\delta,1}(x,t_0)}dx,$$

$$J_{3,l} = \int_\Omega (s\varphi_{\delta,1}(x,t_0))^2|\partial_t L^{(1)} u_l(x,t_0)|^2 e^{2s\psi_{\delta,1}(x,t_0)}dx,$$

$$J_{4,l} = \int_\Omega |\nabla \partial_t u_l(x,t_0)|^2 e^{2s\psi_{\delta,1}(x,t_0)}dx,$$

$$J_{5,l} = \int_\Omega |\nabla \partial_t^2 u_l(x,t_0)|^2 e^{2s\psi_{\delta,1}(x,t_0)}dx,$$

$$J_{6,l} = \int_\Omega |\nabla \partial_t L^{(1)} u_l(x,t_0)|^2 e^{2s\psi_{\delta,1}(x,t_0)}dx$$

for $l = 1, 2$.

Henceforth we assume that $s > 1$ is large enough to satisfy $s\varphi_{\delta,1} > 1$ in Q. We may estimate $J_{1,l}$ through $J_{6,l}$ for $l = 1, 2$, via the same argument which we used in the proof of Theorems 4.3 and 4.7 by using the inequality (4.141) derived by the Carleman estimate

$$J_{1,l} = \int_{t_0-\delta}^{t_0}\int_\Omega \partial_t(s^2\varphi_{\delta,1}^2|y_l|^2 e^{2s\psi_{\delta,1}})dxdt \le$$

$$\le C\int_{t_0-\delta}^{t_0}\int_\Omega [s^2\varphi_{\delta,1}^3|y_l|^2 + s^2\varphi_{\delta,1}^2|\partial_t y_l||y_l| + s^3\varphi_{\delta,1}^4|y_l|^2]e^{2s\psi_{\delta,1}}dxdt \le$$

$$\le C\int_{Q_\delta} s^3\varphi_{\delta,1}^4(|y_l|^2 + |z_l|^2)e^{2s\psi_{\delta,1}}dxdt$$

for $l = 1, 2$. Combining this with (4.141), we may estimate the right-hand side of the above inequality and we obtain

$$J_{1,l} \le C \int_{Q_\delta} \frac{\varphi_{\delta,1}^2}{s^3} (|\nabla \Delta a|^2 + \sum_{|\alpha|\le 2} |\partial_x^\alpha a|^2 + \sum_{|\alpha|\le 2} |\partial_x^\alpha c|^2) e^{2s\psi_{\delta,1}} dxdt + C(s)B^2 \tag{4.145}$$

for $l = 1, 2$. Similarly, we obtain

$$J_{2,l} = \int_{t_0-\delta}^{t_0} \int_\Omega \partial_t (s^2 \varphi_{\delta,1}^2 |\partial_t y_l|^2 e^{2s\psi_{\delta,1}}) dxdt \le$$

$$\le C \int_{t_0-\delta}^{t_0} \int_\Omega [s^2 \varphi_{\delta,1}^3 |\partial_t y_l|^2 + s^2 \varphi_{\delta,1}^2 |\partial_t^2 y_l||\partial_t y_l| + s^3 \varphi_{\delta,1}^4 |\partial_t y_l|^2] e^{2s\psi_{\delta,1}} dxdt \le$$

$$\le C \int_{Q_\delta} s^3 \varphi_{\delta,1}^4 (|\partial_t y_l|^2 + |\partial_t z_l|^2) e^{2s\psi_{\delta,1}} dxdt$$

for $l = 1, 2$. Putting this together with (4.141), we see that

$$J_{2,l} \le C \int_{Q_\delta} s\varphi_{\delta,1}^2 (|\nabla \Delta a|^2 + \sum_{|\alpha|\le 2} |\partial_x^\alpha a|^2 + \sum_{|\alpha|\le 2} |\partial_x^\alpha c|^2) e^{2s\psi_{\delta,1}} dxdt + C(s)B^2 \tag{4.146}$$

for $l = 1, 2$. Moreover we have

$$J_{3,l} = \int_{t_0-\delta}^{t_0} \int_\Omega \partial_t (s^2 \varphi_{\delta,1}^2 |L^{(1)} y_l|^2 e^{2s\psi_{\delta,1}}) dxdt \le$$

$$\le C \int_{t_0-\delta}^{t_0} \int_\Omega [s^2 \varphi_{\delta,1}^3 |L^{(1)} y_l|^2 + s^2 \varphi_{\delta,1}^2 |\partial_t L^{(1)} y_l||L^{(1)} y_l| +$$

$$s^3 \varphi_{\delta,1}^4 |L^{(1)} y_l|^2] e^{2s\psi_{\delta,1}} dxdt \le C \int_{Q_\delta} s^3 \varphi_{\delta,1}^4 (|L^{(1)} y_l|^2 + |L^{(1)} z_l|^2) e^{2s\psi_{\delta,1}} dxdt \le$$

$$\le C \int_{Q_\delta} s^3 \varphi_{\delta,1}^4 (\sum_{|\alpha|\le 2} |\partial_x^\alpha y_l|^2 + \sum_{|\alpha|\le 2} |\partial_x^\alpha z_l|^2) e^{2s\psi_{\delta,1}} dxdt$$

for $l = 1, 2$. This together with (4.141) gives

$$J_{3,l} \le C \int_{Q_\delta} s\varphi_{\delta,1}^2 (|\nabla \Delta a|^2 + \sum_{|\alpha|\le 2} |\partial_x^\alpha a|^2 + \sum_{|\alpha|\le 2} |\partial_x^\alpha c|^2) e^{2s\psi_{\delta,1}} dxdt + C(s)B^2 \tag{4.147}$$

for $l = 1, 2$.

In the same way as getting the estimates of $\widetilde{J}_4$, $\widetilde{J}_5$, $\widetilde{J}_6$ in the proof of Theorem 4.7, we obtain

$$J_{4,l} \le \frac{C}{s^3}\int_{Q_\delta} \varphi_{\delta,1}^2\Big(|\nabla\Delta a|^2 + \sum_{|\alpha|\le 2}|\partial_x^\alpha a|^2 + \sum_{|\alpha|\le 2}|\partial_x^\alpha c|^2\Big)e_{2s\psi_{\delta,1}}dxdt + C(s)B^2, \tag{4.148}$$

$$J_{5,l} \le C\int_{Q_\delta} s\varphi_{\delta,1}^2\Big(|\nabla\Delta a|^2 + \sum_{|\alpha|\le 2}|\partial_x^\alpha a|^2 + \sum_{|\alpha|\le 2}|\partial_x^\alpha c|^2\Big)e_{2s\psi_{\delta,1}}dxdt + C(s)B^2, \tag{4.149}$$

$$J_{6,l} \le C\int_{Q_\delta} s\varphi_{\delta,1}^2\Big(|\nabla\Delta a|^2 + \sum_{|\alpha|\le 2}|\partial_x^\alpha a|^2 + \sum_{|\alpha|\le 2}|\partial_x^\alpha c|^2\Big)e_{2s\psi_{\delta,1}}dxdt + C(s)B^2 \tag{4.150}$$

for $l = 1, 2$.

Summing up the estimate of (4.144) through (4.150), we have

$$\sum_{k=1}^{6} J_{k,l} \le C\int_{Q_\delta} s\varphi_{\delta,1}^2\Big(|\nabla\Delta a|^2 + \sum_{|\alpha|\le 2}|\partial_x^\alpha a|^2 + \sum_{|\alpha|\le 2}|\partial_x^\alpha c|^2\Big)e_{2s\psi_{\delta,1}}dxdt + C(s)B^2$$

for $l = 1, 2$.

This together with (4.144) gives

$$\int_\Omega \big((s\varphi_{\delta,1})^2|F_l(x,t_0)|^2 + |\nabla F_l(x,t_0)|^2\big)e^{2s\psi_{\delta,1}(x,t_0)}dx \le$$

$$\le C\int_{Q_\delta} s\varphi_{\delta,1}^2\Big(|\nabla\Delta a|^2 + \sum_{|\alpha|\le 2}|\partial_x^\alpha a|^2 + \sum_{|\alpha|\le 2}|\partial_x^\alpha c|^2\Big)e_{2s\psi_{\delta,1}}dxdt +$$

$$+ C\int_\Omega \sum_{|\alpha|\le 5}(s\varphi_{\delta,1}(x,t_0))^2|\partial_x^\alpha u_l(x,t_0)|^2 e_{2s\psi_{\delta,1}(x,t0)}dx + C(s)B^2 \tag{4.151}$$

for $l = 1, 2$.

Let us estimate the left-hand side of the inequality (4.151) from below. By (4.137) at $t = t_0$, we have

$$a^{(1)}(x)\nabla r_l(x,t_0)\cdot\partial\Delta a(x) - r_l(x,t_0)div(a^{(1)}(x))\nabla c(x) =$$

$$= F_l(x,t_0) - L_{1,l}a(x) - L_{2,l}c(x),\ x \in \Omega, \tag{4.152}$$

where

$$L_{1,l}a(x) = 2a^{(1)}(x)\sum_{i,j=1}^{n}(\partial_i\partial_j r_l(x,t_0))(\partial_i\partial_j a(x)) +$$

$$+ a^{(1)}(x)\Delta r_l(x,t_0)\Delta a(x) + \nabla a^{(1)}(x)\cdot(\nabla r_l(x,t_0)\cdot\nabla)\nabla a(x) +$$

$$+\big[(\rho_2\partial_t^{\frac{1}{2}}-\rho_1\partial_t)\nabla r_l(x,t_0)+3a^{(1)}(x)\nabla\Delta r_l(x,t_0)+(\Delta r_l(x,t_0))\nabla a^{(1)}(x)-$$

$$-c^{(1)}(x)\nabla r_l(x,t_0)+\frac{\rho_2\nabla r_l(x,0)}{\sqrt{\pi t_0}}\big]\cdot\nabla a(x)+$$

$$+\nabla a^{(1)}(x)\cdot(\nabla a(x)\cdot\nabla)\nabla r_l(x,t_0)+$$

$$+\big[(\rho_2\partial_t^{\frac{1}{2}}-\rho_1\partial_t)\Delta r_l(x,t_0)+\big(\nabla a^{(1)}(x)\cdot\nabla\Delta r_l(x,t_0)\big)+$$

$$+a^{(1)}(x)\Delta^2 r_l(x,t_0)-c^{(1)}(x)\Delta r_l(x,t_0)+\frac{\rho_2\Delta r_l(x,0)}{\sqrt{\pi t_0}}\big]a(x),\ x\in\Omega,$$

$$L_{2,l}c(x)=-2a^{(1)}(x)\nabla r_l(x,t_0)\cdot\nabla c(x)-$$

$$-\big[(\rho_2\partial_t^{\frac{1}{2}}-\rho_1\partial_t)r_l(x,t_0)+div(a^{(1)}(x)\nabla r_l(x,t_0))-$$

$$-c^{(1)}(x)r_l(x,t_0)+\frac{\rho_2 r_l(x,0)}{\sqrt{\pi t_0}}\big]c(x),\ x\in\Omega,$$

for $l=1,2$. Multiplying (4.152) for $l=1$ by $r_2(x,t_0), x\in\Omega$, and multiplying (4.152) for $l=2$ by $r_1(x,t_0), x\in\Omega$, and then by the subtraction of new equations, we obtain

$$a^{(1)}(x)\big(r_2(x,t_0)\nabla r_1(x,t_0)-r_1(x,t_0)\nabla r_2(x,t_0)\big)\cdot\nabla\Delta a(x)=$$

$$=r_2(x,t_0)F_1(x,t_0)-r_1(x,t_0)F_2(x,t_0)-$$

$$-r_2(x,t_0)L_{1,1}a(x)+r_1(x,t_0)L_{1,2}a(x)-$$

$$-r_2(x,t_0)L_{2,1}c(x)+r_1(x,t_0)L_{2,2}c(x),\ x\in\Omega. \tag{4.153}$$

Note that

$$\big|\big(r_2(x,t_0)\nabla r_1(x,t_0)-r_1(x,t_0)\nabla r_2(x,t_0)\big)\cdot\nabla d_1(x)\big|\geq m_3>0,\ x\in\Omega,$$

and $a\in H^4(\Omega)$ satisfies $a(x)=0, x\in D$. Let us apply Lemma 4.8 to (4.153) in Ω. Then we obtain

$$\int_\Omega\big[s\varphi_{\delta,1}(x,t_0)\sum_{i,j,k=1}^n|\partial_i\partial_j\partial_k a(x)|^2+(s\varphi_{\delta,1}(x,t_0))^2|\nabla\Delta a(x)|^2+$$

$$+ (s\varphi_{\delta,1}(x,t_0))^3 \sum_{i,j=1}^{n} |\partial_i \partial_j a(x)|^2 +$$

$$+ (s\varphi_{\delta,1}(x,t_0))^5 (|\nabla a(x)|^2 + |a(x)|^2)\big] e^{2s\psi_{\delta,1}(x,t_0)} dx \le$$

$$\le C \sum_{l=1}^{2} \int_\Omega (|F_l(x,t_0)|^2 + |\nabla F_l(x,t_0)|^2) e^{2s\psi_{\delta,1}(x,t_0)} dx +$$

$$+ C \int_\Omega \left(\sum_{|\alpha|\le 3} |\partial_x^\alpha a(x)|^2 + \sum_{|\alpha|\le 2} |\partial_x^\alpha c(x)|^2 \right) e^{2s\psi_{\delta,1}(x,t_0)} dx. \quad (4.154)$$

On the other hand, by Eq. (4.152) for $l = 1$ and $|r_1(x,t_0)| > 0$, $x \in \Omega$, we have

$$div(a^{(1)}(x))\nabla c(x)) - \frac{L_{2,1}c(x)}{r_1(x,t_0)} =$$

$$= \frac{1}{r_1(x,t_0)} a^{(1)}(x) \nabla r_1(x,t_0) \cdot \nabla \Delta a(x) + \frac{L_{1,1}a(x)}{r_1(x,t_0)} - \frac{F_1(x,t_0)}{r_1(x,t_0)}, \ x \in \Omega. \quad (4.155)$$

Using Eq. (4.155) and the Carleman estimate for the elliptic equation stated in Lemma 4.6 ($p = 2$) for c, we get

$$\int_\Omega s\varphi_{\delta,1}^2(x,t_0) \sum_{|\alpha|\le 2} |\partial_x^\alpha c|^2 e^{2s\psi_{\delta,1}(x,t_0)} dx \le$$

$$\le C \int_\Omega s\varphi_{\delta,1}(x,t_0) \sum_{|\alpha|\le 2} |\partial_x^\alpha c|^2 e^{2s\psi_{\delta,1}(x,t_0)} dx \le$$

$$\le C \int_\Omega \big[s\varphi_{\delta,1}(x,t_0) \sum_{i,j=1}^{n} |\partial_i \partial_j c|^2 +$$

$$+ (s\varphi_{\delta,1}(x,t_0))^3 |\nabla c|^2 + (s\varphi_{\delta,1}(x,t_0))^5 |c|^2 \big] e^{2s\psi_{\delta,1}(x,t_0)} dx \le$$

$$\le C \int_\Omega (s\varphi_{\delta,1}(x,t_0))^2 |F_1(x,t_0)|^2 e^{2s\psi_{\delta,1}(x,t_0)} dx +$$

$$+ C \int_\Omega (s\varphi_{\delta,1}(x,t_0))^2 (|\nabla \Delta a(x)|^2 + \sum_{|\alpha|\le 2} |\partial_x^\alpha a(x)|^2) e^{2s\psi_{\delta,1}(x,t_0)} dx. \quad (4.156)$$

By (4.154), we may estimate the second term on the right-hand side of (4.156) and we see that

$$\int_\Omega s\varphi_{\delta,1}^2(x,t_0)\sum_{|\alpha|\le 2}|\partial_x^\alpha c|^2 e^{2s\psi_{\delta,1}(x,t_0)}\le$$

$$\le C\sum_{l=1}^2\int_\Omega \left((s\varphi_{\delta,1}(x,t_0))^2|F_l(x,t_0)|^2+|\nabla F_l(x,t_0)|^2\right)e^{2s\psi_{\delta,1}(x,t_0)}dx+$$

$$+C\int_\Omega\left(\sum_{|\alpha|\le 3}|\partial_x^\alpha a(x)|^2+\sum_{|\alpha|\le 2}|\partial_x^\alpha c(x)|^2\right)e^{2s\psi_{\delta,1}(x,t_0)}dx.$$

This together with (4.154) gives

$$\int_\Omega\Big[s\varphi_{\delta,1}(x,t_0)\sum_{i,j,k=1}^n|\partial_i\partial_j\partial_k a(x)|^2+(s\varphi_{\delta,1}(x,t_0))^2|\nabla\Delta a(x)|^2+$$

$$+(s\varphi_{\delta,1}(x,t_0))^3\sum_{|\alpha|\le 2}|\partial_x^\alpha a(x)|^2+s\varphi_{\delta,1}^2(x,t_0)\sum_{|\alpha|\le 2}|\partial_x^\alpha c|^2\Big]e^{2s\psi_{\delta,1}(x,t_0)}dx\le$$

$$\le C\sum_{l=1}^2\int_\Omega \left((s\varphi_{\delta,1}(x,t_0))^2|F_l(x,t_0)|^2+|\nabla F_l(x,t_0)|^2\right)e^{2s\psi_{\delta,1}(x,t_0)}dx+$$

$$+C\int_\Omega\Big(\sum_{|\alpha|\le 3}|\partial_x^\alpha a(x)|^2+\sum_{|\alpha|\le 2}|\partial_x^\alpha c(x)|^2\Big)e^{2s\psi_{\delta,1}(x,t_0)}dx.$$

Combining this with (4.151), we obtain

$$\int_\Omega\Big[s\varphi_{\delta,1}(x,t_0)\sum_{i,j,k=1}^n|\partial_i\partial_j\partial_k a(x)|^2+(s\varphi_{\delta,1}(x,t_0))^2|\nabla\Delta a(x)|^2+$$

$$+(s\varphi_{\delta,1}(x,t_0))^3\sum_{|\alpha|\le 2}|\partial_x^\alpha a(x)|^2+s\varphi_{\delta,1}^2(x,t_0)\sum_{|\alpha|\le 2}|\partial_x^\alpha c|^2\Big]e^{2s\psi_{\delta,1}(x,t_0)}dx\le$$

$$\le C\int_\Omega\left(\sum_{|\alpha|\le 3}|\partial_x^\alpha a(x)|^2+\sum_{|\alpha|\le 2}|\partial_x^\alpha c(x)|^2\right)e^{2s\psi_{\delta,1}(x,t_0)}dx+$$

$$+C\int_{Q_\delta}s\varphi_{\delta,1}^2\Big(|\nabla\Delta a|^2+\sum_{|\alpha|\le 2}|\partial_x^\alpha a|^2+\sum_{|\alpha|\le 2}|\partial_x^\alpha c|^2\Big)e^{2s\psi_{\delta,1}}dxdt+$$

$$+C\sum_{l=1}^2\int_\Omega\sum_{|\alpha|\le 5}(s\varphi_{\delta,1}(x,t_0))^2|\partial_x^\alpha u_l(x,t_0)|^2e^{2s\psi_{\delta,1}(x,t_0)}dx+C(s)B^2\le$$

$$\le C\int_\Omega\Big[\sum_{i,j,k=1}^n|\partial_i\partial_j\partial_k a(x)|^2+s\varphi_{\delta,1}^2(x,t_0)\big(|\nabla\Delta a|^2+\sum_{|\alpha|\le 2}|\partial_x^\alpha a|^2\big)\Big]e^{2s\psi_{\delta,1}(x,t_0)}dx+$$

$$+C\int_\Omega\sum_{|\alpha|\le 2}|\partial_x^\alpha c|^2e^{2s\psi_{\delta,1}(x,t_0)}dx+C\int_{Q_\delta}s\varphi_{\delta,1}^2\sum_{|\alpha|\le 2}|\partial_x^\alpha c|^2e^{2s\psi_{\delta,1}}dxdt+$$

$$+C\sum_{l=1}^2\int_\Omega\sum_{|\alpha|\le 5}(s\varphi_{\delta,1}(x,t_0))^2|\partial_x^\alpha u_l(x,t_0)|^2e^{2s\psi_{\delta,1}(x,t_0)}dx+C(s)B^2. \tag{4.157}$$

In the last inequality, we used

$$\varphi_{\delta,1}^2(x,t)e^{2s\psi_{\delta,1}(x,t)}\le\varphi_{\delta,1}^2(x,t_0)e^{2s\psi_{\delta,1}(x,t_0)},\ (x,t)\in Q_\delta,$$

for large $s>0$. As we have seen in the proof of Theorem 4.3, we may estimate the third term on the right-hand side of (4.157).

$$\int_{Q_\delta}s\varphi_{\delta,1}^2\sum_{|\alpha|\le 2}|\partial_x^\alpha c|^2e^{2s\psi_{\delta,1}}dxdt$$

$$=\int_\Omega s\varphi_{\delta,1}^2(x,t_0)\sum_{|\alpha|\le 2}|\partial_x^\alpha c|^2e^{2s\psi_{\delta,1}(x,t_0)}\widetilde{h}_s(x)dx,$$

where

$$\widetilde{h}_s(x)=\frac{1}{\varphi_{\delta,1}^2(x,t_0)}\int_{t_0-\delta}^{t_0+\delta}\varphi_{\delta,1}^2e^{-2s(\psi_{\delta,1}(x,t_0)-\psi_{\delta,1}(x,t))}dt,\ x\in\Omega,$$

and $\widetilde{h}_s$ converges uniformly to 0 in Ω.

Taking sufficiently large $s>0$ and absorbing the first term through the third term on the right-hand side of (4.157) into the left-hand side, then we obtain

$$\int_\Omega\Big[s\varphi_{\delta,1}(x,t_0)\sum_{i,j,k=1}^n|\partial_i\partial_j\partial_k a(x)|^2+(s\varphi_{\delta,1}(x,t_0))^2|\nabla\Delta a(x)|^2+$$

$$+(s\varphi_{\delta,1}(x,t_0))^3\sum_{|\alpha|\le 2}|\partial_x^\alpha a(x)|^2+s\varphi_{\delta,1}^2(x,t_0)\sum_{|\alpha|\le 2}|\partial_x^\alpha c|^2\Big]e^{2s\psi_{\delta,1}(x,t_0)}dx\le$$

$$\le C\sum_{l=1}^2\int_\Omega\sum_{|\alpha|\le 5}(s\varphi_{\delta,1}(x,t_0))^2|\partial_x^\alpha u_l(x,t_0)|^2e^{2s\psi_{\delta,1}(x,t_0)}dx+C(s)B^2. \tag{4.158}$$

Fix $s>0$. Then we obtain

$$\|a\|_{H^3(\Omega)} + \|c\|_{H^2(\Omega)} \le C\|u(\cdot, t_0)\|_{H^5(\Omega)} + CB.$$

Thus we obtain the stability estimate (4.86).

Proof of Theorem 4.10. We may obtain Theorem 4.10 in the same way as Theorem 4.9.

Appendix A.

Lemma A.1. *Let* $u \in H^n(0, T)$*. Then there exists a constant* $C > 0$ *such that*

$$\|\partial_t^\beta u\|_{L^2(0,T)} \le C\|\partial_t^n u\|_{L^2(0,T)}$$

for $n - 1 < \beta < n$ $(n \in N)$.

Proof Let us consider the Caputo derivative defined as

$$\partial_t^\beta u(t) = \frac{1}{\Gamma(n-\beta)} \int_0^t \frac{\partial_\tau^n u(\tau)}{(t-\tau)^{\beta+1-n}} d\tau.$$

We note that $0 < n - \beta < 1$. We have

$$|\partial_t^\beta u|^2 \le \frac{1}{\Gamma(n-\beta)^2} \left[\int_0^t \frac{|\partial_\tau^n u|}{(t-\tau)^{\tau+1-n}} d\tau \right]^2.$$

Hence,

$$\int_0^T |\partial_t^\beta u|^2 dt \le \frac{1}{\Gamma(n-\beta)^2} \int_0^T \left[\int_0^t \frac{1}{(t-\tau)^{\beta+1-n}} d\tau \right] \left[\int_0^t \frac{|\partial_\tau^n u|^2}{(t-\tau)^{\beta+1-n}} d\tau \right] dt =$$

$$= \frac{1}{\Gamma(n-\beta)^2} \int_0^T \frac{t^{n-\beta}}{n-\beta} \left[\int_0^t \frac{|\partial_\tau^n u|^2}{(t-\tau)^{\beta+1-n}} d\tau \right] dt \le$$

$$\le \frac{1}{\Gamma(n-\beta)^2} \frac{T^{n-\beta}}{n-\beta} \int_0^T \int_0^t \frac{|\partial_\tau^n u|^2}{(t-\tau)^{\beta+1-n}} d\tau dt =$$

$$= \frac{1}{\Gamma(n-\beta)^2} \frac{T^{n-\beta}}{n-\beta} \int_0^T \int_\tau^T \frac{|\partial_\tau^n u|^2}{(t-\tau)^{\beta+1-n}} dt d\tau =$$

$$= \frac{1}{\Gamma(n-\beta)^2} \frac{T^{n-\beta}}{n-\beta} \int_0^T \frac{(T-\tau)^{n-\beta}}{n-\beta} |\partial_\tau^n u|^2 d\tau \le$$

$$\le \frac{1}{\Gamma(n-\beta)^2} \frac{T^{2(n-\beta)}}{(n-\beta)^2} \int_0^T |\partial_\tau^n u|^2 d\tau.$$

Therefore there exists a constant $C > 0$ such that

$$\|\partial_t^{\beta} u\|_{L^2(0,T)} \leq C \|\partial_t^{n} u\|_{L^2(0,T)}.$$

4.3 Conclusion

In this chapter, the inverse problems for the first- and half-order time-fractional equation are considered. In inverse source problem of determining the time-independent source factor the conditional stability estimates of Hölder and Lipschitz types are investigated. Also Lipschitz stability estimates in inverse coefficient problems for this equation are established by the Carleman estimate method. The Carleman estimate for the first- and half-order time-fractional diffusion equation is derived.

The first- and half-order Eq. (4.1) is one of the parabolic equations with multiple time-fractional terms, i.e., the time-derivative part in the equation is given by $\sum_{j=0}^{l} p_j \partial_t^{\alpha_j}$, where $0 < \alpha_l < \ldots < \alpha_1 < \alpha_0 \leq 1$ and coefficients p_j generally depend on x. Initial-boundary value problems for multiterm time-fractional diffusion equations were considered in [25]. In the case that all time derivatives are noninteger order and the time-derivative part is given by $\sum_{j=1}^{l} p_j \partial_t^{\alpha_j}$, the well-posedness was investigated [26] and moreover the uniqueness in inverse boundary value problems was proven [27]. An exact solution was obtained in the special case of a two-term time-fractional diffusion equation [28]. The uniqueness for two kinds of inverse problems of identifying fractional orders in diffusion equations with multiple time-fractional derivatives was proved [29]. The uniqueness in determining the spatial component of the source term from interior observation was established [30]. The maximum principle and uniqueness was considered in [31] for the determination of the temporal component of the source term from a single point observation. Also the unique continuation was considered for multiterm time-fractional diffusion equations [32].

References

1. Bukhgeim, A.L., Klibanov, M.V.: Uniqueness in the large of a class of multidimensional inverse problems. Dokl. Akad. Nauk SSSR **260**, 269–272 (1981)
2. Kawamoto, A.: Hölder stability estimate in an inverse source problem for a first and half order time fractional diffusion equation. Inverse Probl. Imaging **12**, 315–330 (2018)
3. Kawamoto, A., Machida, M.: Lipschits stability in inverse source and inverse coefficient problems for a first- and half-order time fractional diffusion equation. SIAM J. Math. Anal. **52**(1), 967–1005 (2020)
4. Carleman, T.: Sur un problème d'unicité pour les systèmes d'équations aux dérivées partielles à deux varlables indépendantes. Ark. Mat. Astr. Fys. **2B**, 1–9 (1939)
5. Hörmander, L.: Linear Partial Differential Operators. Springer, Berlin (1963)
6. Isakov, V.: Inverse Problems for Partial Differential Equations, 2nd edn. Springer, Berlin (2006)

7. Klibanov, M.V., Timonov, A.A.: Carleman Estimates for Coefficient Inverse Problems and Numerical Applications. VSP, Utrecht (2004)
8. Kawamoto, A.: Lipschitz stability estimates in inverse source problems for a fractional diffusion equation of half order in time by Carleman estimates. UTMS Preprint Series, UTMS (2016)
9. Yamamoto, M., Zhang, Y.: Conditional stability in determining a zeroth-order coefficient in a half-order fractional diffusion equation by a Carleman estimate. Inverse Prob. **28** (2012), 105010 (10pp) (2012)
10. Xu, X., Cheng, J., Yamamoto, M.: Carleman estimate for fractional diffusion equation with half order and application. Appl. Anal. **90**, 1355–1371 (2011)
11. Kilbas, A.A., Srivastava, H.M., Trujillo, J.J.: Theory and application of fractional differential equations. Elsevier, North-Holland Mathematical Studies, Amsterdam (2006)
12. Eller, M., Isakov, V.: Carleman estimates with two large parameters and applications. Proc. Conf. Differential-Geometric Methods in the Control of Partial Differential Equations (Boulder, CO, July 1999). Contemp. Math. **268**, 117–136 (2000)
13. Isakov, V., Kim, N.: Carleman estimates with two large parameters for second order operators and applications to elasticity with residual stress. Appl. Math. **35**, 447–465 (2008)
14. Isakov, V., Kim, N.: Carleman estimates with second large parameter for second-order operators, Some Applications of Sobolev Spaces to PDEs. International Mathematical Series, vol. 10. Springer, pp. 135–159 (2009)
15. Yamamoto, M.: Carleman estimates for parabolic equations and applications. Inverse Prob. **25**, 123013 (2009)
16. Fursikov, A.V., Imanuvilov, O.Y.: Controllability of Evolution Equations. (Lecture Notes Series vol. 34) Seoul National University, Seoul (Korea) (1996)
17. Imanuvilov, O.Y., Yamamoto, M.: Lipschitz stability in inverse parabolic problems by the Carleman estimate. Inverse Prob. **14**, 1229–1245 (1998)
18. Kawamoto, A.: Lipschitz stability estimates in inverse source problems for a fractional diffusion equation of half order in time by Carleman estimates. J. Inverse Ill-Posed Probl. **26**, 647–672 (2018)
19. Xu, X., Cheng, J., Yamamoto, M.: Carleman estimate for fractional diffusion equation with half order and application. Appl. Anal. **90**, 1355–1371 (2011)
20. Yamamoto, M., Zhang, Y.: Conditional stability in determining a zeroth-order coefficient in a half-order fractional diffusion equation by a Carleman estimate. Inverse Prob. **28**, 105010 (2012)
21. Sakamoto, K., Yamamoto, M.: Inverse source problem with a final overdetermination for a fractional diffusion equation. Math. Control Relat. Fields **1**, 509–518 (2011)
22. Fursikov, A.V., Imanuvilov, O.Y.: Controllability of Evolution Equations. Lecture Notes Series 34, Seoul National University, Seoul, Korea (1996)
23. Imanuvilov, O.Y.: Controllability of parabolic equations. Sb. Math. **186**, 879–900 (1995)
24. Ren, C., Xu, U.: Local stability for an inverse coefficient problem of a fractional diffusion equation. Chin. Ann. Math. Ser. B **35**, 429–446 (2014)
25. Luchko, Y.: Initial-boundary-value problems for the generalized multi-term time-fractional diffusion equation. J. Math. Anal. Appl. **374**, 538–548 (2011)
26. Li, Z., Liu, Y., Yamamoto, M.: Initial-boundary value problems for multi-term time fractional diffusion equations with positive constant coefficients. Appl. Math. Comput. **257**, 381–397 (2015)
27. Li, Z., Imanuvilov, O.Y., Yamamoto, M.: Uniqueness in inverse boundary value problems for fractional diffusion equations. Inverse Prob. **32**, 015004 (2016)
28. Bazhlekova, E., Dimovski, I.: Exact solution of two-term time-fractional Thornleys problem by operational method. Integral Transf. Spec. Funct. **25**, 61–74 (2014)
29. Li, Z., Yamamoto, M.: Uniqueness for inverse problems of determining orders of multi-term time-fractional derivatives of diffusion equation. Appl. Anal. **94**, 570–579 (2015)

30. Jiang, D., Li, Z., Liu, Y., Yamamoto, M.: Weak unique continuation property and a related inverse source problem for time-fractional diffusion-advection equations. Inverse Prob. **33**, 055013 (2017)
31. Liu, Y.: Strong maximum principle for multi-term time-fractional diffusion equations and its application to an inverse source problem. Comput. Math. Appl. **73**, 96–108 (2017)
32. Lin, C.L., Nakamura, G.: Unique continuation property for multi-terms time fractional diffusion equations. Math. Ann. **373**(2), 1–24 (2019)

Chapter 5
Determination of Source and Initial Functions

This chapter is devoted to the inverse problems (linear problems) of determining source and initial functions in initial-boundary value problems for time-fractional diffusion equation.

In Sects. 5.1 and 5.2, for the solution of the inverse problems two bases for the space $L^2(0, 1)$, which form the bi-orthogonal system are used. Thanks to this bi-orthogonal system, it becomes possible to expand the solution in terms of the functions of the bi-orthogonal system. The existence and uniqueness of the solutions of the inverse problems are showed using properties of the Mittag-Leffler function and using the overdetermination conditions of final time observation in Sect. 5.1 and integral type in Sect. 5.2.

Section 5.3 concerns with the existence and uniqueness of solutions to inverse initial and inverse source problems for a time-fractional differential equation with the Bessel operator in the elliptic part. For both cases of inverse problems, as overdetermination information is used the condition of final time observation. Solutions are obtained in the form of series expansion using an orthogonal set of Bessel functions.

Section 5.4 is devoted to the study of inverse source problem for a mixed type equation with a fractional diffusion equation in the parabolic part and a wave equation in the hyperbolic part of a cylindrical domain. Solution of this problem is obtained in the form of Fourier-Bessel series expansion based on an orthogonal set of Bessel functions. Thus, the theorems of uniqueness and the existence of a solution are proved.

5.1 Determination of an Unknown Source Term

We consider the inverse problem of finding the temperature distribution and the heat source whenever the temperatures at the initial time and the final time are given. The problem considered is one dimension and the unknown heat source is supposed

D. K. Durdiev, *Inverse Problems for Fractional Diffusion Equations*,
Industrial and Applied Mathematics, https://doi.org/10.1007/978-981-96-5338-6_5

to be space dependent only. The existence and uniqueness results are obtained. The statement of this section corresponds to article [1].

We are concerned with the problem of finding $u(x, t)$ (the temperature distribution) and $f(x)$ (the source term) in the domain $Q_T = (0, 1) \times (0, T]$ for the following system

$$\partial_t^\alpha u(x, t) - u_{xx}(x, t) = f(x), \ \ (x, t) \in Q_T, \tag{5.1}$$

$$u(x, 0) = \varphi(x), u(x, T) = \psi(x), x \in [0, 1], \tag{5.2}$$

$$u(1, t) = 0, \ u_x(0, t) = u_x(1, t), t \in [0, T], \tag{5.3}$$

where ∂_t^α stands for the Caputo fractional derivative of order $0 < \alpha < 1, \varphi(x)$ and $\psi(x)$ are the initial and final temperatures respectively.

Now we recall some known facts, which can be found in [2].

The Laplace transform of the Riemann-Liouville integral of order $0 < \alpha < 1$ of a function with at most exponentially growth is

$$\mathcal{L}\left\{I_t^\alpha f(t) : s\right\} = \frac{\mathcal{L}\{f(t) : s\}}{s^\alpha}.$$

Also, for $0 < \alpha < 1$ and an integrable function $f(t)$ we have (see [2, Lemma 2.22])

$$I_t^\alpha \partial_t^\alpha f(t) = f(t) - f(0). \tag{5.4}$$

Let $e_\xi(t, \mu) := E_\xi\left(-\mu t^\xi\right)$, where $E_\xi(t)$ is the Mittag-Leffler function with one parameter ξ as defined in (1.27) and μ is a positive real number. Let us state and prove the

Lemma 5.1 *The solution of equation*

$$v(t) + \mu I_t^\alpha v(t) = g(t) \tag{5.5}$$

for $\mu \in \mathbb{R}$ *satisfies the integral equation*

$$v(t) = \int_0^t g'(t - \tau) e_\alpha(\tau, \mu) d\tau + g(0) e_\alpha(t, \mu). \tag{5.6}$$

Proof Taking the Laplace transform of both sides of the equation (5.5), it follows

$$\mathcal{L}\{v(t) : s\} = \frac{s^\alpha \mathcal{L}\{g(t) : s\}}{s^\alpha + \mu}. \tag{5.7}$$

There are different ways for getting the solution from the equation (5.7) by the inverse Laplace transform. Writing the equation (5.7) as

$$\mathcal{L}\{v(t) : s\} = \frac{s^{\alpha-1}}{s^\alpha + \mu}\left(s\mathcal{L}\{g(t) : s\} - g(0)\right) + \frac{s^{\alpha-1}}{s^\alpha + \mu} g(0)$$

and using the inverse Laplace transform, we obtain (5.6).

The key factor in the determination of the temperature distribution and the unknown source term for the system (5.1)–(5.3) is the specially chosen basis for the space $L^2(0, 1)$ which is

$$\left\{2(1-x),\ \{4(1-x)\cos 2\pi nx\}_{n=1}^{\infty},\ \{4sin2\pi nx\}_{n=1}^{\infty}\right\}. \tag{5.8}$$

In [3], it is proved that the set of functions in (5.8) forms a Riesz basis for the space $L^2(0, 1)$, hence the set is closed, minimal and spans the space $L^2(0, 1)$. The set of functions (5.8) is not orthogonal and in [4] a bi-orthogonal set of functions given by

$$\left\{1,\ \{\cos 2\pi nx\}_{n=1}^{\infty},\ \{x\ \sin 2\pi nx\}_{n=1}^{\infty}\right\} \tag{5.9}$$

is constructed.

The main result of this section is the following statement:

Theorem 5.1 *Let the conditions*

$\varphi(x) \in C^4[0, 1],\ \varphi(0) = \varphi(1) = \varphi''(0) = 0,\ \varphi'(0) = \varphi'(1) = 0,\ \varphi'''(0) = \varphi'''(1) = 0;$

$\psi(x) \in C^4[0, 1],\ \psi(0) = \psi(1) = \psi''(0) = 0,\ \psi'(0) = \psi'(1) = 0,\ \psi'''(0) = \psi'''(1) = 0;$

be satisfied. Then, there exists a unique solution $\{u(x, t), f(x)\} \in C^{2,\alpha}(Q_T) \cap C^{1,0}(\overline{Q}_T) \times C[0, 1]$ *of the problem* (5.1)–(5.3) *(see Sect. 1.2 for definition of* $C^{2,\alpha}(Q_T)$*).*

This theorem is proved in the next three subsections.

5.1.1 Formal Construction of the Inverse Problem Solution

The solution of the inverse problem for the linear system (5.1)–(5.3) can be written in the form

$$u(x, t) = 2u_0(t)(1-x)$$

$$+\sum_{n=1}^{\infty} u_{1n}(t)4(1-x)\cos 2\pi nx + \sum_{n=1}^{\infty} u_{2n}(t)4sin2\pi nx, \tag{5.10}$$

$$f(x, t) = 2f_0(t)(1-x)$$

$$+\sum_{n=1}^{\infty} f_{1n}(t)4(1-x)\cos 2\pi nx + \sum_{n=1}^{\infty} f_{2n}(t)4sin2\pi nx, \tag{5.11}$$

where $u_0(t),\ u_{1n}(t),\ u_{2n}(t),\ f_0,\ f_{1n}$ and f_{2n} are to be determined. By plugging the expressions of $u(x,t)$ and $f(x)$ from (5.10) and (5.11) into the equation (5.1), the following system of fractional differential equations is obtained

$$\partial_t^\alpha u_{2n}(t) - 4\pi n u_{1n}(t) + 4\pi^2 n^2 u_{2n}(t) = f_{2n}, \tag{5.12}$$

$$\partial_t^\alpha u_{1n}(t) + 4\pi^2 n^2 u_{1n}(t) = f_{1n}, \tag{5.13}$$

$$\partial_t^\alpha u_0(t) = f_0. \tag{5.14}$$

Since $I_t^\alpha(1) = \dfrac{t^\alpha}{\Gamma(1+\alpha)}$ and for $0<\alpha<1,\ I_t^\alpha \partial_t^\alpha u_0(t) = u_0(t) - u_0(0)$, then the solution of equation (5.14) is

$$u_0(t) = \frac{f_0 t^\alpha}{\Gamma(1+\alpha)} + u_0(0). \tag{5.15}$$

Denote $\lambda := 4\pi^2 n^2,\ F(t) := \dfrac{f_{1n} t^\alpha}{\Gamma(1+\alpha)} + u_{1n}(0)$, then equation (5.13) can be written as

$$u_{1n}(t) + \lambda I_t^\alpha u_{1n}(t) = F(t),$$

and using Lemma 5.1, the solution of the equation (5.13) is given by

$$u_{1n}(t) = \frac{f_{1n}\alpha}{\Gamma(1+\alpha)} \int_0^t (t-\tau)^{\alpha-1} e_\alpha(\tau,\lambda) d\tau + u_{1n}(0) e_\alpha(t,\lambda), \tag{5.16}$$

Noticing that equation (5.12) can be written as

$$u_{2n}(t) + \lambda I_t^\alpha u_{2n}(t) = h(t), \tag{5.17}$$

where $h(t) := I_t^\alpha \left(f_{2n} + 4\pi n u_{1n}(t)\right) + u_{2n}(0)$, so its solution by virtue of Lemma 5.1 is

$$u_{2n}(t) = \int_0^t h'(t-\tau) e_\alpha(\tau,\lambda) d\tau + h(0) e_\alpha(t,\lambda). \tag{5.18}$$

As $h(0) = u_{2n}(0)$ and

$$h'(t) = \frac{d}{\mathrm{dt}}\left[I_t^{1-\alpha}\left(f_{2n} + 4\pi n u_{1n}(t)\right) - u_{2n}(0)\right] = \frac{\partial^\alpha}{\partial t^\alpha}\left(f_{2n} + 4\pi n u_{1n}(t)\right),$$

where $\frac{\partial^\alpha}{\partial t^\alpha}$ is the Riemann-Liouville fractional derivative. From equation (5.16), we have

$$\frac{\partial^\alpha}{\partial t^\alpha} u_{1n}(t) = \frac{f_{1n}\alpha}{\Gamma(1+\alpha)} \int_0^t \frac{\partial^\alpha}{\partial \tau^\alpha} \tau^{\alpha-1} e_\alpha(t-\tau, \lambda) d\tau + u_{1n}(0) \frac{\partial^\alpha}{\partial t^\alpha} e_\alpha(t, \lambda),$$

since $\frac{\partial^\alpha}{\partial t^\alpha} t^{\alpha-1} = 0$ and $\frac{\partial^\alpha}{\partial t^\alpha} e_\alpha(t, \lambda) = e_{\alpha,1-\alpha}(t, \lambda)$, where $e_{\alpha,\beta}(t, \lambda) := t^{\beta-1} E_{\alpha,\beta}(-\lambda t^\alpha)$, consequently the equation (5.18) becomes

$$u_{2n}(t) = \int_0^t s(t-\tau) e_{\alpha,\alpha}(\tau, \lambda) d\tau + u_{2n}(0) e_\alpha(t, \lambda), \tag{5.19}$$

where we have set

$$s(t) := 4\pi n u_{1n}(0) e_{\alpha,1-\alpha}(t, \lambda) + \frac{f_{2n} t^{-\alpha}}{\Gamma(1-\alpha)}.$$

Remind the unknown constants are $u_{2n}(0),\ u_{1n}(0),\ u_0(0),\ f_{2n},\ f_{1n}$ and f_0. In order to find these unknowns, we use the initial and final temperatures as given in (5.2):

$$2u_0(0)(1-x) + \sum_{n=1}^{\infty} u_{1n}(0) 4(1-x) \cos 2\pi n x + \sum_{n=1}^{\infty} u_{2n}(0) 4 \sin 2\pi n x =$$

$$= 2\varphi_0(1-x) + \sum_{n=1}^{\infty} \varphi_{1n}(0) 4(1-x) \cos 2\pi n x + \sum_{n=1}^{\infty} \varphi_{2n}(0) 4 \sin 2\pi n x,$$

$$2u_0(T)(1-x) + \sum_{n=1}^{\infty} u_{1n}(T) 4(1-x) \cos 2\pi n x + \sum_{n=1}^{\infty} u_{2n}(T) 4 \sin 2\pi n x =$$

$$= 2\psi_0(1-x) + \sum_{n=1}^{\infty} \psi_{1n}(0) 4(1-x) \cos 2\pi n x + \sum_{n=1}^{\infty} \psi_{2n}(0) 4 \sin 2\pi n x.$$

By identification we get

$$u_{2n}(0) = \varphi_{2n},\ u_{1n}(0) = \varphi_{1n},\ u_0(0) = \varphi_0, \tag{5.20}$$

$$u_{2n}(T) = \psi_{2n},\ u_{1n}(T) = \psi_{1n},\ u_0(T) = \psi_0, \tag{5.21}$$

where $\{\varphi_0,\ \varphi_{1n},\ \varphi_{2n}\}$ and $\{\psi_0,\ \psi_{1n},\ \psi_{2n}\}$ are the coefficients of the series expansion in the basis (5.8) of the functions $\varphi(x)$ and $\psi(x)$, respectively. In terms of the bi-orthogonal basis (5.9) these are

$$\varphi_0 = \int_0^1 \varphi(x)dx, \ \varphi_{1n} = \int_0^1 \varphi(x)\cos 2\pi n dx, \ \varphi_{2n} = \int_0^1 \varphi(x) x \sin 2\pi n dx,$$

$$\psi_0 = \int_0^1 \psi(x)dx, \ \psi_{1n} = \int_0^1 \psi(x)\cos 2\pi n dx, \ \psi_{2n} = \int_0^1 \psi(x) x \sin 2\pi n dx.$$

By virtue of (5.15), (5.16), (5.19), and conditions (5.21) we have the expressions for unknowns f_0, f_{1n} and f_{2n} as

$$f_0 = \Gamma(1+\alpha)\left[\frac{\psi_0 - \varphi_0}{T^\alpha}\right], \tag{5.22}$$

$$f_{1n} = \Gamma(1+\alpha)\left[\frac{\psi_{1n} - \varphi_{1n} e_\alpha(T,\lambda)}{\alpha \int_0^T (T-\tau)^{\alpha-1} e_\alpha(\tau,\lambda)d\tau}\right], \tag{5.23}$$

$$f_{2n} = \Gamma(1-\alpha)$$
$$\times \left[\frac{\psi_{2n} - \varphi_{2n} e_\alpha(T,\lambda) - \int_0^T \varphi_{1n} e_{\alpha,1-\alpha}(T-\tau,\lambda) e_\alpha(\tau,\lambda)d\tau}{\int_0^T (T-\tau)^{-\alpha} e_\alpha(\tau,\lambda)d\tau}\right]. \tag{5.24}$$

The unknown source term and the temperature distribution for the problem (5.1)–(5.3) are given by the series (5.11) and (5.10), where the unknowns $u_0(0)$, $u_{1n}(0)$, $u_{2n}(0)$ are calculated from (5.20), while f_0, f_{1n}, f_{2n} are given by (5.22)–(5.24). In the next subsection we will show the uniqueness of the solution.

5.1.2 *Uniqueness of the Inverse Problem Solution*

Suppose $\{u_1(x,t), f_1(x)\}$, $\{u_2(x,t), f_2(x)\}$ are two solution sets of the inverse problem for the system (5.1)–(5.3). Define $\overline{u}(x,t) = u_1 - u_2$ and $\overline{f}(x) = f_1 - f_2$, then the function $\overline{u}(x,t)$ satisfies the following equation and boundary conditions

$$\partial_t^\alpha \overline{u}(x,t) - \overline{u}_{xx}(x,t) = \overline{f}(x), \ (x,t) \in Q_T, \tag{5.25}$$

$$\overline{u}(x,0) = 0, \ \overline{u}(x,T) = 0, \ x \in [0,1], \tag{5.26}$$

$$\overline{u}(1,t) = 0, \ \overline{u}_x(0,t) = \overline{u}_x(1,t), \ t \in [0,T]. \tag{5.27}$$

Using the bi-orthogonal basis (5.9), we have

$$\overline{u}_0(t) = \int_0^1 \overline{u}(x,t)dx, \tag{5.28}$$

$$\overline{u}_{1n}(t) = \int_0^1 \overline{u}(x,t)\cos 2\pi nx dx, \quad n = 1, 2, ..., \tag{5.29}$$

and

$$\overline{u}_{2n}(t) = \int_0^1 \overline{u}(x,t)x\sin 2\pi nx dx, \quad n = 1, 2, ... \tag{5.30}$$

Alike,

$$\overline{f}_0 = \int_0^1 \overline{f}(x)dx, \tag{5.31}$$

$$\overline{f}_{1n} = \int_0^1 \overline{f}(x)\cos 2\pi nx dx, \quad n = 1, 2, ..., \tag{5.32}$$

and

$$\overline{f}_{2n} = \int_0^1 \overline{f}(x)x\sin 2\pi nx dx, \quad n = 1, 2, ... \tag{5.33}$$

Taking the fractional derivative $\frac{\partial^\alpha}{\partial t^\alpha}$ under the integral sign of equation (5.28), using (5.25), (5.31) and integration by parts, we get

$$\frac{\partial^\alpha}{\partial t^\alpha}\overline{u}_0(t) = \overline{f}_0, \tag{5.34}$$

with the boundary conditions

$$\overline{u}_0(0) = 0, \quad \overline{u}_0(T) = 0. \tag{5.35}$$

The solution of the problem (5.34) is

$$\overline{u}_0(t) = \frac{t^\alpha}{\Gamma(1+\alpha)}\overline{f}_0;$$

taking into account the boundary conditions (5.35), we get

$$\overline{f}_0 = 0, \text{ consequently } \overline{u}_0(t) = 0. \tag{5.36}$$

Taking the fractional derivative $\frac{\partial^\alpha}{\partial t^\alpha}$ under the integral sign of equation (5.29) then by using (5.25), (5.26), (5.32) and integration by parts, we obtain

$$\frac{\partial^\alpha}{\partial t^\alpha}\overline{u}_{1n}(t) + \lambda\overline{u}_{1n}(t) = \overline{f}_{1n}, \tag{5.37}$$

the associated boundary conditions are

$$\overline{u}_{1n}(0) = 0, \overline{u}_{1n}(T) = 0. \tag{5.38}$$

The solution of the problem (5.37) is

$$\overline{u}_{1n}(t) = \int_0^t e_{\alpha,\alpha}(t-\tau,\lambda)\overline{f}_{1n}(\tau)d\tau.$$

Using (5.38) we have

$$\overline{f}_{1n} = 0, \text{ consequently } \overline{u}_{1n}(t) = 0, \; n = 1, 2, \ldots.$$

In the same manner from (5.30), we obtain the equation

$$\frac{\partial^\alpha}{\partial t^\alpha}\overline{u}_{2n}(t) + 4\pi^2 n^2 \overline{u}_{2n}(t) = 4\pi n \overline{u}_{1n}(t) + \overline{f}_{2n}. \tag{5.39}$$

The associated boundary conditions are

$$\overline{u}_{2n}(0) = 0, \overline{u}_{2n}(T) = 0. \tag{5.40}$$

In view of $\overline{u}_{1n}(t) = 0$, the solution of the equation (5.39) satisfies

$$\overline{u}_{2n}(t) = \int_0^t e_{\alpha,\alpha}(t-\tau,\lambda)\overline{f}_{2n}(\tau)d\tau,$$

and using (5.40) we have

$$\overline{f}_{2n} = 0, \text{ consequently } \overline{u}_{2n}(t) = 0, \; n = 1, 2, \ldots.$$

For every $t \in [0, T]$, the functions $\overline{u}(x, t)$ and $\overline{f}(x)$ are orthogonal to the system of functions given in (5.9), which form a basis of the space $L^2(0, 1)$, consequently,

$$\overline{u}(x, t) = 0, \quad \overline{f}(x) = 0.$$

5.1.3 Existence of the Inverse Problem Solution

Suppose $\varphi(x) \in C^4[0, 1]$ be such that $\varphi'(0) = \varphi'(1), \varphi'''(0) = \varphi'''(1)$. As φ_{1n} is the coefficient of the cosine Fourier expansion of the function $\varphi(x)$ with respect to basis (5.9), from (5.20) the expression for $u_{1n}(0)$ can be written as

$$u_{1n}(0) = \int_0^1 \varphi(x)\cos 2\pi n x dx,$$

which integrated by parts four times gives

$$u_{1n}(0) = \frac{1}{16\pi^4 n^4}\int_0^1 \varphi^{(4)}(x)\cos 2\pi n x dx = \frac{1}{16\pi^4 n^4}\varphi_{1n}^4,$$

where φ_{1n}^4 is the coefficient of the cosine Fourier expansion of the function $\varphi^{(4)}(x)$ with respect to the basis (5.9). Alike, we obtain

$$u_{2n}(0) = \frac{1}{16\pi^4 n^4}\varphi_{2n}^4, \tag{5.41}$$

where φ_{2n}^4 is the coefficient of sine Fourier transform of the function $\varphi^{(4)}(x)$ with respect to the basis (5.9).

The set of functions $\left\{\varphi_{1n}^4,\ \varphi_{2n}^4,\ n = 1, 2, ...\right\}$ is bounded as by supposition we have $\varphi(x) \in C^4\ [0, 1]$, and due to the inequality

$$\sum_{n=1}^{\infty}\left(\varphi_{\text{in}}^4\right)^2 \le C \parallel \varphi_i^{(4)}(x) \parallel_{L^2(0,1)}^2,\ i = 1, 2,$$

which is true by the Bessel inequality of the trigonometric series. Similarly, we have for $\psi(x) \in C^4[0, 1]$ the set of functions $\left\{\psi_{1n}^4,\ \psi_{2n}^4,\ n = 1, 2, ...\right\}$ is bounded and

$$\sum_{n=1}^{\infty}\left(\psi_{\text{in}}^4\right)^2 \le C \parallel \psi_i^{(4)}(x) \parallel_{L^2(0,1)}^2,\ i = 1, 2.$$

The Mittag-Leffler functions $e_\alpha(t, \mu)$ and $e_{\alpha,\beta}(t, \mu)$ for $\mu > 0$ and $0 < \alpha \le 1,\ 0 < \alpha \le \beta \le 1$ respectively, are completely monotone functions (see [5, p. 268]). Furthermore, we have

$$E_{\alpha,\beta}\left(At^\alpha\right) \le M,\quad t \in [a, b], \tag{5.42}$$

where $[a, b]$ is a finite interval with $a \ge 0$, and

$$\int_0^t (t-\tau)^{\alpha-1}E_{\alpha,\beta}(\lambda t^\alpha)d\tau < \infty$$

(see [6], p. 9). The solution $u(x, t)$ becomes

$$u(x,t) = 2(1-x)\left(\frac{f_0 t^\alpha}{\Gamma(1+\alpha)} + \varphi_0\right)$$

$$+\sum_{n=1}^{\infty}\left(\frac{f_{1n}\alpha}{\Gamma(1+\alpha)}\int_0^t \tau^{\alpha-1}e_\alpha(t-\tau,\lambda)d\tau+\frac{\varphi_{1n}^4}{16\pi^4n^4}e_\alpha(t,\lambda)\right)$$

$$\times 4(1-x)\cos 2\pi nx+\sum_{n=1}^{\infty}\left(\frac{\varphi_{1n}^4}{4\pi^3n^3}e_{\alpha,1-\alpha}(t,\lambda)\right.$$

$$\left.+\frac{f_{2n}}{\Gamma(1+\alpha)}\int_0^t \tau^{-\alpha}e_\alpha(t-\tau,\lambda)d\tau+\frac{\varphi_{2n}^4}{16\pi^4n^4}e_\alpha(t,\lambda)\right)4\sin 2\pi nx. \tag{5.43}$$

Set $M=\max\{M_1,\ M_2\}$, where $|e_\alpha|\le M_1,\ \left|e_{\alpha,1-\alpha}\right|\le M_2,\ t\in[0,T],\ T$ is a positive fixed number. Taking into account the values of $f_0,\ f_{1n},\ f_{2n}$ from (5.22)–(5.24) and the properties of Mittag-Leffler function, the series (5.43) is bounded above by

$$2(1-x)\left(|\psi_0|+\left(1-\frac{t^\alpha}{T^\alpha}\right)|\varphi_0|\right)+$$

$$+\sum_{n=1}^{\infty}\left(\frac{\psi_{1n}^4-\varphi_{1n}^4M}{16\pi^4n^4}\frac{t^\alpha}{T^\alpha}+\frac{\varphi_{1n}^4M}{16\pi^4n^4}\right)4(1-x)+$$

$$+4\sum_{n=1}^{\infty}\left(\frac{\varphi_{1n}^4M}{4\pi^3n^3}+\frac{\psi_{2n}^4-\varphi_{2n}^4M}{16\pi^4n^4}\frac{t^{1-\alpha}}{T^{1-\alpha}}-\frac{\varphi_{1n}^4M}{16\pi^4n^4}+\frac{\varphi_{2n}^4M}{16\pi^4n^4}\right). \tag{5.44}$$

For every $(x,t)\in Q_T$ the series (5.44) is bounded above by the uniformly convergent series

$$\sum_{n=1}^{\infty}\left(\frac{\left|\psi_{1n}^4\right|+\left|\varphi_{1n}^4\right|2M}{16\pi^4n^4}\right)+\sum_{n=1}^{\infty}\left(\frac{(4\pi n-1)\left|\varphi_{1n}^4\right|M}{16\pi^4n^4}+\frac{\left|\psi_{2n}^4\right|-\left|\varphi_{2n}^4\right|2M}{16\pi^4n^4}\right). \tag{5.45}$$

By the Weierstrass M-test, the series (5.43) is uniformly convergent in the domain Q_T. Hence, the solution $u(x,t)$ is continuous in the domain Q_T.

The uniformly convergent series doesn't provide any information about the convergence of the series obtained from its term-by-term differentiation. Now take the $\frac{\partial^\alpha}{\partial t^\alpha}$ derivative from the series expression of $u(x,t)$ given by (5.43):

$$\sum_{n=1}^{\infty}\frac{\partial^\alpha}{\partial t^\alpha}u_n(x,t)=2(1-x)\left(f_0+\varphi_0\frac{t^{-\alpha}}{\Gamma(1-\alpha)}\right)+$$

$$+\sum_{n=1}^{\infty}\left(\frac{\varphi_{1n}^4}{16\pi^4n^4}e_{\alpha,1-\alpha}(t,\lambda)\right)4(1-x)\cos 2\pi nx+$$

$$+\sum_{n=1}^{\infty}\left(\frac{\varphi_{1n}^4}{16\pi^4n^4}e_{\alpha,1-2\alpha}(t,\lambda)+\frac{f_{2n}}{\Gamma(1-\alpha)}\int_0^t \tau^{-\alpha}e_{\alpha,1-\alpha}(t-\tau,\lambda)d\tau\right.$$

$$\left.+\frac{\varphi_{2n}^4}{16\pi^4n^4}e_{\alpha,1-\alpha}(t,\lambda)\right)4\sin 2\pi nx. \tag{5.46}$$

We have used $\frac{\partial^\alpha}{\partial t^\alpha}t^{\alpha-1}=0$, $\frac{\partial^\alpha}{\partial t^\alpha}e_\alpha(t,\lambda)=e_{\alpha,1-\alpha}(t,\lambda)$, $\frac{\partial^\alpha}{\partial t^\alpha}e_{\alpha,1-\alpha}(t,\lambda)=e_{\alpha,1-2\alpha}$ (t,λ) and $u_n(x,t)$ is defined by the right-hand side of the series (5.10). The series (5.46) is bounded above by the uniformly convergent series

$$\sum_{n=1}^{\infty}\left(\frac{\left|\psi_{2n}^4\right|+(4\pi n+1)\left|\varphi_{1n}^4\right|M}{16\pi^4n^4}\right),$$

hence it converges uniformly, so $\frac{\partial^\alpha}{\partial t^\alpha}u(x,t)=\sum_{n=1}^{\infty}\frac{\partial^\alpha}{\partial t^\alpha}u_n(x,t)$.

Take the x-derivative of the series expression of $u(x,t)$ from (5.43), we obtain

$$\sum_{n=1}^{\infty}u_{n,x}(x,t)=-2\left(f_0+\varphi_0\frac{t^{-\alpha}}{\Gamma(1-\alpha)}\right)$$

$$-\sum_{n=1}^{\infty}\left(\frac{f_{1n}\alpha}{\Gamma(1+\alpha)}\int_0^t\tau^{\alpha-1}e_\alpha(t-\tau,\lambda)d\tau+\frac{\varphi_{1n}^4}{16\pi^4n^4}e_\alpha(t,\lambda)\right)$$
$$\times\left(8\pi n(1-x)\sin 2\pi nx+4\cos 2\pi nx\right)$$

$$-\sum_{n=1}^{\infty}\left(\frac{\varphi_{1n}^4}{4\pi^3n^3}e_{\alpha,1-\alpha}(t,\lambda)+\frac{f_{2n}}{\Gamma(1-\alpha)}\int_0^t\tau^{-\alpha}e_\alpha(t-\tau,\lambda)d\tau\right.$$

$$\left.+\frac{\varphi_{2n}^4}{16\pi^4n^4}e_\alpha(t,\lambda)\right)8\pi n\cos 2\pi nx, \tag{5.47}$$

$$\sum_{n=1}^{\infty}u_{n,xx}(x,t)=\sum_{n=1}^{\infty}\left(\frac{f_{1n}\alpha}{\Gamma(1+\alpha)}\int_0^t\tau^{\alpha-1}e_\alpha(t-\tau,\lambda)d\tau+\frac{\varphi_{1n}^4}{16\pi^4n^4}e_\alpha(t,\lambda)\right)$$
$$\times\left(16\pi n\sin 2\pi nx-16\pi^2n^2(1-x)\cos 2\pi nx\right)$$

$$+\sum_{n=1}^{\infty}\left(\frac{\varphi_{1n}^4}{4\pi^3 n^3}e_{\alpha,1-\alpha}(t,\lambda)+\frac{f_{2n}}{\Gamma(1-\alpha)}\int_0^t \tau^{-\alpha}e_{\alpha}(t-\tau,\lambda)d\tau+\right.$$

$$\left.\frac{\varphi_{2n}^4}{16\pi^4 n^4}e_{\alpha}(t,\lambda)\right)16\pi^2 n^2 \sin 2\pi n x. \tag{5.48}$$

The series (5.47) and (5.48) are bounded above by the uniformly convergent series

$$\sum_{n=1}^{\infty}\frac{(1+4\pi n)\left|\psi_{1n}^4\right|+2\pi n(1+M)\left|\varphi_{2n}^4\right|+\left(4\pi^2 n^2+6\pi n+2\right)\left|\varphi_{1n}^4\right|M}{4\pi^4 n^4},$$

$$\sum_{n=1}^{\infty}\frac{\left|\psi_{1n}^4\right|+2\left|\psi_{2n}^4\right|+(TM+1)\left|\varphi_{1n}^4\right|M+\left|\varphi_{2n}^4\right|M}{\pi^2 n^2},$$

respectively, then $\sum_{n=1}^{\infty} u_{n,x}(x,t)$ and $\sum_{n=1}^{\infty} u_{n,xx}(x,t)$ converges uniformly; consequently

$$u_{xx}(x,t)=\sum_{n=1}^{\infty}u_{n,xx}(x,t)$$

is uniformly convergent.

Hence the term-by-term differentiation of the series (5.43) with respect to time t and x is valid. Similar we can show that $f(x)$ obtained by series (5.11) is continuous.

Remark 3.2. *The above analysis remains valid if the boundary conditions given by* (5.3) *are replaced by*

$$u(0,t)=u(1,t),\ \ u_x(0,t)=u_x(1,t),\ \ t\in[0,T],$$

or

$$u(1,t)=0,\ \ u_x(0,t)=0,\ \ t\in[0,T].$$

5.2 An Inverse Source Problem for a Generalized Time-Fractional Diffusion Equation

This section is devoted to the study of the inverse problem of finding the time-dependent factor of the source function in a generalized time fractional diffusion equation, in the case of nonlocal boundary and integral overdetermination conditions. The existence and uniqueness of the solution of the considered inverse problem are obtained by a method based on the expansion of the solution by using a bi-orthogonal system of functions and the fractional calculus. Moreover, we show its continuous dependence on the data. At the end, two examples are presented to illustrate the obtained results. Many articles were addressed for inverse problems of determination of various unknown source functions in time-fractional diffusion equations (see [8]-[14].

The materials of this section correspond to the work [15].

5.2.1 *Statement of the Problem and Preliminaries*

In this subsection, we are concerned with the inverse source coefficient problem for the following generalized time-fractional diffusion equation

$$\partial_t^{\alpha,\beta} u(x,t) = u_{xx}(x,t) + a(t)F(x,t), \ (x,t) \in D_T \tag{5.49}$$

with the initial integral condition

$$I_t^{1-\gamma} u(x,0) = \psi(x), \ \gamma = \alpha + \beta(1-\alpha), \ 0 \le x \le 1, \tag{5.50}$$

the nonlocal boundary conditions

$$u(1,t) = 0, \ u_x(0,t) = u_x(1,t), \ 0 < t \le T, \tag{5.51}$$

and the integral overdetermination data

$$\int_0^1 u(x,t)dx = E(t), \ t \in [0,T], \tag{5.52}$$

where $E(t)$ is a given function, $D_T = \{0 < x < 1, \ 0 < t \le T\}$, $\partial_t^{\alpha,\beta}$ stands for the generalized Riemann-Liouville (Hilfer) time-fractional derivative of order $0 < \alpha < 1$ and type $0 \le \beta \le 1$.

A classical solution of the inverse problem (5.49)–(5.52) consists of determining the pair $\{u(x, \ t), \ a(t)\}$ such that

$$a \in C[0,T], \ t^{1-\gamma} u \in C\left(\overline{D}_T\right), \ t^{1-\gamma}\partial_t^{\alpha,\beta} u \in C\left(\overline{D}_T\right).$$

First, we introduce some notations and definitions of fractional calculus concerned with Hilfer fractional derivative and some results.

New definition of the fractional derivative, known as the left side Hilfer fractional derivative of order $0 < \alpha < 1$ and parameter $0 \le \beta \le 1$ for $f \in L^1(0,b)$ (for some $b > 0$) and $I_t^\alpha f(t) \in W^{1,1}(0,b)$, where $W^{1,1}(0,b)$ is a Sobolev space defined as $W^{1,1}(0,b) = \left\{f(t) \in L_1(0,b) : \ f'(t) \in L_1(0,b)\right\}$, is introduced in [16] as

$$\partial_t^{\alpha,\beta} f(t) := \left(I_t^{\beta(1-\alpha)} \frac{d}{dt}\left(I_t^{(1-\beta)(1-\alpha)} f\right)\right)(t), \tag{5.53}$$

where I_t^α is the fractional integral defined by (1.5). The Hilfer fractional derivative corresponds to the Riemann-Liouville fractional derivative (1.6) when $\beta = 0$ and to the Caputo fractional derivative (1.4) when $\beta = 1$,

$$\partial_t^{\alpha,\beta} f(t) := \begin{cases} \frac{d}{dt} I_t^{1-\alpha} f(t) = \frac{\partial^\alpha}{\partial t^\alpha} f(t), & \beta = 0; \\ I_t^{1-\alpha} \frac{d}{dt} f(t) = \partial_t^\alpha f(t), & \beta = 1. \end{cases}$$

For $0 < \alpha < 1$, $0 \le \beta \le 1$, $\mu > 0$, $t > 0$, the following fractional derivative formula holds:

$$\partial_t^{\alpha,\beta} t^{\mu-1} = \frac{\Gamma(\mu)}{\Gamma(\mu-1)} t^{\mu-\alpha-1}. \tag{5.54}$$

In addition, we have the following relationship between Hilfer and Riemann-Liouville fractional derivatives (see [17]). Let $\alpha \in (0,1)$, $\beta \in [0,1]$, $\gamma = \alpha + \beta(1-\alpha)$ and $f \in L^1(0,b)$, then the left-sided Hilfer derivative of order α and a type β of f can be expressed equivalently as

$$\partial_t^{\alpha,\beta} f(t) := \partial_t^\alpha \left(f(t) - \frac{\left(I_t^{1-\gamma} f\right)(0)}{\Gamma(\gamma)} t^{\gamma-1} \right), \quad t \in [0,b]. \tag{5.55}$$

A three-parameter Mittag-Leffler function is defined by the series expansion [18]:

$$E_{\alpha,\beta}^\rho(z) := \sum_{n=0}^\infty \frac{(\rho)_n z^n}{\Gamma(\alpha n + \beta) n!}, \quad z, \beta, \rho \in \mathbb{C}, \ R(\alpha) > 0, \tag{5.56}$$

where $(\rho)_n$ is the Pochhammer symbol defined by

$$\rho_0 = 1, \ (\rho)_n = \rho(\rho+1)\ldots(\rho+n-1) = \frac{\Gamma(\rho+n)}{\Gamma(\rho)}.$$

For $\rho = 1$, we recover the two-parametric Mittag-Leffler function

$$E_{\alpha,\beta}(z) := \sum_{n=0}^\infty \frac{z^n}{\Gamma(\alpha n + \beta)}, \quad z, \beta \in \mathbb{C}, \ \Re(\alpha) > 0$$

and for $\rho = \beta = 1$, we recover the classical Mittag-Leffler function

$$E_\alpha(z) := E_{\alpha,1}(z) = \sum_{n=0}^\infty \frac{z^n}{\Gamma(\alpha n + 1)}, \quad z \in \mathbb{C}, \ \Re(\alpha) > 0.$$

Let $0 < \alpha < 1$, $0 \le \beta \le 1$, $\nu, \mu, \gamma, \sigma, \rho > 0$, the following results hold:

$$I_t^\alpha \left[t^{\nu-1} E_{\mu,\nu}(\lambda t^\mu) \right] = t^{\nu+\alpha-1} E_{\mu,\nu+\alpha}(\lambda t^\mu), \ t > 0, \tag{5.57}$$

$$\partial_t^{\alpha,\beta}\left[t^{\upsilon-1}E_{\mu,\upsilon}\left(\lambda t^{\mu}\right)\right]=t^{\upsilon-\alpha-1}E_{\mu,\upsilon-\alpha}\left(\lambda t^{\mu}\right),\ t>0, \tag{5.58}$$

and

$$\int_0^t (t-s)^{\upsilon-1}E_{\mu,\upsilon}^{\gamma}\left(\lambda(t-s)^{\mu}\right)s^{\sigma-1}E_{\mu,\upsilon}^{\rho}\left(\lambda s^{\mu}\right)ds=t^{\upsilon+\sigma-1}E_{\mu,\upsilon+\sigma}^{\gamma+\rho}\left(\lambda t^{\mu}\right). \tag{5.59}$$

The solution of the linear Hilfer fractional differential equation with constant coefficient on a bounded interval can be obtained by the same technique used for the linear Riemann-Liouville fractional differential equation which is the method of successive approximations or the operational method (see [16, 19]).

Lemma 5.2 *The solution of the Cauchy-type problem for fractional differential problem of order* $0<\alpha<1$ *and type* $0\le\beta\le 1$ *with initial condition*

$$\begin{cases}\partial_t^{\alpha,\beta}y(t)=\lambda y(t)+f(t),\ t\in(0,T],\ \lambda\in R,\\ I_t^{1-\gamma}y\left(0^{+}\right)=b,\ \gamma=\alpha+\beta(1-\alpha),\ b\in R\end{cases}$$

for $t^{1-\gamma}f\in C[0,\ T]$ *is given by*

$$y(t)=t^{\gamma-1}E_{\alpha,\gamma}\left(\lambda t^{\alpha}\right)b+\int_0^t (t-s)^{\alpha-1}E_{\alpha,\alpha}\left(\lambda(t-s)^{\alpha}\right)f(s)ds,$$

with $t^{1-\gamma}y\in C[0,\ T]$.

Due to relation (5.55) and the linearity of fractional differential operators, we can extend Lemma 15.2 of [20] to the following lemma:

Lemma 5.3 *Let* $(f_i(t))_{i\ge 0}$ *be a sequence of functions defined on* $(0,T]$ *for each* $i\in N$. *Suppose the following conditions are fulfilled:*

I. the fractional derivative $\partial_t^{\alpha,\beta}f_i(t),\ i\ge 0$ *exists for all* $i\in N,\ t\in(0,T]$;
II. both series $\sum_{i=1}^{\infty}f_i(t)$ *and* $\sum_{i=1}^{\infty}\partial_t^{\alpha,\beta}f_i(t)$ *are uniformly convergent series on the interval* $[\varepsilon,T]$ *for any* $\varepsilon>0$.

Then,

$$\partial_t^{\alpha,\beta}\sum_{i=1}^{\infty}f_i(t)=\sum_{i=1}^{\infty}\partial_t^{\alpha,\beta}f_i(t),\ 0<\alpha\le\beta<1,\ 0<t<T.$$

In the next subsection, the existence and uniqueness of the solution of inverse problem (5.49)–(5.52) will be proved based on the fractional calculus and Fourier analysis method. Moreover, the continuous dependence of the solution upon the data of the inverse problem will be shown.

5.2.2 Well-Posedness of the Source Inverse Problem

Here, just like in Sect. 5.1, we use a Riesz basis (5.8) and a bi-orthogonal set of functions (5.9). These two sets of functions are constructed following the fact, that the operator of the spectral problem associated to (5.49)–(5.51) with the eigenvalues λ_k^2 is not self-adjoint. We denote the inner product in $L^2(0, 1)$ by

$$\langle f, g\rangle = \int_0^1 f(x)g(x)dx.$$

Now, let us assume the following assumptions on the data of the problem (5.49)–(5.52):

(A_1) $\varphi \in C^4[0, 1]$ such that $\psi(1) = 0,\ \psi'(0) = \psi'(1),\ \psi''(1) = 0,\ \psi^{(3)}(0) = \psi^{(3)}(1)$;

(A_2) The function $t^{1-\gamma}F(x, .) \in C[0, T]$ for $x \in [0, 1]$ and for $t \in [0, T],\ F(., t) \in C^4[0, 1]$ satisfies the following conditions:

1. $F(1, t) = 0,\ F_x(0, t) = F_x(1, t),\ F_{xx}(1) = 0,\ F_{xxxx}(0, t) = F_{xxxx}(1, t)$.
2. There exists a positive constant M_F such that $t^{1-\gamma}\left|\int_0^1 F(x, t)dx\right| > M_F$ for each $t \in [0, T]$;

(A_3) $t^{1-\gamma}E(t) \in C[0, T],\quad t^{1-\gamma}\partial_t^{\alpha,\beta}E(t) \in C[0, T]$ with $\int_0^1 \psi(x)dx = I_t^{1-\gamma}E(0)$.

From the properties of Mittag-Leffler function, there exists some positive constant M such that

$$M = \max\left(M_{\alpha,\gamma},\ M_{\alpha,\alpha+\gamma},\ M_{\alpha,2\alpha},\ M_{\alpha,\alpha}\right),$$

where

$$M_{\alpha,\gamma} := \max_{0\le t\le T}\left|E_{\alpha,\gamma}\left(-\lambda_k^2 t^\alpha\right)\right|,\quad M_{\alpha,\alpha+\gamma} := \max_{0\le t\le T}\left|E_{\alpha,\alpha+\gamma}^2\left(-\lambda_k^2 t^\alpha\right)\right|,$$

$$M_{\alpha,2\alpha} := \max_{0\le t\le T}\left|E_{\alpha,2\alpha}^2\left(-\lambda_k^2 (t-s)^\alpha\right)\right|,\quad M_{\alpha,\alpha} := \max_{0\le t\le T}\left|E_{\alpha,\alpha}\left(-\lambda_k^2 t^\alpha\right)\right|.$$

The following assertion is valid:

Lemma 5.4 *If conditions (A_1), (A_2) are satisfied, then for some positive constants C, G, we have*

$$\sum_{k=1}^{\infty}\lambda_k\left|\psi_{2k-1}\right| \le C\|\psi\|_{C^4[0,1]}; \tag{5.60}$$

$$\sum_{k=1}^{\infty}\lambda_k\left|F_{2k-1}(t)\right| \le G\ \|F(.,\ t)\|_{C^4[0,1]}. \tag{5.61}$$

Proof Since $\psi \in C^4[0, 1]$, by integration by parts we have

$$\psi_{2k-1} = \langle \psi(x), cos(\lambda_k x) \rangle = \frac{\left\langle \psi^{(4)}(x), cos(\lambda_k x) \right\rangle}{\lambda_k^4}$$

which yields, by the Cauchy-Schwartz and Bessel inequalities

$$\sum_{k=1}^{\infty} \lambda_k \left| \psi_{2k-1} \right| = \sum_{k=1}^{\infty} \frac{\left| \psi_{2k-1}^{(4)} \right|}{\lambda_k^3} \le C \| \psi \|_{C^4[0,1]}.$$

Similarly, for $F(., t) \in C^4[0, 1]$ we find (5.61).

Remark 5.1 In the same way, we can deduce uniform convergence of the series $\sum_{k=1}^{\infty} \lambda_k^2 |\psi_{2k}|$, $\sum_{k=1}^{\infty} \lambda_k^2 |F_{2k}(t)|$, $\sum_{k=1}^{\infty} \lambda_k^n |\psi_{2k-1}|$, $\sum_{k=1}^{\infty} \lambda_k^2 |F_{2k-1}(t)|$ *for* $n = 2, 3$.

Theorem 5.2 *Let conditions* (A_1)-(A_3) *be satisfied. Then, inverse problem* (5.49)–(5.52) *is well-posed.*

Proof We carry out the proof in three steps.

Step 1 (the existence). For arbitrary $a(t) \in C[0, T]$ and by applying the standard procedure of the Fourier method, the solution $u(x, t)$ of problem (5.49)–(5.51) can be expanded to uniformly convergent series form using the basis (5.8) in $L^2(0, 1)$ as follows:

$$u(x,t) = u_0(t)2(1-x) + \sum_{k=1}^{\infty} u_{2k-1}(t)4(1-x)\cos(\lambda_k x) + \sum_{k=1}^{\infty} u_{2k}(t)4\sin(\lambda_k x), \tag{5.62}$$

where the first coefficient

$$u_0(t) = \int_0^1 u(x,t)dx = E(t), \tag{5.63}$$

is known and for $k \ge 1$

$$u_{2k-1}(t) = \int_0^1 u(x,t)\cos(\lambda_k x)dx, \quad u_{2k}(t) = \int_0^1 u(x,t)x\sin(\lambda_n x)dx$$

have to be determined. These coefficients satisfy the following system of linear Cauchy-type problems

$$\partial_t^{\alpha,\beta} u_0(t) = a(t)F_0(t), \ I_{0+}^{1-\gamma} u_0(0) = \psi_0, \tag{5.64}$$

$$\partial_t^{\alpha,\beta} u_{2k-1}(t) + \lambda_k^2 u_{2k-1}(t) = a(t)F_{2k-1}(t), \ I_{0+}^{1-\gamma} u_{2k-1}(0) = \psi_{2k-1}, \tag{5.65}$$

$$\partial_t^{\alpha,\beta} u_{2k}(t) + \lambda_k^2 u_{2k}(t) - 2\lambda_k u_{2k-1}(t) = a(t)F_{2k}(t),\ I_{0+}^{1-\gamma} u_{2k}(0) = \psi_{2k}, \quad (5.66)$$

respectively, $F_k(t)$ and ψ_k, $k = 0, 1, 2, \ldots$ are the coefficients of the series expansion of $F(x, t)$ and $\psi(x)$ respectively, in the basis (5.9). In view of Lemma 5.2, the solutions of problems (5.65)–(5.66) satisfy the following integrals equations:

$$u_{2k-1}(t) = t^{\gamma-1} E_{\alpha,\gamma}\left(-\lambda_k^2 t^\alpha\right) \psi_{2k-1}$$
$$+ \int_0^t (t-s)^{\alpha-1} E_{\alpha,\alpha}\left(-\lambda_k^2 (t-s)^\alpha\right) a(s) F_{2k-1}(s) ds, \quad (5.67)$$

and

$$u_{2k}(t) = t^{\gamma-1} E_{\alpha,\gamma}\left(-\lambda_k^2 t^\alpha\right) \psi_{2k}$$
$$+ 2\lambda_k \int_0^t (t-s)^{\alpha-1} E_{\alpha,\alpha}\left(-\lambda_k^2 (t-s)^\alpha\right) u_{2k-1}(s) ds$$
$$+ \int_0^t (t-s)^{\alpha-1} E_{\alpha,\alpha}\left(-\lambda_k^2 (t-s)^\alpha\right) a(s) F_{2k}(s) ds, \quad (5.68)$$

respectively. By using (5.67), we can transform the first integral in (5.68) as follows

$$\int_0^t (t-s)^{\alpha-1} E_{\alpha,\alpha}\left(-\lambda_k^2 (t-s)^\alpha\right) u_{2k-1}(s) ds$$
$$= \psi_{2k-1} \int_0^t (t-s)^{\alpha-1} E_{\alpha,\alpha}\left(-\lambda_k^2 (t-s)^\alpha\right) s^{\gamma-1} E_{\alpha,\gamma}\left(-\lambda_k^2 s^\alpha\right) ds$$
$$+ \int_0^t (t-s)^{\alpha-1} E_{\alpha,\alpha}\left(-\lambda_k^2 (t-s)^\alpha\right) \left[\int_0^s (s-\tau)^{\gamma-1} \right.$$
$$\left. \times E_{\alpha,\alpha}\left(-\lambda_k^2 (s-\tau)^\alpha\right) a(\tau) F_{2k-1}(\tau) d\tau \right] ds$$
$$= \psi_{2k-1} \int_0^t (t-s)^{\alpha-1} E_{\alpha,\alpha}\left(-\lambda_k^2 (t-s)^\alpha\right) s^{\gamma-1} E_{\alpha,\gamma}\left(-\lambda_k^2 s^\alpha\right) ds$$
$$+ \int_0^t a(\tau) F_{2k-1}(\tau) \left[\int_\tau^t (t-s)^{\alpha-1} E_{\alpha,\alpha}\left(-\lambda_k^2 (t-s)^\alpha\right) (s-\tau)^{\alpha-1} \right.$$
$$\left. \times E_{\alpha,\alpha}\left(-\lambda_k^2 (t-s)^\alpha\right) \right] ds d\tau.$$

Then, taking into account (5.59), we represent (5.68) in the following form

$$u_{2k}(t) = t^{\gamma-1} E_{\alpha,\gamma}\left(-\lambda_k^2 t^\alpha\right) \psi_{2k} + 2\lambda_k t^{\alpha+\gamma-1} E^2_{\alpha,\alpha+\gamma}\left(-\lambda_k^2 t^\alpha\right) \psi_{2k-1} +$$

$$+ 2\lambda_k \int_0^t (t-s)^{2\alpha-1} E^2_{\alpha,2\alpha}\left(-\lambda_k^2 (t-s)^\alpha\right) a(s) F_{2k-1}(s) ds +$$

$$+ \int_0^t (t-s)^{\alpha-1} E_{\alpha,\alpha}\left(-\lambda_k^2 (t-s)^\alpha\right) a(s) F_{2k}(s) ds. \tag{5.69}$$

Next, from (5.63) and (5.64) we get

$$a(t) = \frac{\partial_t^{\alpha,\beta} u_0(t)}{\int_0^1 F(x,t)dx} = \frac{\partial_t^{\alpha,\beta} E(t)}{F_0(t)}. \tag{5.70}$$

Then, by (A_3) we have

$$\|a\|_{C[0,T]} \leq \frac{1}{M_F} \left\| t^{1-\gamma} \partial_t^{\alpha,\beta} E \right\|_{C[0,T]} < \infty, \tag{5.71}$$

so, we deduce that $a(t) \in C[0,\ T]$.

For the existence of a classical solution, we will show that the series corresponding to $t^{1-\gamma} u(x,t)$, $t^{1-\gamma} u_{xx}(x,t)$, $t^{1-\gamma} \partial_t^{\alpha,\beta} u(x,t)$ are uniformly convergent in $[0, 1] \times (\epsilon,\ T]$, for any $\epsilon > 0$. Before we proceed further, (5.63) implies that

$$t^{1-\gamma} |u_0(t)| \leq \left\| t^{1-\gamma} E \right\|_{C[0,T]}. \tag{5.72}$$

Next, from (5.67) and using (5.54), we obtain

$$t^{1-\gamma} |u_{2k-1}(t)| \leq M_{\alpha,\gamma} |\psi_{2k-1}|$$

$$+ t^{1-\gamma} M_{\alpha,\alpha} \|a\|_{C[0,T]} \int_0^t (t-s)^{\alpha-1} s^{1-\gamma} s^{\gamma-1} |F_{2k-1}(s)| ds \leq$$

$$\leq M |\psi_{2k-1}| + \frac{\Gamma(\alpha)\Gamma(\gamma)T^\alpha}{\Gamma(\alpha+\gamma)} M \|a\|_{C[0,T]} \left\| t^{1-\gamma} F_{2k-1} \right\|_{C[0,T]}.$$

Taking the sum of the last inequality and by Bessel's inequality, we find

$$t^{1-\gamma} \sum_{k=1}^{\infty} |u_{2k-1}(t)|$$

$$\leq M \|\psi\|_{C^4[0,1]} + \frac{\Gamma(\alpha)\Gamma(\gamma)T^\alpha}{\Gamma(\alpha+\gamma)} M \|a\|_{C[0,T]} \left\| t^{1-\gamma} F \right\|_{C([0,T];\, C^4[0,1])}. \tag{5.73}$$

The same for (5.69), according to (5.54) and Lemma 5.4, we find after summation

$$t^{1-\gamma}\sum_{k=1}^{\infty}|u_{2k}(t)| \le M_{\alpha,\gamma}\sum_{k=1}^{\infty}|\psi_{2k}(t)| + 2t^{\alpha}M_{\alpha,\alpha+\gamma}\sum_{k=1}^{\infty}\lambda_k\,|\psi_{2k-1}(t)| +$$

$$+\,2t^{1-\gamma}M_{\alpha,2\alpha}\,\|a\|_{C[0,T]}\int_0^t (t-s)^{\alpha-1}s^{1-\gamma}s^{\gamma-1}\sum_{k=1}^{\infty}\lambda_k\,|F_{2k-1}(s)|ds +$$

$$+\,t^{1-\gamma}M_{\alpha,\alpha}\,\|a\|_{C[0,T]}\int_0^t (t-s)^{\alpha-1}s^{1-\gamma}s^{\gamma-1}\sum_{k=1}^{\infty}|F_{2k}(s)|\,ds.$$

Thus,

$$t^{1-\gamma}\sum_{k=1}^{\infty}|u_{2k}(t)| \le L_{\psi}\,\|\psi\|_{C^4[0,1]} + L_F\,\|a\|_{C[0,T]}\left\|t^{1-\gamma}F\right\|_{C([0,T];\,C^4[0,1])}. \tag{5.74}$$

Since, the series expression (5.62) of the solution $u(x,\ t)$ gives

$$t^{1-\gamma}\,|u(x,t)| \le 2t^{1-\gamma}\,|u_0(t)| + 4t^{1-\gamma}\sum_{k=1}^{\infty}|u_{2k-1}(t)| + 4t^{1-\gamma}\sum_{k=1}^{\infty}|u_{2k}(t)|\,, \tag{5.75}$$

we conclude by (5.72)–(5.74) that the series $t^{1-\gamma}u(x,\ t)$ is uniformly convergent on $[0,\ 1]\ \times\ (\epsilon,\ T\ ]$ for any $\epsilon >\ 0$. Similarly, we can obtain the uniform convergence of $t^{1-\gamma}u_{xx}(x,\ t)$. From (5.63), (5.65) and (5.66), we get the following inequalities

$$t^{1-\gamma}\left|\partial_t^{\alpha,\beta}u_0(t)\right| \le \left\|t^{1-\gamma}\partial_t^{\alpha,\beta}E\right\|_{C[0,T]};$$

$$t^{1-\gamma}\left|\partial_t^{\alpha,\beta}u_{2k-1}(t)\right| \le t^{1-\gamma}\lambda_k^2\,|u_{2k-1}(t)| + \|a\|_{C[0,T]}\left\|t^{1-\gamma}F_{2k}\right\|_{C[0,T]};$$

$$t^{1-\gamma}\left|\partial_t^{\alpha,\beta}u_{2k}(t)\right| \le t^{1-\gamma}\lambda_k^2\,|u_{2k}(t)| + 2t^{1-\gamma}\lambda_k\,|u_{2k-1}(t)| + \|a\|_{C[0,T]}\left\|t^{1-\gamma}F_{2k-1}\right\|_{C[0,T]}.$$

Due to Lemma 5.4 and Remark 5.1, we deduce the uniform convergence of the series

$$t^{1-\gamma}\sum_{k=1}^{\infty}\lambda_k\,|u_{2k-1}(t)|,\ \ t^{1-\gamma}\sum_{k=1}^{\infty}\lambda_k^2\,|u_{2k-1}(t)|,\ \ t^{1-\gamma}\sum_{k=1}^{\infty}\lambda_k^2\,|u_{2k}(t)|,$$

the Weierstrass M-test will ensure the uniform convergence of the series

$$t^{1-\gamma}\sum_{k=1}^{\infty}\left|\partial_t^{\alpha,\beta}u_{2k-1}(t)\right|,\ t^{1-\gamma}\sum_{k=1}^{\infty}\left|\partial_t^{\alpha,\beta}u_{2k}(t)\right|$$

on $(\epsilon,\ T]$, for each $\epsilon > 0$.

In view of (5.75) and Lemma 5.3, the series corresponding to $t^{1-\gamma}\left|\partial_t^{\alpha,\beta}u(x,t)\right|$ is uniformly convergent.

Now, recall that we have for $t = 0$

$$I_t^{1-\gamma}u(x,0) = \lim_{t\to 0^+} I_t^{1-\gamma}u(x,t).$$

From (5.63) and (A_3), we get

$$I_t^{1-\gamma}u_0(0) = I_t^{1-\gamma}E(0) = \psi_0.$$

The same, by using (5.57) we obtain

$$I_t^{1-\gamma}u_{2k-1}(0) = \psi_{2k-1},\ \ I_t^{1-\gamma}u_{2k}(0) = \psi_{2k}.$$

Therefore, we deduce that

$$\begin{aligned}&I_t^{1-\gamma}u(x,0)\\ =&\psi_0 2(1-x)+\sum_{k=1}^{\infty}\psi_{2k-1}4(1-x)\cos(\lambda_k x)+\sum_{k=1}^{\infty}\psi_{2k}4\sin(\lambda_k x)=\psi(x).\end{aligned}$$

Step 2 (the uniqueness). To prove the uniqueness of the solution, we suppose that $\{u_1(x,\ t),\ a_1(t)\}$ and $\{u_2(x,\ t),\ a_2(t)\}$ are two solution sets of inverse problem (5.49)–(5.52).

Let $\overline{u}(x,\ t) = u_1(x,\ t) - u_2(x,\ t)$ and $\overline{a}(t) = a_1(t) - a_2(t)$, then $\overline{u}(x,\ t)$ satisfies the problem

$$\partial_t^{\alpha,\beta}\overline{u}(x,t) = \overline{u}_{xx}(x,t) + \overline{a}(t)F(x,t),\ 0 < x < 1, 0 < t < T, \tag{5.76}$$

$$I_t^{1-\gamma}\overline{u}(x,0) = 0,\ \gamma = \alpha + \beta(1-\alpha),\ 0 \le x \le 1, \tag{5.77}$$

$$\overline{u}(1,t) = 0,\ \overline{u}_x(0,t) = \overline{u}_x(1,t),\ 0 < t \le T, \tag{5.78}$$

and the integral overdetermination data

$$\int_0^1 \overline{u}(x,t)dx = 0,\ t \in [0,T]. \tag{5.79}$$

Using the basis (5.8), we can write

$$\overline{u}(x,t) = \overline{u}_0(t)2(1-x) + \sum_{k=1}^{\infty} \overline{u}_{2k-1}(t)4(1-x)\cos(\lambda_k x) + \sum_{k=1}^{\infty} \overline{u}_{2k}(t)4\sin(\lambda_k x),$$

where the coefficient $\overline{u}_0(t) = 0, \overline{u}_{2k-1}(t)$ and $\overline{u}_{2k}(t)$ are defined by the formulas (5.67) and (5.69) respectively with $\overline{a}(t)$ instead of $a(t)$, according to condition (5.77), we get

$$\psi_0 = \psi_{2k-1} = \psi_{2k} = 0, \; k \geq 1.$$

Consequently, $\overline{u}_{2k-1}(t) = \overline{u}_{2k}(t) = 0$ and by virtue of (5.70) and (5.79) we deduce that $\overline{a}(t) = 0$, therefore we conclude the uniqueness of the solution of the inverse problem.

Step 3 (the stability). Next, to show the continuous dependence of the solution upon the data of the inverse problem, let $\{u(x,\ t),\ a(t)\}$ and $\{\overline{u}(x,\ t),\ \overline{a}(t)\}$ be two solutions of inverse problem (5.49)–(5.52), corresponding to two sets of the data $\Phi(t) = \{\psi(x),\ F(x,\ t),\ E(t)\}$ and $\overline{\Phi}(t) = \left\{\overline{\psi}(x),\ \overline{F}(x,\ t),\ \overline{E}(t)\right\}$ respectively. Denote

$$\|\Phi\| = \|\psi\|_{C^4[0,1]} + \left\|t^{1-\gamma}E\right\|_* + \left\|t^{1-\gamma}F\right\|_{C([0,T];\, C^4[0,1])},$$

where

$$\left\|t^{1-\gamma}E\right\|_* = \max\left(\left\|t^{1-\gamma}E\right\|_{C[0,T]},\ \left\|t^{1-\gamma}\partial_t^{\alpha,\gamma}E\right\|_{C[0,T]}\right). \tag{5.80}$$

First, in view of (5.63) we have

$$t^{1-\gamma}\left|u_0(t) - \overline{u}_0(t)\right| \leq \left\|t^{1-\gamma}\left(E - \overline{E}\right)\right\|_{C[0,T]}.$$

Next, by similar previous computations, we have from (5.67) for $k \geq 1$

$$\begin{gathered} t^{1-\gamma}\left|u_{2k-1}(t) - \overline{u}_{2k-1}(t)\right| \leq M_{\alpha,\gamma}\left|\psi_{2k-1} - \overline{\psi}_{2k-1}\right| + \\ + t^{1-\gamma}M_{\alpha,\alpha}\|a - \overline{a}\|_{C[0,T]}\int_0^t (t-s)^{\alpha-1}s^{1-\gamma}s^{\gamma-1}\left|F_{2k-1}(s)\right|ds + \\ + t^{1-\gamma}M_{\alpha,\alpha}\|\overline{a}\|_{C[0,T]}\int_0^t (t-s)^{\alpha-1}s^{1-\gamma}s^{\gamma-1}\left|F_{2k-1}(s) - \overline{F}_{2k-1}(s)\right|ds \leq \end{gathered}$$

$$\begin{aligned} \leq M\left|\psi_{2k-1} - \overline{\psi}_{2k-1}\right| + \frac{\Gamma(\alpha)\Gamma(\gamma)T^\alpha}{\Gamma(\alpha+\gamma)}M\|a - \overline{a}\|_{C[0,T]}\left\|t^{1-\gamma}F_{2k-1}\right\|_{C[0,T]} + \\ + \frac{\Gamma(\alpha)\Gamma(\gamma)T^\alpha}{\Gamma(\alpha+\gamma)}M\|\overline{a}\|_{C[0,T]}\left\|t^{1-\gamma}\left(F_{2k-1} - \overline{F}_{2k-1}\right)\right\|_{C[0,T]}. \end{aligned}$$

Then, there exist positive constants $\beta_i,\ i = 1,\ 2,\ 3$ such that

$$t^{1-\gamma}\sum_{k=1}^{\infty}|u_{2k-1}(t)-\overline{u}_{2k-1}(t)|=$$
$$\beta_1\left\|\psi-\overline{\psi}\right\|_{C^4[0,1]}+\beta_2\left\|a-\overline{a}\right\|_{C[0,T]}+\beta_3\left\|t^{1-\gamma}\left(F-\overline{F}\right)\right\|_{C([0,T];\,C^4[0,1])}.$$

By the same procedure, for some positive constants $\beta_i,\ i\ =\ 4,\ 5,\ 6$, we have from (5.69)

$$t^{1-\gamma}\sum_{k=1}^{\infty}|u_{2k}(t)-\overline{u}_{2k}(t)|=$$
$$=\beta_4\left\|\psi-\overline{\psi}\right\|_{C^4[0,1]}+\beta_5\left\|a-\overline{a}\right\|_{C[0,T]}+\beta_6\left\|t^{1-\gamma}\left(F-\overline{F}\right)\right\|_{C([0,T];\,C^4[0,1])}.$$

Thus, we deduce for some constants $\rho_i>\ 0,\ i\ =\ 1,\ 2,\ 3$

$$t^{1-\gamma}\left|u(x,t)-\overline{u}(x,t)\right|\leq\left\|t^{1-\gamma}\left(E-\overline{E}\right)\right\|_{C[0,T]}+\rho_1\left\|\psi-\overline{\psi}\right\|_{C^4[0,1]}+$$
$$+\rho_2\left\|a-\overline{a}\right\|_{C[0,T]}+\rho_3\left\|t^{1-\gamma}\left(F-\overline{F}\right)\right\|_{C([0,T];C^4[0,1])}.$$

Consequently, we get

$$\left\|t^{1-\gamma}\left(u-\overline{u}\right)\right\|_{C([0,T]\times[0,1])}\leq\Pi_0\left\|\Phi-\overline{\Phi}\right\|+\rho_2\left\|a-\overline{a}\right\|_{C[0,T]},\tag{5.81}$$

where $\Pi_0=\max\left(1,\ \rho_1,\ \rho_3\right)$. On the other hand, from (5.70) we have

$$|a-\overline{a}|=\left|\frac{\partial_t^{\alpha,\beta}E(t)}{\int_0^1 F(x,t)dx}-\frac{\partial_t^{\alpha,\beta}\overline{E}(t)}{\int_0^1\overline{F}(x,t)dx}\right|\leq$$

$$\leq\frac{1}{M_F^2}\left(\left\|t^{1-\gamma}\partial_t^{\alpha,\beta}\left(E-\overline{E}\right)\right\|_{C[0,T]}\left\|t^{1-\gamma}\overline{F}\right\|_{C([0,T];\,C^4[0,1])}+\right.$$
$$\left.\left\|t^{1-\gamma}\partial_t^{\alpha,\beta}\overline{E}\right\|_{C[0,T]}\left\|t^{1-\gamma}\left(F-\overline{F}\right)\right\|_{C([0,T];\,C^4[0,1])}\right).$$

Then, from (5.80) we obtain for some positive constants $\rho_i,\ i\ =\ 4,\ 5$

$$\left\|a(t)-\overline{a}(t)\right\|_{C[0,T]}\leq\rho_4\left\|t^{1-\gamma}\left(E-\overline{E}\right)\right\|_*+\rho_5\left\|t^{1-\gamma}\left(F-\overline{F}\right)\right\|_{C([0,T];\,C^4[0,1])}.$$

Therefore, we find

$$\left\|a-\overline{a}\right\|_{C[0,T]}\leq\Pi_1\left\|\Phi-\overline{\Phi}\right\|,\quad\Pi_1=\ max(\rho_4,\ \rho_5).$$

In view of (5.81), we get

$$\left\|t^{1-\gamma}\left(u-\overline{u}\right)\right\|_{C([0,T]\times[0,1])}\leq\Pi_2\left\|\Phi-\overline{\Phi}\right\|$$

for positive constant $\Pi_2 = max(\Pi_0, \rho_2, \Pi_1)$.

Hence, $\left\|t^{1-\gamma}(u-\overline{u})\right\|_{C([0,T]\times[0,1])} \to 0$ and $\|a-\overline{a}\|_{C[0,T]} \to 0$, when $\left\|\Phi-\overline{\Phi}\right\| \to 0$. This induces the desired result and completes the proof. □

5.2.3 Examples for Illustration

Example 1. Consider problem (5.49)–(5.52) with $0 < \alpha + \gamma \le \mu < 1$ and

$$\psi(x) = \sin(2\pi x),\ \ F(x,\ t) = 2t^{\gamma-1}(1-x),\ \ E(t) = t^{\mu-1} = u_0(t).$$

Then, we get

$$\psi_0 = 0,\ \ F_0(t) = t^{\gamma-1},\ \ \partial_t^{\alpha,\beta} E(t) = \frac{\Gamma(\mu)}{\Gamma(\mu-\alpha)} t^{\mu-\alpha-1}$$

and

$$\psi_{2k-1} = 0,\ \ \psi_{2k} = \begin{cases} \frac{1}{4}, & k=1; \\ 0, & k \ne 1 \end{cases},\ \ F_{2k-1}(t) = F_{2k}(t) = 0,\ \ k = 1, 2, \ldots$$

- Using (5.67)–(5.69), we obtain

$$u_{2k-1} = 0,\ \ u_{2k} = t^{1-\gamma} E_{\alpha,\gamma}\left(-\lambda_k^2 t^{\alpha}\right)\psi_{2k},\ k = 1, 2, \ldots.$$

Hence, by (5.62)–(5.70), we obtain the solution of this inverse problem

$$\left\{u(x,t) = 2t^{\mu-1}(1-x) + t^{1-\gamma} E_{\alpha,\gamma}\left(-\lambda_k^2 t^{\alpha}\right)\sin(2\pi x),\ a(t) = \frac{\Gamma(\mu)}{\Gamma(\mu-\alpha)} t^{\mu-\alpha-1}\right\}.$$

Example 2. Consider problem (5.49)–(5.52) with $0 < \alpha + \gamma \le \mu < 1$, and

$$\psi(x) = 0,\ \ F(x,t) = t^{\mu-\alpha-1}(1-x),\ \ E(t) = \frac{t^{\mu-1}}{2} E_{\alpha,\mu}(t^{\alpha}) = u_0(t).$$

Clearly, we get

$$\psi_0 = \psi_{2k-1} = \psi_{2k} = 0,\ \ k = 1, 2, \ldots$$

and $F_0(t) = \frac{1}{2} t^{\mu-\alpha-1}$ and $F_{2k-1}(t) = F_{2k}(t) = 0,\ k = 1,\ 2, \ldots.$

In the light of formula (5.58), we have

$$\partial_t^{\alpha,\beta} E(t) = \frac{t^{\mu-\alpha-1}}{2} E_{\alpha,\mu-\alpha}(t^{\alpha}).$$

By using (5.67)–(5.69), we obtain

$$u_{2k-1}(t) = u_{2k}(t) = 0, \ k = 1, \ 2, \dots$$

Hence, from (5.62)–(5.70), the solution is this pair of functions

$$\left\{u(x,t) = (1-x)t^{\mu-1}E_{\alpha,\mu}\left(t^{\alpha}\right), \ a(t) = E_{\alpha,\mu-\alpha}\left(t^{\alpha}\right)\right\}.$$

We emphasize that in [26, 27, 29, 30], inverse source problems involving Hilfer fractional derivative in diffusion (and wave) equations with the other type boundary and overdetermination conditions are discussed.

5.3 Inverse Problems for a Fractional Differential Equation with Bessel Operator

In this section, inverse initial and inverse source problems of a time-fractional differential equation with Bessel operator are considered. Results on existence and uniqueness of solutions to these problems are presented. The solution method is based on series expansions using a set of Bessel functions of order zero. The results of this section correspond to the work [21].

5.3.1 Inverse Initial Problem for the Fractional Diffusion Equation

Here, we consider the following inverse initial problem: Find a pair of functions $\{u(x,t), g(x)\}$, in the domain Q_T, which satisfies the fractional differential equation

$$\partial_t^{\alpha} u = \frac{\partial^2 u}{\partial x^2} + \frac{1}{x}\frac{\partial u}{\partial x}, \tag{5.82}$$

along with the boundary conditions

$$\lim_{x\to 0} x\frac{\partial u}{\partial x} = 0, \quad u(1,t) = 0, \tag{5.83}$$

the initial condition

$$u(x,0) = g(x), \tag{5.84}$$

and the overdetermining condition

$$u(x,T) = f(x). \tag{5.85}$$

Here $f(x)$ is a given function and ∂_t^α stands for the Caputo fractional derivative of order $0 < \alpha < 1$.

In studying the problem under consideration, we need Bessel's function and the conditions for the convergence of the Fourier-Bessel series. Linear differential Bessel equation (or equation of cylindrical functions) with parameter λ of order or index $\nu \geq 0$ with respect to the function z of the real variable x has the form ([22], ch. 8)

$$z'' + \frac{1}{x}z' + \left(\lambda^2 - \frac{\nu^2}{x^2}\right) z = 0. \tag{5.86}$$

Solution of equation (5.86), except for very particular values ν, not expressed in terms of elementary functions (in the final form) and leads to the so-called Bessel functions. When ν is an integer number, then equation (5.86) has the following solution:

$$z(x) = C_1 J_\nu(\lambda x) + C_2 Y_\nu(\lambda x),$$

where J_ν and Y_ν are the Bessel functions of the first and second kind of order ν, respectively. Bessel functions of the second kind are not bounded near the point $x = 0$, so for a bounded solution near zero it is necessary that $C_2 = 0$, i.e., solution (5.86) has the following form:

$$z(x) = C J_\nu(\lambda x).$$

Moreover, if the boundary condition $z(1) = 0$ is imposed, then the parameter λ must satisfy $J_\nu(\lambda) = 0$, i.e., the values of λ are the zeros of the Bessel function $J_\nu(x)$, which has the following asymptotic representation ([22], p. 213)

$$J_\nu(x) = \sqrt{\frac{2}{\pi x}} sin\left(x - \frac{\nu\pi}{2} + \frac{\pi}{4}\right) + \frac{r_\nu(x)}{x\sqrt{x}},$$

where the function $r_\nu(x)$ is bounded for $x \to \infty$. Therefore, for any large k, the zeros of $J_\nu(x)$ are given by the expression ([22], p. 214)

$$k\pi + \frac{\nu\pi}{2} - \frac{\pi}{4}.$$

We define the Fourier-Bessel expansion of the given function $f(x)$ as follows: for any function $f(x)$, absolutely integrable on $[0, 1]$, one can compose a Fourier series in the system $J_\nu(\lambda_k x)$, $k = 1, 2, \ldots,$ or, in briefly, the Fourier-Bessel series

$$f(x) = \sum_{k=1}^{\infty} c_k J_\nu(\lambda_k x), \tag{5.87}$$

where the constants c_k are determined by the formula:

$$c_k = \frac{2}{J_{\nu+1}^2(\lambda_k)} \int_0^1 x f(x) J_\nu(\lambda_k x) dx, k = 1, 2, \ldots$$

and are called the Fourier-Bessel coefficients.

Let us give without proof the most important criteria for the convergence of the Fourier-Bessel series to the function for which it is composed. These criteria are similar to the ones known to us for the convergence of trigonometric Fourier series.

Theorem 5.3 ([22, p. 225]) *If* $\nu \geq 0$ *and for all sufficiently large* k*, we have the estimate*

$$|c_k| \leq \frac{M}{\lambda_k^{1+\varepsilon}},$$

where $\varepsilon > 0$ *and* $M > 0$ *are constants, then series* (5.87) *converges absolutely and uniformly on* $[0, 1]$.

Theorem 5.4 ([22, pp. 230–232]) *Let the function* $f(x)$ *is defined and* $2s$ *times continuously differentiable on the interval* $[0, 1]$ $(s \geq 1)$*, and*

(1) $f(0) = f'(0) = \ldots = f^{(2s-1)}(0) = 0,$
(2) $f^{(2s)}(x)$ *is bounded (this derivative may not exist at some points),*
(3) $f(1) = f'(1) = \ldots = f^{(2s-2)}(1) = 0.$

Then, for the Fourier-Bessel coefficients of the function $f(x)$*, the inequality*

$$|c_k| \leq \frac{M}{\lambda_k^{2s-(1/2)}} \quad (M = const) \tag{5.88}$$

is valid.

5.3.2 Solution Method: Fourier-Bessel Series

Using the separation of variables method, we obtain, from (5.82) and (5.83), the following spectral problem

$$X'' + \frac{1}{x}X' + \lambda^2 X = 0, \quad 0 < x < 1, \tag{5.89}$$

$$\lim_{x \to 0} xX'(x) = 0, \quad X(1) = 0, \tag{5.90}$$

which is a self-adjoint problem. Note that the above spectral equation (5.89) represents Bessel's equation of order zero and hence one can show that the eigenfunctions, X_k, for the problem (5.89), (5.90) are the zero-order Bessel functions of the first kind, i.e.,

$$X_k = J_0(\lambda_k x), \quad k = 1, 2, 3, \ldots.$$

The corresponding eigenvalues, λ_k, are the positive zero of $J_0(x)$, i.e., they satisfy $J_0(\lambda_k) = 0$ and for large k, they are given by

$$\lambda_k = k\pi - \frac{\pi}{4}$$

as discussed earlier. The set of eigenfunctions $\{X_k\}$ forms a complete orthogonal basis in $L^2(0, 1)$ ([23], p. 40). Hence, the functions $u(x, t)$, $g(x)$ and $f(x)$ can be expanded as follows:

$$u(x,t) = \sum_{k=1}^{\infty} u_k(t) J_0(\lambda_k x), \tag{5.91}$$

$$g(x) = \sum_{k=1}^{\infty} g_k J_0(\lambda_k x), \quad f(x) = \sum_{k=1}^{\infty} f_k J_0(\lambda_k x), \tag{5.92}$$

where $u_k(t)$, g_k are unknowns and the coefficient f_k is given by

$$f_k = \frac{2}{J_1^2(\lambda_k)} \int_0^1 x f(x) J_0(\lambda_k x) dx. \tag{5.93}$$

Substituting (5.91) and (5.92) into (5.82), (5.84) and (5.85) gives the following linear fractional differential equation:

$$\partial_t^\alpha u_k(t) + \lambda_k^2 u_k(t) = 0, \tag{5.94}$$

along with the conditions:

$$u_k(0) = g_k, \quad u_k(T) = f_k.$$

Applying Laplace transform to equation (5.94), we obtain

$$U_k(s) = \frac{s^{\alpha-1}}{s^\alpha + \lambda_k^2} g_k,$$

and consequently, we get

$$u_k(t) = g_k E_\alpha(-\lambda_k^2 t^\alpha).$$

Next, using the condition $u_k(T) = f_k$, we get

$$g_k = \frac{f_k}{E_\alpha(-\lambda_k^2 T^\alpha)}.$$

Therefore, $u(x,t)$ and $g(x)$ can now be written as

$$u(x,t) = \sum_{k=1}^{\infty} \frac{f_k}{E_\alpha(-\lambda_k^2 T^\alpha)} E_\alpha(-\lambda_k^2 t^\alpha) J_0(\lambda_k x), \tag{5.95}$$

$$g(x) = \sum_{k=1}^{\infty} \frac{f_k}{E_\alpha(-\lambda_k^2 T^\alpha)} J_0(\lambda_k x). \tag{5.96}$$

To complete the existence of solution, we need to show that the series representations of $u(x,t)$, $g(x)$, $\partial_t^\alpha u(x,t)$ and $u_{xx}(x,t)$ converge uniformly. We start by estimating the coefficient of the Fourier-Bessel series of $u(x,t)$. Since $E_\alpha(-\lambda t^\alpha)$ is bounded, we have

$$\left| \frac{f_k E_\alpha(-\lambda_k^2 t^\alpha)}{E_\alpha(-\lambda_k^2 T^\alpha)} \right| \leq M|f_k|,$$

whereupon using (5.88) for $s = 1$, we obtain

$$\left| \frac{f_k E_\alpha(-\lambda_k^2 t^\alpha)}{E_\alpha(-\lambda_k^2 T^\alpha)} \right| \leq \frac{M}{\lambda_k^{3/2}}.$$

Therefore, by Theorem 5.3, we conclude that the series representation of $u(x,t)$ converges absolutely and uniformly in Ω. Similarly, one can show that the series

$$g(x) = \sum_{k=1}^{\infty} \frac{f_k}{E_\alpha(-\lambda_k^2 T^\alpha)} J_0(\lambda_k x)$$

converges absolutely and uniformly on $[0, 1]$.

Now, the series expansion of $\partial_t^\alpha u(x,t)$ is given by

$$\partial_t^\alpha u(x,t) = \sum_{k=1}^{\infty} \frac{-\lambda_k^2 f_k}{E_\alpha(-\lambda_k^2 T^\alpha)} E_\alpha(-\lambda_k^2 t^\alpha) J_0(\lambda_k x),$$

Then, for convergence we have the following estimate

$$\left| \frac{\lambda_k^2 f_k}{E_\alpha(-\lambda_k^2 T^\alpha)} E_\alpha(-\lambda_k^2 t^\alpha) \right| \leq M\lambda_k^2 |f_k|,$$

and by using Theorem 5.4 for $s = 2$, we get

$$M\lambda_k^2 |f_k| \leq \frac{M\lambda_k^2}{\lambda_k^{7/2}} = \frac{M}{\lambda_k^{3/2}}.$$

Therefore, the uniform and absolute convergence of the series expansion of $\partial_t^\alpha u(x,t)$ is ensured by Theorem 5.3. Finally, it remains to show the uniform convergence of the series representation of $u_{xx}(x,t)$, which is given by

$$u_{xx}(x,t) = \sum_{k=1}^{\infty} \frac{f_k}{E_\alpha(-\lambda_k^2 T^\alpha)} E_\alpha(-\lambda_k^2 t^\alpha) \frac{d^2}{dx^2} \left(J_0(\lambda_k x)\right).$$

Using properties of Bessel functions, namely, (see [22])

$$J_0'(x) = -J_1(x),$$

$$2J_1'(x) = J_0(x) - J_2(x),$$

the above representation can be rewritten as

$$u_{xx}(x,t) = \sum_{k=1}^{\infty} \frac{f_k}{2E_\alpha(-\lambda_k^2 T^\alpha)} E_\alpha(-\lambda_k^2 t^\alpha) \left[J_2(\lambda_k x) - J_0(\lambda_k x)\right],$$

and hence, for convergence, we have the following estimate:

$$\left|\lambda_k^2 f_k \frac{E_\alpha(-\lambda_k^2 t^\alpha)}{2E_\alpha(-\lambda_k^2 T^\alpha)}\right| \le M\lambda_k^2 |f_k| \le \frac{M\lambda_k^2}{\lambda_k^{7/2}} = \frac{M}{\lambda_k^{3/2}}.$$

Therefore, by Theorem 5.3, the series expansion of $u_{xx}(x,t)$ is uniformly and absolutely convergent in Ω. Note that the uniform convergence of $u_x(x,t)$ follows from the convergence $u_{xx}(x,t)$.

Uniqueness of solution

Suppose $\{u_1(x,t), g_1(x)\}$ and $\{u_2(x,t), g_2(x)\}$ are two solution sets of the inverse problem (5.82)–(5.85), then $\widehat{u}(x,t) = u_1(x,t) - u_2(x,t)$ and $\widehat{g}(x) = g_1(x) - g_2(x)$ satisfy the following boundary value problem:

$$\partial_t^\alpha \widehat{u} = \frac{\partial^2 \widehat{u}}{\partial x^2} + \frac{1}{x}\frac{\partial \widehat{u}}{\partial x}, \quad (x,t) \in \Omega_T, \tag{5.97}$$

$$\lim_{x\to 0} x\frac{\partial \widehat{u}}{\partial x} = 0, \quad \widehat{u}(1,t) = 0, \quad 0 \le t \le T, \tag{5.98}$$

$$\widehat{u}(x,0) = \widehat{g}(x), \quad \widehat{u}(x,T) = 0, \quad 0 \le x \le 1. \tag{5.99}$$

Using the completeness property of system (5.89), one can then show that $\widehat{u}(x,t) = 0$ and $\widehat{g}(x) = 0$, in Ω_T, which in turn implies the uniqueness of solution of the inverse problem (5.82)–(5.85).

The main result of this section on existence and uniqueness of solution can be summarized in the following theorem:

Theorem 5.5 *Assume that the function f is differentiable 4 times such that:*

- $f(0) = f'(0) = f''(0) = f'''(0) = 0$;
- $f(1) = f'(1) = f''(1) = 0$;
- $f^{(4)}(x)$ *is bounded;*

then the inverse problem

$$\partial_t^\alpha u = \frac{\partial^2 u}{\partial x^2} + \frac{1}{x}\frac{\partial u}{\partial x}, \qquad (x,t) \in \Omega_T,$$

$$\lim_{x\to 0} x\frac{\partial u}{\partial x} = 0, \quad u(1,t) = 0, \quad 0 \le t \le T,$$

$$u(x,0) = g(x), \quad u(x,T) = f(x), \quad 0 \le x \le 1$$

has a unique solution given by

$$u(x,t) = \sum_{k=1}^{\infty} \frac{f_k}{E_\alpha(-\lambda_k^2 T^\alpha)} E_\alpha(-\lambda_k^2 t^\alpha) J_0(\lambda_k x),$$

$$g(x) = \sum_{k=1}^{\infty} \frac{f_k}{E_\alpha(-\lambda_k^2 T^\alpha)} J_0(\lambda_k x),$$

where

$$f_k = \frac{2}{J_1^2(\lambda_k)} \int_0^1 x f(x) J_0(\lambda_k x) dx.$$

5.3.3 Inverse Source Problem for the Fractional Diffusion Equation

In this subsection, we consider the following inverse source problem:
Find a pair of functions $\{u(x,t), h(x)\}$, in the domain Ω_T, which satisfies the boundary value problem:

$$\partial_t^\alpha u = \frac{\partial^2 u}{\partial x^2} + \frac{1}{x}\frac{\partial u}{\partial x} + h(x), \tag{5.100}$$

$$\lim_{x\to 0} x\frac{\partial u}{\partial x} = 0, \quad u(1,t) = 0, \tag{5.101}$$

$$u(x,0) = g(x), \quad u(x,T) = f(x). \tag{5.102}$$

Here $g(x)$ and $f(x)$ are given functions and ∂_t^α as usual stands for the Caputo fractional derivative of order $0 < \alpha < 1$.

Existence of solution

Using the solution method described in the previous section, we seek solutions of the form

$$u(x,t) = \sum_{k=1}^{\infty} u_k(t) J_0(\lambda_k x), \quad h(x) = \sum_{k=1}^{\infty} h_k J_0(\lambda_k x),$$

where the coefficients $u_k(t)$ and h_k are the unknowns to be determined. Substituting these representations into (5.100) and (5.102), we obtain the following nonhomogeneous fractional differential equation:

$$\partial_t^\alpha u_k(t) + \lambda_k^2 u_k(t) = h_k, \tag{5.103}$$

with the boundary conditions:

$$u_k(0) = g_k, \quad u_k(T) = f_k, \tag{5.104}$$

where g_k and f_k are the coefficients of Fourier-Bessel series for $g(x)$ and $f(x)$, respectively, which are given by

$$I_k = \frac{2}{J_1^2(\lambda_k)} \int_0^1 x I(x) J_0(\lambda_k x) dx, \quad I(x) = g(x),\ f(x).$$

Solving (5.103) we obtain ([2], p. 243)

$$u_k(t) = C_k E_\alpha(-\lambda_k^2 t^\alpha) + \frac{h_k}{\lambda_k^2},$$

and using the conditions (5.104), we deduce the following:

$$C_k = \frac{g_k - f_k}{1 - E_\alpha(-\lambda_k^2 T^\alpha)}, \quad h_k = \lambda_k^2 (g_k - C_k).$$

Consequently, we have

$$u(x,t) = \sum_{k=1}^{\infty} \left(C_k E_\alpha(-\lambda_k^2 t^\alpha) + g_k - C_k\right) J_0(\lambda_k x) =$$

$$= g(x) - \sum_{k=1}^{\infty} \frac{1 - E_\alpha(-\lambda_k^2 t^\alpha)}{1 - E_\alpha(-\lambda_k^2 T^\alpha)} (g_k - f_k) J_0(\lambda_k x),$$

and

$$h(x) = \sum_{k=1}^{\infty} \lambda_k^2 (g_k - C_K) J_0(\lambda_k x) = g''(x) - \sum_{k=1}^{\infty} \frac{\lambda_k^2 (g_k - f_k)}{1 - E_\alpha(-\lambda_k^2 T^\alpha)} J_0(\lambda_k x).$$

Using appropriate conditions on $f(x)$ and $g(x)$, as given in the theorem below, the convergence of the series expansions of $u(x,t)$, $u_x(x,t)$, $u_{xx}(x,t)$, $\partial_t^\alpha u(x,t)$ and $h(x)$ can be shown in a similar approach as discussed in the inverse initial problem in the previous section.

The following theorem summarizes the main result of this section:

Theorem 5.6 *Assume that the functions g and f is differentiable* 4 *times such that*

- $g^{(i)}(0) = f^{(i)}(0) = 0, \quad (i = 0, 1, 2, 3)$;
- $g^{(j)}(1) = f^{(j)}(1) = 0, \quad (j = 0, 1, 2)$;
- $g^{(4)}(x)$ *and* $f^{(4)}(x)$ *are bounded.*

Then the inverse problem

$$\partial_t^\alpha u = \frac{\partial^2 u}{\partial x^2} + \frac{1}{x}\frac{\partial u}{\partial x} + h(x), \quad 0 < x < 1, \ 0 < t < T,$$

$$\lim_{x \to 0} x \frac{\partial u}{\partial x} = 0, \quad u(1,t) = 0, \quad 0 < t < T,$$

$$u(x,0) = g(x), \quad u(x,T) = f(x), \quad 0 < x < 1,$$

has a unique solution given by

$$u(x,t) = g(x) - \sum_{k=1}^{\infty} \frac{1 - E_\alpha(-\lambda_k^2 t^\alpha)}{1 - E_\alpha(-\lambda_k^2 T^\alpha)} (g_k - f_k) J_0(\lambda_k x),$$

$$h(x) = g''(x) - \sum_{k=1}^{\infty} \frac{\lambda_k^2 (g_k - f_k)}{1 - E_\alpha(-\lambda_k^2 T^\alpha)} J_0(\lambda_k x),$$

where f_k *and* g_k *are given by*

$$f_k = \frac{2}{J_1^2(\lambda_k)} \int_0^1 x f(x) J_0(\lambda_k x) dx \quad and \quad g_k = \frac{2}{J_1^2(\lambda_k)} \int_0^1 x g(x) J_0(\lambda_k x) dx.$$

5.4 Inverse Source Problem for a Mixed Type Equation Involving Time-Fractional Derivative

This section is devoted to the study of inverse source problem for a mixed type equation with a fractional diffusion equation in the parabolic part and a wave equation in the hyperbolic part of a cylindrical domain. Solution is obtained in the form of Fourier-Bessel series expansion using an orthogonal set of Bessel functions. Thus, the theorems of uniqueness and existence of a solution are proved.

The results of this section are obtained in [25].

Consider in a cylinder $G := \{(x, y, t) : 0 < r < 1, \ -a < t < b\}, r = \sqrt{x^2 + y^2}$ the equation of mixed type

$$Lu = \begin{cases} \partial_t^\alpha u - \Delta u = f(r), & t > 0, \\ u_{tt} - \Delta u = f(r), & t < 0, \end{cases} \tag{5.105}$$

where $a > 0$, $b > 0$, α $(0 < \alpha \leq 1$ are given numbers, $\partial_t^\alpha u$ is the Caputo fractional derivative, Δ is the Laplace operator in variables x and y.

We pose the following problem: Find in the domain G the functions $u(x, y, t)$ and $f(r)$ satisfying the equation (5.105) and conditions

$$((x, y), \nabla u)\,|_{r=0} = u|_{r=1} = 0, \quad -a \leq t \leq b, \tag{5.106}$$

$$u|_{t=-a} = \varphi(r), \quad 0 \leq r \leq 1, \tag{5.107}$$

$$u|_{t=b} = \psi(r), \quad 0 \leq r \leq 1. \tag{5.108}$$

Here $((x, y), \nabla u)$ is scalar product of vectors (x, y) and ∇u, $\varphi(\cdot)$ and $\psi(\cdot)$ are given sufficiently smooth functions.

Denote $G_+ = G \cap \{t > 0\}, \quad G_- = G \cap \{t < 0\}$.

Definition 5.1 The solution of problem (5.105)–(5.108) are the functions $u(x, y, t)$ and $f(r)$ from the classes $C^{2,\alpha}_{x,y;t}(G_+ \cup \{t = b\}) \cap C^2(G_- \cup \{t = -a\})$ and $C[0, 1]$, respectively, satisfying relations (5.105)–(5.108) and the following conjugation conditions:

$$u(x, y, +0) = u(x, y, -0), \quad \lim_{t \to +0} \partial_t^\alpha u(x, y, t) = u_t(x, y, -0), \quad r \in (0, 1). \tag{5.109}$$

Here $C^{2,\alpha}_{x,y;t}(\Omega) := \left\{v(x, y, t) : v \in C(\Omega), \left(\Delta v, \partial_t^\alpha v\right) \in C(\Omega)\right\}$.

If $\alpha = 1$, conditions (5.109) mean the continuity of the solution $u(x, y, t)$ and its derivative with respect to t on the line of change of equation type $t = 0$.

We now turn to the study of the problem (5.105)–(5.108).

5.4.1 Formal Construction of the Inverse Problem Solution

Note that since the right side of equation (5.105) and the functions in (5.107) and (5.108) depend on the distance r, then $u(x, y, t) = u(r, t)$, i.e., we have an axisymmetric case. Then the operator Laplace on the function $u(x, y, t)$ in polar coordinate systems will not depend on the angle and has the form:

$$\Delta u(x, y, t) = \frac{\partial^2 u}{\partial r^2} + \frac{1}{r}\frac{\partial u}{\partial r}.$$

Therefore, equation (5.105) in these coordinate systems is written as follows:

$$\begin{aligned} \partial_t^\alpha u &= \frac{1}{r}\frac{\partial u}{\partial r} + \frac{\partial^2 u}{\partial r^2} + f(r), \quad t > 0, \\ \frac{\partial^2 u}{\partial t^2} &= \frac{1}{r}\frac{\partial u}{\partial r} + \frac{\partial^2 u}{\partial r^2} + f(r), \quad t < 0. \end{aligned} \tag{5.110}$$

Conditions (5.106)–(5.108) take the following forms:

$$\left[r\frac{\partial}{\partial r}u(r, t) \right]_{r=0} = 0, \quad u|_{r=1} = 0, \quad -a \le t \le b, \tag{5.111}$$

$$u(r, -a) = \varphi(r), \quad 0 \le r \le 1, \tag{5.112}$$

$$u(r, b) = \psi(r), \quad 0 \le r \le 1. \tag{5.113}$$

Thus, the inverse problem (5.106)–(5.108) is reduced to the problem definitions of the functions $u(r, t)$, $f(r)$ from equalities (5.110)–(5.113).

According to the Fourier method, searching partial solutions of equation (5.110) for the case $f = 0$ in form

$$u(r, t) = R(r)T(t),$$

we get the following relations:

$$\partial_t^\alpha T(t)R(r) = \frac{1}{r}T(t)R'(r) + T(t)R''(r), \quad t > 0,$$

$$T''(t)R(r) = \frac{1}{r}T(t)R'(r) + T(t)R''(r), \quad t < 0.$$

Hence, by separating the variables, we have

$$\frac{\partial_t^\alpha T(t)}{T(t)} = \frac{1}{r}\frac{R'(r)}{R(r)} + \frac{R''(r)}{R(r)} = -\lambda^2, t > 0,$$

$$\frac{T''(t)}{T(t)} = \frac{1}{r}\frac{R''(r)}{R(r)} + \frac{R''(r)}{R(r)} = -\lambda^2, \quad t < 0,$$

where λ is an arbitrary real parameter. Hence, for finding the function $R(r)$ we get the problem for the equation

$$R''(r) + \frac{1}{r}R'(r) + \lambda^2 R(r) = 0 \tag{5.114}$$

with boundary conditions

$$\lim_{r\to 0}(rR'(r)) = 0, \quad R(1) = 0, \tag{5.115}$$

which is a self-adjoint problem.

The solutions of equation (5.111) are the following zero-order Bessel functions of the first kind:

$$R_k(r) = J_0(\lambda_k r), k = 1, 2, 3, \ldots$$

They also are eigenfunctions. We find the eigenvalues using the second boundary condition of (5.115) (the validity of the first boundary condition in (5.115) is obvious), positive roots of the equation $J_0(\lambda_k) = 0$. As noted in the previous section, they look like:

$$\lambda_k = k\pi - \frac{\pi}{4} = (4k-1)\frac{\pi}{4}.$$

Expand now all functions in a Fourier-Bessel series in terms of eigenfunctions $J_0(\lambda_k r)$, i.e.,

$$u(r,t) = \sum_{k=1}^{\infty} u_k(t) J_0(\lambda_k r), \tag{5.116}$$

$$f(r) = \sum_{k=1}^{\infty} f_k J_0(\lambda_k r), \tag{5.117}$$

where $u_k(t) = \frac{2}{J_1^2(\lambda_k)}\int_0^1 ru(r,t)J_0(\lambda_k r)dr$, $f_k = \frac{2}{J_1^2(\lambda_k)}\int_0^1 rf(r)J_0(\lambda_k r)dr$.

Substituting (5.116), (5.117) into (5.110), we obtain

$$\partial_t^\alpha u_k(t) = -\lambda_k^2 u_k(t) + f_k, \quad t > 0,$$

$$u_k''(t) = -\lambda_k^2 u_k(t) + f_k, \quad t < 0. \tag{5.118}$$

It is not difficult to find that these differential equations have general solutions:

$$u_k(t) = c_k E_\alpha\left(-\lambda_k^2 t^\alpha\right) + \frac{f_k}{\lambda_k^2}, \quad t > 0,$$
$$u_k(t) = d_k \cos(\lambda_k t) + e_k \sin(\lambda_k t) + \frac{f_k}{\lambda_k^2}, \quad t < 0, \tag{5.119}$$

where $E_\alpha(\cdot)$ is the Mittag-Leffler function, $d_k,\ e_k,\ c_k$ are arbitrary constants.

In order to find the coefficients $d_k,\ e_k,\ c_k$, we use conditions

$$u_k(0+0) = u_k(0-0), \quad \partial_t^\alpha u_k(0+0) = u_k'(0-0),$$

which follow from the conditions (5.109). In view of this, from (5.119) we have

$$d_k = c_k, \ e_k = -\lambda_k c_k.$$

From the initial and additional conditions (5.112), (5.113), we get

$$d_k \cos(\lambda_k a) - e_k \sin(\lambda_k a) + \frac{f_k}{\lambda_k^2} = \varphi_k,$$

$$c_k E_\alpha\left(-\lambda_k^2 b^\alpha\right) + \frac{f_k}{\lambda_k^2} = \psi_k,$$

where $\varphi_k,\ \psi_k$ are Fourier-Bessel coefficients of functions $\varphi,\ \psi$, respectively:

$$\varphi(r) = \sum_{k=1}^{\infty} \varphi_k J_0(\lambda_k r), \ \psi(r) = \sum_{k=1}^{\infty} \psi_k J_0(\lambda_k r). \tag{5.120}$$

Substituting the values $d_k,\ e_k$ found through c_k, into the previous equations and solving the resulting system with respect to c_k and f_k, we find

$$c_k = \frac{\psi_k - \varphi_k}{E_\alpha\left(-\lambda_k^2 b^\alpha\right) - \left(\cos(\lambda_k a) + \lambda_k \sin(\lambda_k a)\right)},$$
$$f_k = \lambda_k^2\left(\psi_k - c_k E_\alpha\left(-\lambda_k^2 b^\alpha\right)\right). \tag{5.121}$$

Introduce the notation

$$A_{ab}(k) = E_\alpha\left(-\lambda_k^2 b^\alpha\right) - \left(\cos(\lambda_k a) + \lambda_k \sin(\lambda_k a)\right). \tag{5.122}$$

5.4.2 Existence and Uniqueness of Solution

We find the values of a and b for which (5.122) takes values not equal to zero. To do this, we rewrite (5.122) in the following form:

$$A_{ab}(k) = E_\alpha\left(-\frac{(4k-1)^2\pi^2}{16}b^\alpha\right) - \sqrt{1+\frac{(4k-1)^2\pi^2}{16}}\sin\left(\frac{4k-1}{4}a\pi + \gamma_k\right), \tag{5.123}$$

where $\gamma_k = \arcsin\left(1/\sqrt{1+\frac{(4k-1)^2\pi^2}{16}}\right)$. Obtain the values of a for which $A_{ab}(k) = 0$. It equals

$$a = \frac{4}{(4k-1)\pi}\left(\arcsin\left(\frac{E_\alpha\left(-\frac{(4k-1)^2\pi^2}{16}b^\alpha\right)}{\sqrt{1+\frac{(4k-1)^2\pi^2}{16}}}\right) + \pi n - \gamma_k\right)$$

Now we find the values of a and b, for which the following condition is met:

$$|A_{ab}(k)| \geq C_0 > 0. \tag{5.124}$$

For this, we calculate

$$|A_{ab}(k)| = \left|E_\alpha\left(-\frac{(4k-1)^2\pi^2}{16}b^\alpha\right) - \sqrt{1+\frac{(4k-1)^2\pi^2}{16}}\sin\left(\frac{4k-1}{4}a\pi + \gamma_k\right)\right|.$$

If $a = 4n, n \in \mathbb{N}$, then

$$|A_{ab}(k)| = \left|E_\alpha\left(-\frac{(4k-1)^2\pi^2}{16}b^\alpha\right) - \sqrt{1+\frac{(4k-1)^2\pi^2}{16}}\sin\left(\frac{4k-1}{4}a\pi + \gamma_k\right)\right|$$

$$\geq \left|E_\alpha\left(-\frac{(4k-1)^2\pi^2}{16}b^\alpha\right) \pm 1\right| \geq \left|1 - E_\alpha\left(-\frac{(4k-1)^2\pi^2}{16}b^\alpha\right)\right|.$$

According to Proposition 1.2 we have $0 < E_\alpha\left(-\frac{(4k-1)^2\pi^2}{16}b^\alpha\right) < 1$, for all $k = 1, 2, ...,$ then $|A_{ab}(k)| \geq C_0 > 0$. As C_0, it can be taken $1 - E_\alpha\left(-\frac{9\pi^2}{16}b^\alpha\right)$ as the largest of all possible such constants.

Thus, we have obtained the following uniqueness criterion:

Theorem 5.7 *If there exists a solution to problem* (5.105)–(5.108), *then it is unique for the values* $a = 4n,\ n \in \mathbb{N}$ *for any* $b > 0$.

We investigate the existence of a solution. To this end, we will prove the following assertion:

Theorem 5.8 *Assume that* $\{\varphi(r), \psi(r)\} \in C^6[0, 1]$ *and, in addition, condition* (5.124) *and the equalities*

$$\varphi^{(i)}(0) = 0, \ \psi^{(i)}(0) = 0, \ i = 0, 1, ..., 5,$$

$$\varphi^{(i)}(1) = 0, \ \psi^{(i)}(1) = 0, \ j = 0, 1, ..., 4$$

are satisfied.

Then there is a unique solution to problem (5.105)–(5.108), *which is defined by formulas* (5.125)–(5.127), *where* $\varphi^{(i)}$, $\psi^{(i)}$ *are ith derivatives of the functions* φ, ψ, *and* φ_k, ψ_k *are the Fourier-Bessel coefficients of the functions* φ *and* ψ, *respectively.*

To prove the theorem, substituting the found values of the coefficients d_k, e_k, c_k in (5.119), (5.121), we find $u_k(t)$ e f_k:

$$u_k(t) = \frac{\psi_k - \varphi_k}{A_{ab}(k)} E_\alpha\left(-\lambda_k^2 t^\alpha\right) + \psi_k - \frac{\psi_k - \varphi_k}{A_{ab}} E_\alpha\left(-\lambda_k^2 b^\alpha\right), \quad t > 0,$$

$$u_k(t) = \frac{\psi_k - \varphi_k}{A_{ab}(k)} \cos(\lambda_k t) - \lambda_k \frac{\psi_k - \varphi_k}{A_{ab}(k)} \sin(\lambda_k t) + \psi_k - \frac{\psi_k - \varphi_k}{A_{ab}(k)} E_\alpha\left(-\lambda_k^2 b^\alpha\right), \quad t < 0,$$

$$f_k = \lambda_k^2 \left(\psi_k - \frac{\psi_k - \varphi_k}{A_{ab}(k)} E_\alpha\left(-\lambda_k^2 b^\alpha\right)\right).$$

Taking into account these relations, from (5.116) and (5.117) we obtain the formal solution of problem in the form of series:

$$u(r, t) = \sum_{k=1}^{\infty} u_k(t) J_0(\lambda_k r)$$

$$= \sum_{k=1}^{\infty} \left[\frac{\psi_k - \varphi_k}{A_{ab}(k)} E_\alpha\left(-\lambda_k^2 t^\alpha\right) + \psi_k - \frac{\psi_k - \varphi_k}{A_{ab}(k)} E_\alpha\left(-\lambda_k^2 b^\alpha\right)\right] J_0(\lambda_k r), \quad t > 0, \tag{5.125}$$

$$u(r, t) = \sum_{k=1}^{\infty} u_k(t) J_0(\lambda_k r) = \sum_{k=1}^{\infty} \left[\frac{\psi_k - \varphi_k}{A_{ab}(k)} \cos(\lambda_k t) - \lambda_k \frac{\psi_k - \varphi_k}{A_{ab}(k)} \sin(\lambda_k t) + \psi_k \right.$$
$$\left. - \frac{\psi_k - \varphi_k}{A_{ab}(k)} E_\alpha\left(-\lambda_k^2 b^\alpha\right)\right] J_0(\lambda_k r), \quad t < 0, \tag{5.126}$$

$$f(r) = \sum_{k=1}^{\infty} f_k J_0(\lambda_k r) = \sum_{k=1}^{\infty} \lambda_k^2 \left[\psi_k - \frac{\psi_k - \varphi_k}{A_{ab}(k)} E_\alpha\left(-\lambda_k^2 b^\alpha\right)\right] J_0(\lambda_k r). \tag{5.127}$$

To prove the existence of a solution, we need to show that the series in (5.125)–(5.127) and the series obtained as a result of the action of fractional differentiation ∂_t^α, by differentiating with respect to r twice, in domain G_+ and by differentiating twice in r, t in domain G_-, converge uniformly. To this end, we calculate $\partial_t^\alpha u(r,t)$, $\left(\partial^2/\partial t^2\right) u(r,t)$, $\left(\partial^2/\partial r^2\right) u(r,t)$, performing differentiation formally under the signs of sums. Using properties of Bessel functions, namely, (see [22, 24]) $J_0'(r) = -J_1(r)$, $2J_1'(r) = J_0(r) - J_2(r)$, from formulas (5.125), (5.126) we obtain

$$\partial_t^\alpha u(r,t) = \sum_{k=1}^{\infty} \partial_t^\alpha u_k(t) J_0(\lambda_k r)$$

$$= \sum_{k=1}^{\infty} \left[-\lambda_k^2 E_\alpha\left(-\lambda_k^2 t^\alpha\right)\right] \frac{(\psi_k - \varphi_k)}{A_{ab}(k)} J_0(\lambda_k r), \quad t > 0, \tag{5.128}$$

$$\frac{\partial^2 u(r,t)}{\partial t^2} = \sum_{k=1}^{\infty} \frac{\partial^2 u_k(t)}{\partial t^2} J_0(\lambda_k r)$$

$$= \sum_{k=1}^{\infty} \left[-\lambda_k^2 \cos(\lambda_k t) + \lambda_k^3 \sin(\lambda_k t)\right] \frac{(\psi_k - \varphi_k)}{A_{ab}(k)} J_0(\lambda_k r), \quad t < 0, \tag{5.129}$$

$$\frac{\partial^2 u(r,t)}{\partial r^2} = \sum_{k=1}^{\infty} u_k(t) \frac{d^2 J_0(\lambda_k r)}{dr^2} = \sum_{k=1}^{\infty} \left[\frac{\psi_k - \varphi_k}{A_{ab}(k)} E_\alpha\left(-\lambda_k^2 t^\alpha\right)\right.$$

$$\left. + \psi_k - \frac{\psi_k - \varphi_k}{A_{ab}(k)} E_\alpha\left(-\lambda_k^2 b^\alpha\right)\right] \frac{\lambda_k^2}{2} \left(J_2(\lambda_k r) - J_0(\lambda_k r)\right), \quad t > 0, \tag{5.130}$$

$$\frac{\partial^2 u(r,t)}{\partial r^2} = \sum_{k=1}^{\infty} u_k(t) \frac{d^2 J_0(\lambda_k r)}{dr^2} = \sum_{k=1}^{\infty} \left[\frac{\psi_k - \varphi_k}{A_{ab}(k)} \cos(\lambda_k t) - \lambda_k \frac{\psi_k - \varphi_k}{A_{ab}(k)} \sin(\lambda_k t)\right.$$

$$\left. + \psi_k - \frac{\psi_k - \varphi_k}{A_{ab}(k)} E_\alpha\left(-\lambda_k^2 b^\alpha\right)\right] \frac{\lambda_k^2}{2} \left(J_2(\lambda_k r) - J_0(\lambda_k r)\right), \quad t < 0. \tag{5.131}$$

Let the functions $\varphi(r)$ and $\psi(r)$ satisfy conditions of Theorem 5.4 with some $s \geq 1$ (we define the number s later). Then if the Fourier-Bessel coefficients of these functions are true the following estimates:

$$|\varphi_k| \leq \frac{M_1}{\lambda_k^{2s-(1/2)}}, \quad |\psi_k| \leq \frac{M_2}{\lambda_k^{2s-(1/2)}}.$$

Now we will evaluate the expressions at Bessel functions in the right-hand sides of equalities (5.125)–(5.131). In this case, the expressions in (5.125), (5.126) are estimated as follows:

$$\left|\frac{\psi_k - \varphi_k}{A_{ab}(k)} E_\alpha\left(-\lambda_k^2 t^\alpha\right) + \psi_k - \frac{\psi_k - \varphi_k}{A_{ab}(k)} E_\alpha\left(-\lambda_k^2 b^\alpha\right)\right|$$

$$\leq M\left(\frac{M_1}{\lambda_k^{2s-(1/2)}} + \frac{M_2}{\lambda_k^{2s-(1/2)}}\right) \leq \frac{N_1}{\lambda_k^{2s-(1/2)}},\ t < 0,$$

$$\left|\frac{\psi_k - \varphi_k}{A_{ab}(k)} \cos\left(\lambda_k t\right) - \lambda_k \frac{\psi_k - \varphi_k}{A_{ab}(k)} \sin\left(\lambda_k t\right) + \psi_k - \frac{\psi_k - \varphi_k}{A_{ab}(k)} E_\alpha\left(-\lambda_k^2 b^\alpha\right)\right|$$

$$\leq N_2 \frac{\lambda_k + 1}{\lambda_k^{2s-(1/2)}},\ t > 0,$$

where $M,\ M_1,\ M_2,\ N_1,\ N_2$ are positive constants.

In a similar way, it is established that the expressions in (5.127), (5.128), (5.130) are less than $N_3 \frac{\lambda_k^2}{\lambda_k^{2s-(1/2)}}$, and the expressions in (5.129)–(5.131) are less then $N_4 \frac{\lambda_k^2+\lambda_k^3}{\lambda_k^{2s-(1/2)}}$, where N_3, N_4 are positive constants.

It follows from these estimates that if $s = 3$, then, according to Theorem 5.105, the series in (5.129)–(5.131) and the series obtained as a result of the action of fractional differentiation ∂_t^α, by differentiating with respect to r twice, in domain G_+ and by differentiating twice in r, t, in domain G_-, converge uniformly. Thus, Theorem 5.8 is proved.

5.5 Conclusion

In Sects. 5.1 and 5.2, for the solution of the inverse problems we used two bases for the space $L^2(0, 1)$, which form the bi-orthogonal system. Due to this bi-orthogonal system we are able to expand the solution in terms of the functions of the bi-orthogonal system. We showed the existence and uniqueness of the solutions of the inverse problems using properties of the Mittag-Leffler function and using the overdetermination conditions of final time observation in Sect. 5.1 and integral type in Sect. 5.2.

Section 5.3 concerns with the existence and uniqueness of solutions to inverse initial and inverse source problems for a time-fractional differential equation with the Bessel operator in the elliptic part. For both cases of inverse problems, the condition of final time observation is used. Solutions are obtained in the form of series expansion using an orthogonal set of Bessel functions.

Section 5.4 is devoted to the study of inverse source problem for a mixed type equation with a fractional diffusion equation in the parabolic part and a wave equation in the hyperbolic part of a cylindrical domain. The solution is obtained in the form of Fourier-Bessel series expansion using an orthogonal set of Bessel functions. Thus, the theorems of uniqueness and existence of a solution are proved.

It should be noted that problems close to those considered in this chapter have been also studied in the works [26–31]. We also point out recent papers [32–37] in which new classes of inverse problems of determining the time-dependent source and unknown order of the fractional derivative are investigated for various types of additional conditions.

References

1. Kirane, M., Malik, S.A.: Determination of an unknown source term and the temperature distribution for the linear heat equation involving fractional derivative in time. Appl. Math. Comput. **218**(1), 163–170 (2011)
2. Kilbas, A.A., Srivastava, H.M., Trujillo, J.J.: Theory and Application of Fractional Differential Equations. North-Holland Mathematical Studies, Elsevier, Amsterdam (2006)
3. Ionkin, N.I., Moiseev, E.I.: A two-point boundary value problem for a heat conduction equation. Differentsialnye Uravneniya **15**(7), 1284–1295 (1979)
4. Moiseev, E.I.: The solution of a nonlocal boundary value problem by the spectral method. Differ. Equ. **35**(8), 1105–1112 (1999)
5. Gorenflo, R., Mainardi, R.: Fractional Calculus: integral and differential equations of fractional order. In: Carpinteri, A., Mainardi, F. (eds.) Fractals and Fractional Calculus in Continuum Mechanics. Springer Verlag, Wien, pp. 223–276 (1997)
6. Prabhakar, T.R.: A singular intgeral equation with a generalized Mittag-Leffler function in the kernel. Yokohama Math. J. **19**, 7–15 (1971)
7. Ismailov, M.I., Cicek, M.: Inverse source problem for a time-fractional diffusion equation with nonlocal boundary conditions. Appl. Math. Model. **40**(7–8), 4891–4899 (2016)
8. Nguyen, H.C., Nguyen, H.L., Baleanu, D., Zhou, L., Long, L.D.: Inverse source problem for time fractional diffusion equation with Mittag-Leffler kernel. Adv. Differ. Equ. **2020**(210), 18 (2020)
9. Liao, K., Wei, T.: Identifying a fractional order and a space source term in a time-fractional diffusion-wave equation simultaneously. Inverse Prob. **35**(11), 115002 (2019)
10. Liu, W., Du, Ch., Li, Z.: New type of the unique continuation property for a fractional diffusion equation and an inverse source problem. Boundary Value Prob. **2024**(19), 13 (2025)
11. Derbissaly, B., Sadybekov, M.: Inverse source problem for multi-term time-fractional diffusion equation with nonlocal boundary conditions. AIMS Math. **9**(4), 9969–9988 (2024)
12. Durdiev, D.K., Jumaev, J.J., Turdiev, H.H.: Inverse problem for determining time dependent coefficient and source functions in a time-fractional diffusion equation. J. Math. Sci. (2024). https://doi.org/10.1007/s10958-024-07204-y
13. Tuan, V.: Inverse problem for fractional diffusion equation. Fract. Calculus Appl. Anal. **14**(1), 31–55 (2011)
14. Wei, T., Li, X.L., Li, Y.S.: An inverse time-dependent source problem for a time-fractional diffusion equation. Inverse Prob. **32**(8), 085003 (2011)
15. Faizi, R., Armania, R.: An inverse source problem for a generalized time fractional diffusion equation. Eurasian J. Math. Comput. Appl. **10**(1), 26–39 (2022)
16. Hilfer, R., Luchko, Y., Tomovski, Z.: Operational method for the solution of fractional differential equations with generalized Riemann-Liouville fractional derivatives. Frac. Calc. Appl. Anal. **12**(3), 299–318 (2009)

17. Kamocki, R.: A new representation formula for the Hilfer fractional derivative and its applications. J. Comput. Appl. Math. **308**, 39–45 (2016)
18. Haubold, H.J., Mathai, A.M., Saxena, R.K.: Mittag-Leffler functions and their applications. J. Appl. Math. **2011**, 51 (2011)
19. Chitalkar-Dhaigude, C.P., Bhairat, S.P., Dhaigude, D.B.: Solution of fractional differential equations involving Hilfer fractional derivative: method of successive approximations. Bull. Marathwada Math. Soc. **18**(2), 1–12 (2017)
20. Samko, S.G., Kilbas, A.A., Marichev, O.I.: Fractional Integrals and Derivatives, Theory and Applications. Gordon and Breach Science Publishers, Yveron (1993)
21. AL-Musalhi, F., AL-Salti, N., Kerbal, S.: inverse problems of a fractional differential equation with Bessel operator. Math. Model. Nat. Phenomena **12**(3), 105–113 (2017)
22. Tolstov, G.P.: Fourier Series (translated by Silverman R.A.). Prentice Hall, Inc., Englewood Cliffs, N. J (1962)
23. Higgins, J.R.: Completeness and Basis Properties of Sets of Special Functions. Cambridge University Press (1977)
24. Petrova, T.S.: Application of Bessel's functions in the modelling of chemical engineering processes. Bul. Chem. Commun. **41**, 343–354 (2009)
25. Durdiev, D.K.: Inverse source problem for an equation of mixed parabolic-hyperbolic type with the time fractional derivative in a cylindrical domain. J. Samara State Technical Univ. Ser. Phys. Math. Sci. **26**(2), 355–367 (2022)
26. Ali, M., Malik, S.A.: An inverse problem for a family of two parameters time fractional diffusion equations with nonlocal boundary conditions. Math. Methods Appl. Sci. **40**(18), 7737–7748 (2017)
27. Furati, K.M., Iyiola, O.S., Kirane, M.: An inverse problem for a generalized fractional diffusion. Appl. Math. Comput. **249**, 24–31 (2014)
28. Durdiev, D.K., Sultonov, M.A., Rahmonov, A.A., Nurlanuly, Y.: Inverse problems for a time-fractional diffusion equation with unknown right-hand side. Prog. Fract. Differ. Appl. **9**, 639–653 (2023)
29. Kinash, N., Janno, J.: An inverse problem for a generalized fractional derivative with an application in reconstruction of time-and space-dependent sources in fractional diffusion and wave equations. Mathematics **7**(12), 1138 (2019)
30. Malik, S.A., Ilyas, A., Samreen, A.: Simultaneous determination of a source term and diffusion concentration for a multi-term space-time fractional diffusion equation. Math. Model. Anal. **26**(3), 411–431 (2021)
31. Aleroev, T.S., Kirane, M., Malik, S.A.: Determination of a source term for a time fractional diffusion equation with an integral type over-determination condition. Electron. J. Differ. Equ. **270**, 1–16 (2013)
32. Ashurov, R., Umarov, S.: Determination of the order of fractional derivative for subdiffusion equations. Fract. Calc. Appl. Anal. **23**(6), 1647–1662 (2020)
33. Umarov, S., Ashurov, R., Chen, Ya., Q.: On a method of solution of systems of fractional pseudo-differential equations. Fract. Calc. Appl. Anal. **24**(1), 254–277 (2021)
34. Ashurov, R., Fayziev, Y.: On the nonlocal problems in time for time-fractional subdiffusion equations. Fract. Fract. **6**(1), 41 (2022)
35. Ashurov, R., Vaisova, N.: Backward and non-local problems for the Rayleigh-Stokes equation. Fract. Fract. **6**(10), 587 (2022)
36. Karimov, E., Ruzhansky, M., Tokmagambetov, N.: Cauchy type problems for fractional differential equations. Integral Transf. Spec. Funct. **33**(1), 47–64 (2022)
37. Karimov, E.T., Yuldashev, T.K.: Inverse problem for a mixed type integro-differential equation with fractional order caputo operators and spectral parameters. Axioms **9**(4), 121 (2020)

Chapter 6
Convolution Kernel Determination Problems

This chapter studies two inverse problems of determining the convolution kernel from the fractional diffusion equation.

In Sect. 6.1, an inverse problem of determining a kernel in a one-dimensional integro-differential time-fractional diffusion equation with initial-boundary and overdetermination conditions is investigated. First an auxiliary problem which is equivalent to the original problem is introduced. By the Fourier method, this auxiliary problem is reduced to equivalent integral equations. Then, using estimates of the Mittag-Leffler function and successive approximation method, an a priori estimate for the solution of the direct problem is obtained in terms of the norm of the unknown kernel which will be used in the study of the inverse problem. For solving this equation the contracted mapping principle is applied. The local existence and global uniqueness results are proven.

In Sect. 6.2, we study the inverse problem of determining the convolution kernel in the Cauchy problem for the time-fractional diffusion equation by a single observation at the point $x = 0$ of the fractional diffusion process. To represent the solution of the direct problem, it is used the fundamental solution of the time-fractional diffusion equation, which is expressed through $H-$ Fox function. The inverse problem is reduced to the equivalent integral equation of the Volterra type. By the contracted mapping principle, the local existence and global uniqueness results are proven. Also the stability estimate is obtained.

6.1 Inverse Kernel Problem in a Bounded Domain

Inverse problems for classical parabolic differential equations have been extensively studied. In the literature, the most commonly investigated cases involve linear inverse source problems and nonlinear inverse coefficient problems for the heat equation, formulated with various types of initial-boundary conditions (direct problems) and

D. K. Durdiev, *Inverse Problems for Fractional Diffusion Equations*,
Industrial and Applied Mathematics, https://doi.org/10.1007/978-981-96-5338-6_6

overdetermination conditions (see, for example, [1–5] and references therein). These studies address the unique solvability, stability estimates, and numerical approaches for solving such problems.

The works [6–9] focus on memory recovery problems in second-order parabolic integro-differential equations, where the integral term follows a convolution structure. In [10], it was demonstrated that when the kernel of the convolution integral in a classical integro-differential diffusion equation takes the form of a two-parametric Mittag-Leffler function with a specific argument, the equation models anomalous diffusive transport of solutes in heterogeneous porous media [11].

In this section, we examine an inverse problem for determining the kernel in a one-dimensional integro-differential time-fractional diffusion equation with initial boundary and overdetermination conditions. We first introduce an auxiliary problem equivalent to the original formulation. Applying the Fourier method, we transform this auxiliary problem into an equivalent system of integral equations. Using estimates of the Mittag-Leffler function and the method of successive approximations, we establish an estimate for the solution of the direct problem in terms of the norm of the unknown kernel, which plays a key role in analyzing the inverse problem. The inverse problem is then reformulated as an equivalent integral equation, which is solved using the contraction mapping principle. We prove the local existence and global uniqueness of the solution.

The results presented in this section are based on the findings in [12].

6.1.1 The Problem Formulation and Its Reduction

Let $T > 0, l > 0$ be fixed numbers and $D_T := \{(x, t) : 0 < x < l, 0 < t \leq T\}$. Consider the time-fractional diffusion equation with convolution integral

$$\partial_t^\alpha u - u_{xx} = \int_0^t k(\tau)u(x, t-\tau)d\tau, \quad (x, t) \in D_T, \tag{6.1}$$

with the initial condition

$$u(x, 0) = \varphi(x), \quad x \in [0, l], \tag{6.2}$$

the boundary conditions

$$u|_{x=0} = u|_{x=l} = 0, \quad \varphi(0) = \varphi(l) = 0, \quad t \in [0, T], \tag{6.3}$$

where ∂_t^α is the Caputo fractional derivative of order $0 < \alpha \leq 1$ in the time variable (see (1.4)), $\varphi(x)$ is the initial temperature.

We pose the inverse problem as follows: find the function $k(t)$ in (6.1), if the solution of the problem (6.1)–(6.3) satisfies

$$u_x(x,t)|_{x=0} = h(t), \quad t \in [0,T], \tag{6.4}$$

$h(t)$ is given function of $t \in [0,T]$.

The additional condition (6.4) corresponds to a given heat flow at the left end of a rod of length l.

Definition 6.1 A function $u(x,t)$ is called a classical solution to the initial-boundary problem (6.1)–(6.3) if:

(1) $u(x,t)$ is twice continuously differentiable in D_T;
(2) for each $x \in (0,l)$ $u(x,t)$ is continuous in t on $[0,T]$, and its fractional integral

$$(I_t^{1-\alpha}u)(x,t) = \frac{1}{\Gamma(1-\alpha)} \int_0^t \frac{u(x,\tau)d\tau}{(t-\tau)^\alpha} \tag{6.5}$$

is continuously differentiable in t for $t \in [0,T]$;
(3) $u(x,t)$ satisfies (6.1)−(6.3).

Let $[a,b]$ $(-\infty < a < b < \infty)$ be finite interval and $\gamma \in \mathbb{C}$ $(0 \le \Re(\gamma) < 1)$.

We introduce the weighted space $C_\gamma[a,b]$ of functions $f(t)$ given on $(a,b]$, such that the function $(t-a)^\gamma f(t) \in C[a,b]$, and

$$\|f\|_{C_\gamma} = \|(t-a)^\gamma f(t)\|_C, \quad C_0[a,b] = C[a,b].$$

For $n \in \mathbb{N}$ we denote by $C_\gamma^n[a,b]$ the Banach space of functions $f(t)$ which are continuously differentiable on $[a,b]$ up to order $n-1$ and have the derivative $f^{(n)}(t)$ of order n on $(a,b]$ such that $f^{(n)}(t) \in C_\gamma[a,b]$:

$$C_\gamma^n[a,b] = \left\{ f(t) : \|f\|_{C_\gamma^n[a,b]} = \sum_{k=0}^{n-1} \|f^{(k)}\|_C + \|f^{(n)}\|_{C_\gamma} \right\}, \quad C_\gamma^0[a,b] = C_\gamma[a,b].$$

We also define the class of functions $C_\gamma^{2,\alpha}(D_T)$:

$$C_\gamma^{2,\alpha}(D_T)$$

$$= \left\{ v(x,t) : v(\cdot,t) \in C^2(0,l), \text{ and } \partial_t^\alpha v(x,\cdot) \in C_\gamma(0,T], \ (x,t) \in D_T \right\}.$$

Let the functions φ and h satisfy the following assumptions:

(E1) $\varphi(x) \in \Phi^4[0,l]$, *(see Sect. 1.1.3 for definition of* $\Phi^4[0,l]$*),*
(E2) $h(t) \in C_\gamma^1[0,T]$ *and* $\varphi'(0) = h(0) > 0$.

Now we present some assertions on fractional calculus (see [13], pp. 96–99), which will be used in the future.

Proposition 6.1 (see [13], pp. 83) *Let* $\alpha > 0, \beta > 0$ *and* $\alpha + \beta < \frac{1}{p}$. *If* $f(t) \in L_p(\mathbb{R}^+)$, *then the semiproperties*

$$(I_t^\alpha I_t^\beta f)(t) = (I_t^{\alpha+\beta} f)(t)$$

hold, where I_t^α *is the Riemann-Liouville fractional integral operator (see (1.5)).*

Proposition 6.2 (see [13], pp. 71) *If* ≥ 0 *and* $\beta \in \mathbb{C}(\Re(\beta) > 0)$, *then*

$$(I_t^\alpha t^{\beta-1})(x) = \frac{\Gamma(\beta)}{\Gamma(\alpha+\beta)} x^{\beta+\alpha-1}.$$

Theorem 6.1 (see [13], pp.135–140) *The solution* $y(t) \in L[0, T]$ *of the linear nonhomogeneous fractional problem*

$$\frac{\partial^\alpha}{\partial t^\alpha} y(t) + \lambda y(t) = f(t), \quad t \in (0, T], \quad \lambda > 0,$$

$$I_t^{1-\alpha} y(0+) = c,$$

where $0 < \alpha < 1$, $f(t) \in L^1(0, T)$, $\frac{\partial^\alpha}{\partial t^\alpha}$ *is the Riemann-Liouville fractional derivative; is given by the integral expression*

$$y(t) = ct^{\alpha-1} E_{\alpha,\alpha}(-\lambda t^\alpha) + \int_0^t (t-\tau)^{\alpha-1} E_{\alpha,\alpha}(-\lambda(t-\tau)^\alpha) f(\tau) d\tau.$$

We reduce the inverse problem to equivalent one.

Let $u(x, t), k(t)$ be a classical solution to the problem (6.1)–(6.4) and φ, h be enough smooth functions. The next assertion contains the auxiliary equivalent problem.

Lemma 6.1 *Problem* (6.1)–(6.4) *is equivalent to the auxiliary problem of determining the functions* $\vartheta(x; t), k(t)$ *from the following equations:*

$$\frac{\partial^\alpha}{\partial t^\alpha} \vartheta - \vartheta_{xx} = k(t)\varphi''(x) + \int_0^t k(\tau)\vartheta(x, t-\tau), \quad (x, t) \in D_T, \tag{6.6}$$

$$I_t^{1-\alpha} \vartheta(x, t)\Big|_{t=0} = \varphi^{(IV)}(x), \quad x \in [0, l]. \tag{6.7}$$

$$\vartheta(x, t)\Big|_{x=0} = \vartheta(x, t)\Big|_{x=l} = 0, t \in [0, T], \tag{6.8}$$

$$\vartheta_x(x, t)\Big|_{x=0} = \frac{\partial^\alpha}{\partial t^\alpha} h'(t) - k(t)\varphi'(0) - \int_0^t k(\tau)h'(t-\tau)d\tau. \tag{6.9}$$

where $\vartheta(x, t) := u_{txx}(x, t)$.

Proof Introducing new function $\vartheta^{(1)}(x,t) = u_t(x,t)$, we differentiate the equalities (6.1) and (6.3), (6.4) with respect to t. For this, we use the following equalities

$$\frac{\partial}{\partial t}\partial_t^\alpha u(x,t) = \frac{\partial}{\partial t}\frac{1}{\Gamma(1-\alpha)}\int_0^t \frac{u_\tau(x,\tau)}{(t-\tau)^\alpha}d\tau =$$

$$= \frac{1}{\Gamma(1-\alpha)}\frac{\partial}{\partial t}\int_0^t \frac{\vartheta^{(1)}(x,\tau)}{(t-\tau)^\alpha}d\tau = \frac{\partial^\alpha}{\partial t^\alpha}\vartheta^1(x,t).$$

As result, one has problem of finding functions $\vartheta^{(1)}(x,t)$, $k(t)$ from the following equations:

$$\frac{\partial^\alpha}{\partial t^\alpha}\vartheta^{(1)} - \vartheta^{(1)}_{xx} = k(t)\varphi(x) + \int_0^t k(\tau)\vartheta^{(1)}(x,t-\tau), \quad (x,t)\in D_T, \tag{6.10}$$

$$I_t^{1-\alpha}\vartheta^{(1)}(x,t)\Big|_{t=0} = \varphi''(x), \quad x\in[0,l]. \tag{6.11}$$

$$\vartheta^{(1)}(x,t)\Big|_{x=0} = \vartheta^{(1)}(x,t)\Big|_{x=l} = 0, \quad t\in[0,T], \tag{6.12}$$

$$\vartheta^{(1)}_x(x,t)\Big|_{x=0} = h'(t), \quad t\in[0,T]. \tag{6.13}$$

Here, the initial condition (6.11) was obtained from equality (6.1) by setting $t = 0$.

Now let us denote by $\vartheta(x,t)$ the function $\vartheta^{(1)}_{xx}(x,t)$. Differentiating (6.11) and (6.12) twice with respect to x we obtain the equations (6.6) and (6.7). Taking into account (6.12) fact and substituting in (6.10) $x = 0$ and $x = l$, after some calculus we obtain to the boundary condition (6.8). In order to derive an additional condition for the function $\vartheta(x,t)$, taking into account condition (6.13), we differentiate Eq. (6.10) once with respect to x and, putting $x = 0$, we get (6.9). Thus, problem (6.1)–(6.4) is reduced to problem (6.6)–(6.9). It is not difficult to show that the reverse changes occur. The lemma is proved. □

First, we show that the direct problem is uniquely solvable.

6.1.2 *Solvability of the Initial-Boundary Value Problem*

We will seek the solution of problem (6.6)–(6.8) as before in the form

$$\vartheta(x,t) = \sum_{n=1}^{\infty}\vartheta_n(t)\sin(\lambda_n x), \quad \lambda_n = \frac{\pi n}{l}, \tag{6.14}$$

where

$$\vartheta_n(t) = \sqrt{\frac{2}{l}} \int_0^l \vartheta(x, t) \sin(\lambda_n x) dx.$$

Taking into account equality (6.14), we obtain from (6.6) following equations for v_n:

$$\left(\frac{\partial^\alpha}{\partial t^\alpha} \vartheta_n\right)(t) + \lambda_n^2 \vartheta_n(t) = F_n(t; \vartheta, k, \varphi), n = 1, 2, ..., \tag{6.15}$$

where

$$F_n(t; \vartheta, k, \varphi) := k(t)\varphi_n + \int_0^t k(\tau)\vartheta_n(t-\tau)d\tau,$$

$$\varphi_n = \sqrt{\frac{2}{l}} \int_0^l \varphi(x) \sin(\lambda_n x) dx.$$

The initial condition (6.7) gives

$$I_t^{1-\alpha}\vartheta_n(0) = \int_0^l \vartheta(x, 0) \sin(\lambda_n x) dx = -\lambda_n^2 \int_0^l \varphi(x) \sin(\lambda_n x) dx = -\lambda_n^2 \varphi_n. \tag{6.16}$$

Based Theorem 6.1 we conclude that the initial-value problem (6.15), (6.16) is equivalent in the space $C[0, T]$ to the Volterra integral equation of the second kind

$$\vartheta_n(t) = -\lambda_n^2 \varphi_n t^{\alpha-1} E_{\alpha,\alpha}(-\lambda_n^2 t^\alpha)$$
$$+ \int_0^t (t-\tau)^{\alpha-1} E_{\alpha,\alpha}\left(-\lambda_n^2 (t-\tau)^\alpha\right) k(\tau)\varphi_n(\tau) d\tau$$
$$+ \int_0^t (t-\tau)^{\alpha-1} E_{\alpha,\alpha}\left(-\lambda_n^2 (t-\tau)^\alpha\right) \int_0^\tau k(\alpha)\vartheta_n(\tau-\alpha) d\alpha d\tau. \tag{6.17}$$

We prove the existence of a solution to the integral equation (6.17) in the space $C_\gamma[0, T]$ using the method of successive approximation. Let us denote the sum of the first two summands on the right-hand side of (6.17) by $F_n(t)$. For this equation, we consider in the domain $[0, T]$ the sequence of functions

$$t^\gamma (\vartheta_n)_m(t) =$$
$$= t^\gamma \int_0^t (t-\tau)^{\alpha-1} E_{\alpha,\alpha}\left(-\lambda_n^2 (t-\tau)^\alpha\right) \int_0^\tau k(\alpha)(\vartheta_n)_{m-1}(\tau-\alpha) d\alpha d\tau.$$

where $m = 1, 2, ..., 0 < \gamma < 1, \gamma + \alpha > 1$, and $(\vartheta_n)_0(t) = F_n(t)$ for $t \in [0, T]$.

We introduce following notation: $\|k\|_\gamma = \max_{t\in[0,T]} |t^\gamma k(t)|$. Using the formula (6.17) based on Proposition 6.1, we estimate $t^\gamma \vartheta_n(t)$ in the domain $[0, T]$:

$$|t^\gamma (\vartheta_n)_0(t)| = |t^\gamma F_n(t)| \leq \lambda_n^2 |\varphi_n| t^{\alpha+\gamma-1} E_{\alpha,\alpha}(-\lambda_n^2 t^\alpha)$$

$$+ t^\gamma \left| \varphi_n \int_0^t (t-\tau)^{\alpha-1} \tau^{-\gamma} E_{\alpha,\alpha}\left(-\lambda_n^2 (t-\tau)^\alpha\right) \tau^\gamma k(\tau) d\tau \right|$$

$$\leq \frac{\lambda_n^2 |\varphi_n| t^{\alpha+\gamma-1}}{\Gamma(\alpha)} + t^\gamma |\varphi_n| \|k\|_\gamma I_t^\alpha (t^{-\gamma})$$

$$= \frac{\lambda_n^2 |\varphi_n| t^{\alpha+\gamma-1}}{\Gamma(\alpha)} + \frac{\Gamma(1-\gamma)}{\Gamma(1+\alpha-\gamma)} t^\alpha |\varphi_n| \|k\|_\gamma \leq C_0 \lambda_n^2 |\varphi_n| =: F_n^0;$$

where the constant C_0 depends on $\|k\|_\gamma$, α, T; and $I_t^\alpha (t^{-\gamma})$ is the fractional integral of the weight function $t^{-\gamma}$.

In accordance with Propositions 6.1 and 6.2, we obtain following estimates:

$$|t^\gamma (\vartheta_n)_1(t)| \leq t^\gamma \int_0^t (t-\tau)^{\alpha-1} E_{\alpha,\alpha}\left(-\lambda_n^2 (t-\tau)^\alpha\right)$$

$$\times \int_0^\tau \beta^{-\gamma} (\tau-\beta)^{-\gamma} \left|\beta^\gamma k(\beta)\right| \left|(\tau-\beta)^\gamma (\vartheta_n)_0(\tau-\beta)\right| d\beta d\tau$$

$$\leq t^\gamma \|k\|_\gamma F_n^0 \Gamma(1-\gamma) I_t^\alpha (I_t^{1-\gamma}(t^{-\gamma})) = \|k\|_\gamma F_n^0 \frac{\Gamma^2(1-\gamma)}{\Gamma(2+\alpha-2\gamma)} t^{1+\alpha-\gamma};$$

Similarly

$$|t^\gamma (\vartheta_n)_2(t)| \leq t^\gamma \int_0^t (t-\tau)^{\alpha-1} E_{\alpha,\alpha}\left(-\lambda_n^2 (t-\tau)^\alpha\right)$$

$$\times \int_0^\tau \beta^{-\gamma} (\tau-\beta)^{-\gamma} \left|\beta^\gamma k(\beta)\right| \left|(\tau-\beta)^\gamma (\vartheta_n)_1(\tau-\beta)\right| d\beta d\tau$$

$$\leq t^\gamma \|k\|_\gamma^2 F_n^0 \frac{\Gamma^2(1-\gamma)}{\Gamma(2+\alpha-2\gamma)} I_t^{2(1+\alpha-\gamma)}(t^{-\gamma}) = \|k\|_\gamma^2 F_n^0 \frac{\Gamma^3(1-\gamma)}{\Gamma(3+2\alpha-3\gamma)} t^{2(1+\alpha-\gamma)};$$

For arbitrary $m = 0, 1, 2, \ldots.$, we hold

$$|t^\gamma (\vartheta_n)_m(t)| \leq t^\gamma \int_0^t (t-\tau)^{\alpha-1} E_{\alpha,\alpha}\left(-\lambda_n^2 (t-\tau)^\alpha\right) \times$$

$$\times \int_0^\tau \beta^{-\gamma} (\tau-\beta)^{-\gamma} \left|\beta^\gamma k(\beta)\right| \left|(\tau-\beta)^\gamma (\vartheta_n)_{m-1}(\tau-\beta)\right| d\beta d\tau \leq$$

$$\leq \Gamma(1-\gamma)F_n^0 \frac{(\|k\|_\gamma \Gamma(1-\gamma)t^{1+\alpha-\gamma})^m}{\Gamma((1+\alpha-\gamma)m+1-\gamma)}.$$

It follows from the last estimates that the series

$$t^\gamma \vartheta_n(t) = \sum_{m=1}^{\infty} t^\gamma (\vartheta_n)_m(t)$$

converged uniformly in $[0, T]$, because it can be majorized in $[0, T]$ by the convergent numerical series

$$\Gamma(1-\gamma)F_n^0 \sum_{m=1}^{\infty} \frac{(\|k\|_\gamma \Gamma(1-\gamma)t^{1+\alpha-\gamma})^m}{\Gamma((1+\alpha-\gamma)m+1-\gamma)}.$$

This means the following estimate for the solution of the integral equation (6.17) takes place:

$$|t^\gamma \vartheta_n(t)| \leq \Gamma(1-\gamma)F_n^0 \sum_{m=1}^{\infty} \frac{(\|k\|_\gamma \Gamma(1-\gamma)t^{1+\alpha-\gamma})^m}{\Gamma((1+\alpha-\gamma)m+1-\gamma)} =$$

$$= \Gamma(1-\gamma)F_n^0 E_{1+\alpha-\gamma,1-\gamma}(\|k\|_\gamma \Gamma(1-\gamma)T^{1+\alpha-\gamma}) \leq C_1 \lambda_n^2 |\varphi_n|, \tag{6.18}$$

$$C_1 = \Gamma(1-\gamma)C_0 E_{1+\alpha-\gamma,1-\gamma}(\|k\|_\gamma \Gamma(1-\gamma)T^{1+\alpha-\gamma}),$$

where $E_{1+\alpha-\gamma,1-\gamma}(\cdot)$ is the Mittag-Leffler function of a nonnegative real argument.

Further, using equality (6.15), we obtain the estimate for $\frac{\partial^\alpha}{\partial t^\alpha}\vartheta_n(t)$:

$$\left| t^\gamma \frac{\partial^\alpha}{\partial t^\alpha}\vartheta_n(t) \right| \leq \left(\lambda_n^2 + \frac{\|k\|_\gamma \Gamma^2(1-\gamma)T^{1-\gamma}}{\Gamma(2-2\gamma)} \right)$$

$$\times \Gamma(1-\gamma)F_n^0 E_{1+\alpha-\gamma,1-\gamma}(\|k\|_\gamma \Gamma(1-\gamma)T^{1+\alpha-\gamma}) + |\varphi_n| \|k\|_\gamma \leq C_2 \lambda_n^2 |\varphi_n|,$$

where the constant C_2 depends on C_0, C_1, γ, T.

Thus we have proved the following lemma:

Lemma 6.2 *For any $t \in [0, T]$ and for $n \in \mathbb{N}$, the estimates are valid*

$$|t^\gamma \vartheta_n(t)| \leq \overline{C}_1 n^2 |\varphi_n|,$$

$$|t^\gamma D_{0+}^\alpha \vartheta_n(t)| \leq \overline{C}_2 n^4 |\varphi_n|,$$

here $\overline{C}_i, i = 1, 2$, are positive constants independent on n, φ.

Formally, from (6.14) by term-by-term differentiation we compose the series

$$t^{\gamma}\frac{\partial^{\alpha}}{\partial t^{\alpha}}\vartheta(x,t)=\sum_{n=1}^{\infty}t^{\gamma}\frac{\partial^{\alpha}}{\partial t^{\alpha}}\vartheta_n(t)\sin(\lambda_n x), \tag{6.19}$$

$$t^{\gamma}\vartheta_{xx}(x,t)=\sum_{n=1}^{\infty}t^{\gamma}\lambda_n^2\vartheta_n(t)\sin(\lambda_n x), \tag{6.20}$$

In view of Lemma 6.2, if following series converge then the functional series (6.14), (6.19), and (6.20) to be converge for any $(x,t)\in D_T$:

$$\overline{C}_3\sum_{n=1}^{\infty}n^4|\varphi_n|, \tag{6.21}$$

where the constant $\overline{C}_3$ depends on C_1, C_2, γ, T.

We hold the following auxiliary lemma.

Lemma 6.3 *If the conditions of* $(E1)$ *are true, then there are equalities*

$$\varphi_n=\frac{1}{\lambda_n^5}\varphi_n^{(5)}, \tag{6.22}$$

where

$$\varphi_n^{(5)}=\sqrt{\frac{2}{l}}\int_0^l\varphi^{(5)}(x)\cos(\lambda_n x)dx,$$

with the following estimate:

$$\sum_{n=1}^{\infty}|\varphi_n^{(5)}|^2\le\|\varphi^{(5)}\|^2_{L_2(0,l)}. \tag{6.23}$$

If the function $\varphi(x)$ satisfies the condition $E1$, then due to representations (6.22) and (6.23) series (6.14), (6.19) and (6.20) converge uniformly in domain D_T, therefore, function $\vartheta(x,t)$ satisfies relations (6.6)–(6.8).

Using the above results, we obtain the following assertion.

Lemma 6.4 *Let* $k(t)\in C_{\gamma}[0,T]$ *and* $(E1)$ *are satisfied, then there exists a unique solution of the direct problem* (6.6)–(6.8) *so that* $\vartheta(x,t)\in C_{\gamma}^{2,\alpha}(D_T)\cap C_{\gamma}\left(\overline{D}_T\right)$.

Now we derive an a priori estimate for the norm of the difference between the solution of the original integral equation (6.17) and the solution of this equation with perturbed functions $\widetilde{k}, \widetilde{\varphi}_n$. Let $\widetilde{\vartheta}_n(t)$ be solution of the integral equation (6.17) corresponding to the functions $\widetilde{k}, \widetilde{\varphi}_n$; i.e.,

$$\widetilde{\vartheta}_n(t)=-\lambda_n^2t^{\alpha-1}\widetilde{\varphi}_nE_{\alpha,\alpha}(-\lambda_n^2t^{\alpha})$$
$$+\widetilde{\varphi}_n\int_0^t(t-\tau)^{\alpha-1}E_{\alpha,\alpha}\left(-\lambda_n^2(t-\tau)^{\alpha}\right)\widetilde{k}(\tau)d\tau$$

$$+\int_0^t (t-\tau)^{\alpha-1} E_{\alpha,\alpha}\left(-\lambda_n^2(t-\tau)^{\alpha}\right) \int_0^{\tau} \widetilde{k}(\alpha)\widetilde{\vartheta}_n(\tau-\alpha)d\alpha d\tau. \tag{6.24}$$

Composing the difference $\vartheta - \widetilde{\vartheta}$ with the help of the equations (6.17), (6.24) and introducing the notations $\vartheta - \widetilde{\vartheta} = \overline{\vartheta}_n$, $k - \widetilde{k} = \overline{k}$, $\varphi_n - \widetilde{\varphi}_n = \overline{\varphi}_n$, we obtain the integral equation

$$\overline{\vartheta}_n(t) = -\lambda_n^2 t^{\alpha-1}\overline{\varphi}_n E_{\alpha,\alpha}(-\lambda_n^2 t^{\alpha})$$
$$+\varphi_n \int_0^t (t-\tau)^{\alpha-1} E_{\alpha,\alpha}\left(-\lambda_n^2(t-\tau)^{\alpha}\right)\overline{k}(\tau)d\tau$$
$$+\overline{\varphi}_n \int_0^t (t-\tau)^{\alpha-1} E_{\alpha,\alpha}\left(-\lambda_n^2(t-\tau)^{\alpha}\right)\widetilde{k}(\tau)d\tau$$
$$+\int_0^t (t-\tau)^{\alpha-1} E_{\alpha,\alpha}\left(-\lambda_n^2(t-\tau)^{\alpha}\right) \int_0^{\tau} \overline{k}(\beta)\vartheta_n(\tau-\beta)d\beta d\tau$$
$$+\int_0^t (t-\tau)^{\alpha-1} E_{\alpha,\alpha}\left(-\lambda_n^2(t-\tau)^{\alpha}\right) \int_0^{\tau} \widetilde{k}(\beta)\overline{\vartheta}_n(\tau-\beta)d\beta d\tau.$$

From the last relation is derived the following linear integral inequality for $|t^{\gamma}\overline{\vartheta}_n(t)|$:

$$|t^{\gamma}\overline{\vartheta}_n(t)| \le \frac{\lambda_n^2|\overline{\varphi}_n|t^{\alpha+\gamma-1}}{\Gamma(\alpha)} + \frac{\Gamma(1-\gamma)}{\Gamma(1+\alpha-\gamma)}t^{1+\alpha}\Big(|\varphi_n|\|\overline{k}\|_{\gamma} + |\overline{\varphi}_n|\|\widetilde{k}\|_{\gamma}\Big) +$$
$$+\frac{\Gamma^2(1-\gamma)}{\Gamma(2+\alpha-2\gamma)}t^{1+\alpha-\gamma}\|\overline{k}\|_{\gamma}F_n^0 E_{1+\alpha-\gamma,1-\gamma}(\|k\|_{\gamma}\Gamma(1-\gamma)T^{1+\alpha-\gamma})+$$
$$+t^{\gamma}\|\widetilde{k}\|_{\gamma}\frac{1}{\Gamma(\alpha)}\int_0^t (t-\tau)^{\alpha-1}\int_0^{\tau}(\tau-\beta)^{-\gamma}\beta^{-\gamma}|(\tau-\beta)^{\gamma}\overline{\vartheta}_n(\tau-\beta)|d\beta d\tau. \tag{6.25}$$

Applying the successive approximation method to inequality (6.25) with the help scheme

$$|t^{\gamma}(\overline{\vartheta}_n)_0(t)| \le \frac{\lambda_n^2|\overline{\varphi}_n|t^{\alpha+\gamma-1}}{\Gamma(\alpha)} + \frac{\Gamma(1-\gamma)}{\Gamma(1+\alpha-\gamma)}t^{1+\alpha}\Big(|\varphi_n|\|\overline{k}\|_{\gamma} + |\overline{\varphi}_n|\|\widetilde{k}\|_{\gamma}\Big) +$$
$$+\frac{\Gamma^2(1-\gamma)}{\Gamma(2+\alpha-2\gamma)}t^{1+\alpha-\gamma}\|\overline{k}\|_{\gamma}F_n^0 E_{1+\alpha-\gamma,1-\gamma}(\|k\|_{\gamma}\Gamma(1-\gamma)T^{1+\alpha-\gamma}) \le$$
$$\le \overline{C}_0(|\overline{\varphi}_n| + \|\overline{k}\|_{\gamma}),$$

where

$$\overline{C}_0 = \max\left\{\frac{\lambda_n^2 t^{\alpha+\gamma-1}}{\Gamma(\alpha)} + \frac{\Gamma(1-\gamma)}{\Gamma(1+\alpha-\gamma)} t^{1+\alpha}\|\widetilde{k}\|_\gamma, \frac{\Gamma(1-\gamma)}{\Gamma(1+\alpha-\gamma)} t^{1+\alpha}|\varphi_n| + \right.$$

$$\left. + \frac{\Gamma^2(1-\gamma)}{\Gamma(2+\alpha-2\gamma)} t^{1+\alpha-\gamma} F_n^0 E_{1+\alpha-\gamma,1-\gamma}(\|k\|_\gamma \Gamma(1-\gamma)T^{1+\alpha-\gamma})\right\},$$

and

$$|t^\gamma(\overline{\vartheta}_n)_m(t)|$$

$$\leq t^\gamma \|\widetilde{k}\|_\gamma \frac{1}{\Gamma(\alpha)}\int_0^t (t-\tau)^{\alpha-1}\int_0^\tau (\tau-\beta)^{-\gamma}\beta^{-\gamma}|(\tau-\beta)^\gamma(\overline{\vartheta}_n(\tau-\beta))_{m-1}|d\beta d\tau.$$

We arrive at the estimate:

$$|t^\gamma \overline{\vartheta}_n(t)| \leq \Gamma(1-\gamma)E_{1+\alpha-\gamma,1-\gamma}(\|\widetilde{k}\|_\gamma \Gamma(1-\gamma)T^{1+\alpha-\gamma})\overline{C}_0(|\overline{\varphi}_n| + \|\overline{k}\|_\gamma). \quad (6.26)$$

The estimate (6.26) will be used in studying the inverse problem using Banach's fixed point theorem.

6.1.3 *Local Solvability of the Kernel Determination Inverse Problem*

This subsection studies the inverse problem as the problem of determining of functions $k(t)$ from relations (6.6)–(6.9), using the contraction mapping principle.

Firstly, by differentiating (6.14) with respect to x, we get the following equality

$$\vartheta_x(x,t) = \sum_{n=1}^{\infty} \lambda_n \vartheta_n(t)\cos(\lambda_n x). \quad (6.27)$$

Setting in (6.27) $x = 0$ and using additional condition (6.9), after simplifications converting, we get the following integral equation for determining $k(t)$:

$$k(t) = k_0(t) - \frac{1}{\varphi'(0)}\sum_{n=1}^{\infty}\lambda_n\vartheta_n(t;k) - \frac{1}{\varphi'(0)}\int_0^t k(\tau)h'(t-\tau)d\tau, \quad (6.28)$$

where $k_0(t) = \frac{\partial^\alpha}{\partial t^\alpha}h'(t)\frac{1}{\varphi'(0)}$, and $\vartheta_n(t;k)$ means that the solution of integral equation (6.17) depends on $k(t)$.

We introduce an operator B, defining it by the right-hand side of (6.28)

$$B[k](t) = k_0(t) - \frac{1}{\varphi'(0)}\sum_{n=1}^{\infty}\lambda_n\vartheta_n(t;k) - \frac{1}{\varphi'(0)}\int_0^t k(\tau)h'(t-\tau)d\tau. \quad (6.29)$$

Then, Eq. (6.28) is written in more convenient form as

$$k(t) = B[k](t). \tag{6.30}$$

Let $k_{00} := \max_{t\in[0,T]} |k_0(t)|$. Fix a number $\rho > 0$ and consider the ball

$$\Phi^T(k_0, \rho) := \left\{k(t) : k(t) \in C_\gamma[0, T], \|k - k_0\|_{C_\gamma[0,T]} \le \rho\right\}.$$

Theorem 6.2 *Let (E1), (E2) be satisfied. Then there exists a number $T^* \in (0, T)$, such that there exists a unique solution $k(t) \in C_\gamma[0, T^*]$ of the inverse problem* (6.6)–(6.9).

Proof Let us first prove that for an enough small $T > 0$ the operator B maps the ball $\Phi^T(k_0, \rho)$ implies that $B[k](t) \in \Phi^T(k_0, \rho)$. Indeed, for any continuous function $k(t)$, the function $B[k](t)$ calculated using formula (6.29) will be continuous. Moreover, estimating the norm of the differences, we find that

$$\|t^\gamma(B[k](t) - k_0(t))\| \le \frac{1}{\varphi'(0)}|\sum_{n=1}^{\infty} \lambda_n t^\gamma \vartheta_n(t; k)| + \frac{t^\gamma}{\varphi'(0)}|\int_0^t k(\tau)h'(t-\tau)d\tau|$$

$$\le \frac{\Gamma(1-\gamma)}{\varphi'(0)} E_{1+\alpha-\gamma,1-\gamma}(\|k\|_\gamma \Gamma(1-\gamma)T^{1+\alpha-\gamma}) \sum_{n=1}^{\infty} \lambda_n F_n^0$$

$$+ \frac{\Gamma^2(1-\gamma)\|k\|_\gamma \|h'\|_\gamma}{\varphi'(0)\Gamma(1-2\gamma)} T^{1-\gamma}.$$

Here we have used the estimate (6.18). In view of Lemmas 6.2 and 6.3 last series is a convergent series. Note that the function occurring on the right-hand side in this inequality is monotone increasing with T, and the fact that the function $k(t)$ belongs to the ball $\Phi^T(k_0, \rho)$ implies the inequality

$$\|k\| \le \rho + \|k_0\|. \tag{6.31}$$

Therefore, we only strengthen the inequality if we replace $\|k\|_\gamma$ in this inequality with the expression $\rho + \|k_0\|$. Performing these replacements, we obtain the estimate

$$\|t^\gamma(B[k](t) - k_0(t))\| \le \frac{\Gamma(1-\gamma)}{\varphi'(0)} E_{1+\alpha-\gamma,1-\gamma}((\rho + \|k_0\|)\Gamma(1-\gamma)T^{1+\alpha-\gamma})$$

$$\times \sum_{n=1}^{\infty} \lambda_n F_n^0 + \frac{\Gamma^2(1-\gamma)(\rho + \|k_0\|)\|h'\|_\gamma}{\varphi'(0)\Gamma(1-2\gamma)} T^{1-\gamma}.$$

Let T_1 be a positive root of the equation

$$m_1(T) = \frac{\Gamma(1-\gamma)}{\varphi'(0)} E_{1+\alpha-\gamma,1-\gamma}((\rho + \|k_0\|)\Gamma(1-\gamma)T^{1+\alpha-\gamma}) \sum_{n=1}^{\infty} \lambda_n F_n^0$$

$$+ \frac{\Gamma^2(1-\gamma)(\rho + \|k_0\|)\|h'\|_\gamma}{\varphi'(0)\Gamma(1-2\gamma)} T^{1-\gamma} = \rho.$$

We choose ρ as $\rho > \left[\varphi'(0)\right]^{-1} \sum_{n=1}^{\infty} \lambda_n F_n^0$. Then, the equation $m_1(T) = \rho$ has a unique positive root. This root we denote by T_1. It is clear, for $T \in [0, T_1]$, we have $B[k](t) \in \Phi^T(k_0, \rho)$.

Further we consider two functions $k(t)$ and $\widetilde{k}(t)$ belonging to the ball $\Phi^T(k_0, \rho)$ and estimate the distance between their images $B[k](t)$ and $B[\widetilde{k}](t)$ in the space $C[0, T]$. The function $\widetilde{\vartheta}_n(t)$ corresponding to $\widetilde{k}(t)$ satisfies the integral equation (6.24) with the function $\varphi_n = \widetilde{\varphi}_n$. Composing the difference $B[k](t) - B[\widetilde{k}](t)$ based on equations (6.17), (6.24) and then estimating its norm, we obtain

$$\|t^\gamma (B[k](t) - B[\widetilde{k}](t))\| \leq \frac{1}{\varphi'(0)} \sum_{n=1}^{\infty} \lambda_n |t^\gamma \overline{\vartheta}_n(t; k)| + \frac{t^\gamma}{\varphi'(0)} \left| \int_0^t \overline{k}(\tau) h'(t-\tau) d\tau \right|.$$

Using inequality (6.18) and the estimate (6.26) with $\varphi_n = \widetilde{\varphi}_n$, we continue the previous inequality

$$\|t^\gamma (B[k](t) - B[\widetilde{k}](t))\| \leq \frac{1}{\varphi'(0)} \sum_{n=1}^{\infty} \lambda_n \Big\{ \frac{\Gamma(1-\gamma)}{\Gamma(1+\alpha-\gamma)} t^{1+\alpha} |\varphi_n| \|\overline{k}\|_\gamma$$

$$+ \frac{\Gamma^2(1-\gamma)}{\Gamma(2+\alpha-2\gamma)} t^{1+\alpha-\gamma} \|\overline{k}\|_\gamma F_n^0 E_{1+\alpha-\gamma,1-\gamma}(\|k\|_\gamma \Gamma(1-\gamma) T^{1+\alpha-\gamma}) \Big\}$$

$$\times \Gamma(1-\gamma) E_{1+\alpha-\gamma,1-\gamma}(\|\widetilde{k}\|_\gamma \Gamma(1-\gamma) T^{1+\alpha-\gamma})$$

$$+ \frac{\Gamma^2(1-\gamma)}{\varphi'(0)\Gamma(2+\alpha-2\gamma)} t^{1-\gamma} \|h'\|_\gamma \|\overline{k}\|_\gamma. \tag{6.32}$$

The functions $k(t)$ and $\widetilde{k}(t)$ belong to the ball $\Phi^T(k_0, \rho)$, and hence for each of these functions one has inequality (6.31). Note that the function on the right-hand side in inequality (6.32) at the factor $\|\overline{k}\|_\gamma$ is monotone increasing with $\|k\|_\gamma$, $\|\widetilde{k}\|_\gamma$, and T. Consequently, replacing $\|k\|_\gamma$ and $\|\widetilde{k}\|_\gamma$ in inequality (6.32) by $\rho + \|k\|_\gamma$ will only strengthen the inequality. This, we have

$$\|t^\gamma (B[k](t) - B[\widetilde{k}](t))\| \leq \Big[\frac{1}{\varphi'(0)} \sum_{n=1}^{\infty} \lambda_n \Big\{ \frac{\Gamma(1-\gamma)}{\Gamma(1+\alpha-\gamma)} t^{1+\alpha} |\varphi_n|$$

$$+ \frac{\Gamma^2(1-\gamma)}{\Gamma(2+\alpha-2\gamma)} t^{1+\alpha-\gamma} F_n^0 E_{1+\alpha-\gamma,1-\gamma}(\|k\|_\gamma \Gamma(1-\gamma) T^{1+\alpha-\gamma}) \Big\}$$

$$\times \Gamma(1-\gamma)E_{1+\alpha-\gamma,1-\gamma}\|\widetilde{k}\|_\gamma\Gamma(1-\gamma)T^{1+\alpha-\gamma}$$

$$+\frac{\Gamma^2(1-\gamma)}{\varphi'(0)\Gamma(2+\alpha-2\gamma)}t^{1-\gamma}\|h'\|_\gamma\Big]\|\widetilde{k}\|_\gamma.$$

Let T_2 be a root of the equation

$$m_2(T)=\frac{1}{\varphi'(0)}\sum_{n=1}^{\infty}\lambda_n\Big\{\frac{\Gamma(1-\gamma)}{\Gamma(1+\alpha-\gamma)}T^{1+\alpha}|\varphi_n|$$

$$+\frac{\Gamma^2(1-\gamma)}{\Gamma(2+\alpha-2\gamma)}T^{1+\alpha-\gamma}F_n^0E_{1+\alpha-\gamma,1-\gamma}((\rho+\|k\|_\gamma)\Gamma(1-\gamma)T^{1+\alpha-\gamma})\Big\}$$

$$\times \Gamma(1-\gamma)E_{1+\alpha-\gamma,1-\gamma}((\rho+\|k\|_\gamma)\Gamma(1-\gamma)T^{1+\alpha-\gamma})$$

$$+\frac{\Gamma^2(1-\gamma)}{\varphi'(0)\Gamma(2+\alpha-2\gamma)}T^{1-\gamma}\|h'\|_\gamma=1.$$

Then for $T\in[0,T_2)$ we have that the distance between the functions $B[k](t)$ and $B[\widetilde{k}](t)$ in the function space $C_\gamma[0,T]$ is not greater than the distance between the functions $\|k\|_\gamma$ and $\|\widetilde{k}\|_\gamma$ multiplied by $m_2(T)<1$. Consequently, If we choose $T^*<\min(T_1,T_2)$, then the operator B is a contraction in the ball $\Phi^{T^*}(k_0,\rho)$. However, in accordance with the Banach theorem, the operator B has a unique fixed point in the ball $\Phi^{T^*}(k_0,\rho)$; i.e., there exists a unique solution of Eq. (6.30). Theorem 6.2 is proven. □

By found function $\vartheta(x,t)$, $k(t)$, the function $u(x,t)$ is found by virtue of the notation $\vartheta(x,t)=u_{txx}$ from which follows

$$u(x,t)=\int_0^x(x-\xi)\int_0^t\vartheta(\xi,\tau)d\tau d\xi+\varphi(x)-x\varphi'(0)+xh(t).$$

6.2 Kernel Identification Problem in an Unbounded Domain

Currently, there are still relatively few results on inverse coefficient problems for fractional-order diffusion equations. Among the most common are inverse source problems (see Chap. 5 and the references therein). Regarding inverse coefficient problems for fractional diffusion equations, we highlight the work [14], which establishes a uniqueness theorem for the inverse problem of determining various time-independent smooth coefficients appearing in time-fractional diffusion equations, based on solution measurements at a fixed time over a specific subset.

In [15], the inverse problem of identifying a time-dependent reaction-diffusion coefficient in the Cauchy problem for the time-fractional diffusion equation is studied using a single observation at $x = 0$. The work [16] investigates an inverse reaction coefficient problem for a semilinear generalized fractional diffusion equation with a time-dependent principal part, utilizing additional integral-type measured data. More recently, [17, 18, 24] have examined multidimensional inverse problems related to the identification of the lowest order coefficient in fractional diffusion equations. These studies establish local existence theorems, global uniqueness results, and conditional stability estimates. Additionally, in [19], the inverse problem of determining both the kernel function and the time-dependent right-hand side in the first initial-boundary value problem for an integro-differential fractional diffusion equation, subject to additional integral-type conditions, is investigated.

In this section, we focus on the inverse problem of determining the convolution kernel in the Cauchy problem for the time-fractional diffusion equation, using a single observation at $x = 0$. To represent the solution of the direct problem, we construct the fundamental solution of the time-fractional diffusion equation. By deriving estimates for the fundamental solution and its derivatives, we establish an estimate for the direct problem solution in terms of the norm of the unknown kernel function, which will be used in the analysis of the inverse problem.

The inverse problem is reformulated as an equivalent integral equation of the Volterra type. Using the contraction mapping principle, we prove local existence and global uniqueness results. Additionally, a stability estimate is derived.

The results of this section are obtained from the work [20].

6.2.1 Kernel Identification Problem and Its Reduction

We consider the following one-dimensional anomalously diffusive equation with convolution integral:

$$\partial_t^\alpha u - u_{xx} - \int_0^t k(\tau)u(x, t-\tau)d\tau = f(x,t), \; t > 0, \; x \in \mathbb{R}, \tag{6.33}$$

the solution of which satisfies the initial condition

$$u(x, 0) = \varphi(x), \; x \in \mathbb{R}, \tag{6.34}$$

where $f(x, t)$, $\varphi(x)$ are given enough smooth functions.

We pose *the inverse problem* as follows: find the function $k(t)$, $t > 0$ in (6.33), if the solution of the Cauchy problem (6.33), (6.34) satisfies

$$u(0, t) = g(t), \; t > 0, \tag{6.35}$$

$g(t)$ is a given function.

Definition 6.2 A function $u(x,t)$ is called *a classical solution* to the Cauchy problem (6.33) and (6.34) if:

(1) $u(x,t)$ is twice continuously differentiable in x and $u,\ u_x,\ u_{xx}$ are bounded in $x \in \mathbb{R}$ functions for each $t > 0$;

(2) for each $x \in \mathbb{R}$ $u(x,t)$ is continuous in t on $[0,T]$, and its fractional integral

$$\left(I_t^{1-\alpha}u\right)(x,t) = \frac{1}{\Gamma(1-\alpha)}\int_0^t \frac{u(x,\tau)d\tau}{(t-\tau)^\alpha} \tag{6.36}$$

is continuously differentiable in t for $t > 0$;

(3) $u(x,t)$ satisfies (6.33) and (6.34) in usual sense.

Assuming $u(x,t)$ is a classical solution of the Cauchy problem (6.33), (6.34) and $f,\ \varphi,\ g$ be enough smooth functions, we transform the inverse problem (6.33)–(6.35). Denote for this purpose the second derivative of $u(x,t)$ with respect to x, by $\tilde{u}(x,t)$, i.e., $\tilde{u}(x,t) := u_{xx}(x,t)$. Differentiating (6.33) and (6.34) twice in x, we get

$$\partial_t^\alpha \tilde{u}(x,t) - \tilde{u}_{xx} - \int\limits_0^t k(\tau)\tilde{u}(x,t-\tau)d\tau = f_{xx}(x,t),\ t > 0,\ x \in \mathbb{R}, \tag{6.37}$$

$$\tilde{u}(x,0) = \varphi''(x),\ x \in \mathbb{R}. \tag{6.38}$$

To obtain an additional condition for the function $\tilde{u}(x,t)$, we note that the second term in (6.33) is $\tilde{u}(x,t)$. Setting $x = 0$ in (6.33) and using equality (6.35), we obtain

$$\tilde{u}(0,x) = \partial_t^\alpha g(t) - f(0,t),\ t > 0. \tag{6.39}$$

When the matching condition $\varphi(0) = g(0)$ is fulfilled, it is easy to derive from (6.37)–(6.39) the equations (6.33)–(6.35) (see, for example, [9]).

Denoting $v(x,t) := \frac{\partial}{\partial t}\tilde{u}(x,t)$, we differentiate (6.37) with respect to t. Taking into account the following relations:

$$\frac{\partial}{\partial t}\left(\partial_t^\alpha \tilde{u}\right) = \frac{1}{\Gamma(1-\alpha)}\frac{\partial}{\partial t}\left[\int_0^t \frac{\tilde{u}_\tau(x,\tau)}{(t-\tau)^\alpha}d\tau\right] =: \frac{\partial^\alpha v(x,\tau)}{\partial t^\alpha},$$

where $\frac{\partial^\alpha}{\partial t^\alpha}$ is the Riemann-Liouville fractional differential operator (see (1.6)):

$$\frac{\partial^\alpha v(x,\tau)}{\partial t^\alpha} := \frac{\partial}{\partial t}I_t^{1-\alpha}v(x,t) = \frac{1}{\Gamma(1-\alpha)}\frac{\partial}{\partial t}\int_0^t \frac{v(x,\tau)}{(t-\tau)^\alpha}d\tau.$$

As result we get

$$\frac{\partial^\alpha v(x,\tau)}{\partial t^\alpha} - v_{xx} - \int_0^t k(\tau)v(x,t-\tau)d\tau$$
$$= k(t)\varphi''(x) + f_{txx}(x,t),\ t>0,\ x\in\mathbb{R}. \quad (6.40)$$

In what follows, we will consider this transformed equation. To obtain an initial condition for (6.40) equation, we note that $\partial_t^\alpha \tilde{u}(x,t)$ in (6.37) is equal to $I_t^{1-\alpha}v(x,t)$. In view of this and (6.38), putting in (6.37) $t=0$ we have the following condition:

$$\left(I_t^{1-\alpha}\right)v(x,0) = \varphi''''(x) + f_{xx}(x,0),\ \ x\in\mathbb{R}. \quad (6.41)$$

In order to form an overdetermination condition for the function $v(x,t)$, we differentiate (6.33) in time variable t. At the same time, taking into account $u_{txx}(x,t)=v(x,t)$ in the resulting equality and using the condition (6.35), we get

$$v(0,t) = \frac{\partial^\alpha}{\partial t^\alpha}g'(t) - k(t)\varphi(0) - \int_0^t k(\tau)g'(t-\tau)d\tau - f_t(0,t),\ \ t>0. \quad (6.42)$$

In view of the matching condition $\varphi''(0)=\partial_t^\alpha g(0) - f(0,0)$ from (6.40)–(6.42), the relations (6.37)–(6.39) can be derived using inverse transformations.

Remark 6.1 In fact, we have replaced the original Eq. (6.33) with the Caputo fractional derivative by differentiation by Eq. (6.40) with the Riemann-Liouville fractional derivative. Later we will see that it is convenient to study the inverse problem with Eqs. (6.40)–(6.42), since the additional condition (6.42) contains an unknown function $k(t)$ outside the integral.

The problem of determining the solution to the Cauchy-type problem (6.40) and (6.41) we call as *the direct problem* for Eq. (6.40). In this case, the function $k(t)$ is considered known. *A regular solution* of this problem consists of determining the function $u(x,t)$ such that

$$t^{1-\alpha}u \in C\left(\overline{R_T}\right),\ t^{1-\alpha}\frac{\partial^\alpha}{\partial t^\alpha}u \in C\left(\overline{R_T}\right),$$

where $R_T := \{(x,t) : x\in\mathbb{R}, 0<t\le T\}$, $T>0$ is any fixed number.

We introduce the weighted space

$$C_\alpha[0,T] = \left\{g(t) : \|g\|_\alpha := \left\|t^{1-\alpha}g(t)\right\|_C < \infty\right\}.$$

The weighted space $C_\alpha^n[0,T]$ of functions g is defined by

$$C_\alpha^n[a,b] = \left\{g(t) : [0,T]\to\mathbb{R},\ g^{(n-1)}(t)\in C[a,b],\ g^{(n)}(t)\in C_\alpha[a,b]\right\}$$

with the norm

$$\|g\|_\alpha = \sum_{k=0}^{n-1} \|g^{(k)}\|_{C[0,T]} + \left\|g^{(n)}\right\|_{C_\alpha[0,T]}.$$

Let $C^{1-\alpha,m}(R_T)$ be the class of the m times continuously differentiable with respect to $x \in \mathbb{R}$ variable, continuous in t and its fractional integral of the order $1-\alpha$ is continuously differentiable in t on $[0,T]$, and bounded together with all derivatives in R_T functions. By $C^m(\mathbb{R})$ we denote the class of functions that are bounded in $\mathbb{R}$ together with all derivatives up to the order of m. When $m=0$, $C^0(\mathbb{R}) =: C(\mathbb{R})$.

We also introduce the class $H^l(\mathbb{R})$ of bounded Hölder continuous functions on $\mathbb{R}$ with $l \in (0,1)$. For the space $H^{m+l}(\mathbb{R})$ (m is a nonnegative integer) and norms $|\cdot|^l$, $|\cdot|^{m+l}$ we use the concept of Sect. 3.1.1 (see also ([21], pp. 16–27).

By $C\left([0,\ T],\ H^l(\mathbb{R})\right)$ we denote the class of continuous functions with respect to t variable on the segment $[0,\ T]$ with values in $H^l(\mathbb{R})$. For a fixed t, the norm of the function $f(x,t)$ in the $H^l(\mathbb{R})$ will be denoted by $|f|^l(t)$. The norm of a function $t^{1-\alpha}f(x,t)$ in $C\left([0,\ T],\ H^l(\mathbb{R})\right)$ is defined by the equality

$$\|f\|_\alpha^l := \max_{t\in[0,\ T]}\left[t^{1-\alpha}\,|f|^l(t)\right].$$

Further we will use Definitions and Propositions of Sect. 3.1 in Chap. 3.

6.2.2 *Investigation of the Cauchy Problem: Method of Successive Approximations*

First we consider the following Cauchy-type problem:

$$\frac{\partial^\alpha v(x,t)}{\partial t^\alpha} - \Delta v = F(x,t),\ t>0,\ x \in \mathbb{R}^n \tag{6.43}$$

with the initial condition

$$I_t^\alpha v(x,0) = \Phi(x),\ \ x \in \mathbb{R}^n, \tag{6.44}$$

where Δ is the n-dimensional Laplacian in x variable.

To find the solution of (6.43) and (6.44) we apply the Fourier transform (see Definition 3.4) with respect to space variable x to this equation. As result we have

$$\frac{\partial^\alpha \tilde{v}(\xi,t)}{\partial t^\alpha} + |\xi|^2\tilde{v}(\xi,t) = \tilde{F}(\xi,t),\ t>0,\ \xi \in \mathbb{R}^n \tag{6.45}$$

with the initial condition

$$I_t^{\alpha}\tilde{v}(\xi,0)=\tilde{\Phi}(\xi),\ \ \xi\in\mathbb{R}^n, \tag{6.46}$$

where the symbol $\tilde{}$ over functions denotes the Fourier transform of the corresponding functions.

The solution to problem (6.45) and (6.46) has the following form (see Theorem 1.1 in Sect. 1.1.3):

$$\tilde{v}(\xi,t)=t^{\alpha-1}E_{\alpha,\alpha}\left(-|\xi|^2t^{\alpha}\right)\tilde{\Phi}(\xi)+$$

$$+\int_0^t(t-\tau)^{\alpha-1}E_{\alpha,\alpha}\left(-|\xi|^2(t-\tau)^{\alpha}\right)\tilde{F}(\xi,\tau)d\tau,\ t>0,\ \xi\in\mathbb{R}^n. \tag{6.47}$$

Now we find the inverse Fourier transform (see Definition 3.4) of function $E_{\alpha,\alpha}\left(-|\xi|^2t^{\alpha}\right)$ in terms of Fox's $H-$ function. Based on (3.15) and Proposition 3.3, we have the following relations:

$$F^{-1}\left[E_{\alpha,\alpha}\left(-|\xi|^2t^{\alpha}\right)\right]=\frac{1}{(2\pi)^n}\int_{\mathbb{R}^n}H_{1,2}^{1,1}\left[|\xi|^2t^{\alpha}\left|\begin{matrix}(0,1)\\(0,1),(1-\alpha,\alpha)\end{matrix}\right.\right]e^{i\xi\cdot x}d\xi=$$

$$=\frac{|x|^{1-\frac{n}{2}}}{(2\pi)^n}\int_0^{\infty}y^{\frac{n}{2}}H_{1,2}^{1,1}\left[y^2t^{\alpha}\left|\begin{matrix}(0,1)\\(0,1),(1-\alpha,\alpha)\end{matrix}\right.\right]J_{\frac{n}{2}-1}\left(|x|y\right)dy.$$

Using Proposition 3.4, (3.12)–(3.15), we continue the previous calculations

$$F^{-1}\left[E_{\alpha,\alpha}\left(-|\xi|^2t^{\alpha}\right)\right]=\frac{1}{(2\pi)^{\frac{n}{2}}|x|^n}H_{3,2}^{1,2}\left[\frac{4t^{\alpha}}{|x|^2}\left|\begin{matrix}(1-\frac{n}{2},1),(0,1),(0,1)\\(0,1),(1-\alpha,\alpha)\end{matrix}\right.\right]=$$

$$=\frac{1}{(2\pi)^{\frac{n}{2}}|x|^n}H_{2,1}^{0,2}\left[\frac{4t^{\alpha}}{|x|^2}\left|\begin{matrix}(1-\frac{n}{2},1),(0,1)\\(1-\alpha,\alpha)\end{matrix}\right.\right]=\frac{1}{(2\pi)^{\frac{n}{2}}|x|^n}H_{1,2}^{2,0}\left[\frac{|x|^2}{4t^{\alpha}}\left|\begin{matrix}(\alpha,\alpha)\\(\frac{n}{2},1),(1,1)\end{matrix}\right.\right]. \tag{6.48}$$

Applying the inverse Fourier transform to (6.47), in view of (6.48) and the convolution property, we obtain

$$v(x,t)=\int_{\mathbb{R}^n}G_{\alpha}(x-\xi,t)\Phi(\xi)d\xi$$

$$+\int_0^t\int_{\mathbb{R}^n}G_{\alpha}(x-\xi,t-\tau)F(\xi,\tau)d\xi d\tau,\ t>0,\ \xi\in\mathbb{R}^n, \tag{6.49}$$

where

$$G_\alpha(x,t) = \frac{t^{\alpha-1}}{(2\pi)^{\frac{n}{2}}|x|^n} H^{2,0}_{1,2}\left[\frac{|x|^2}{4t^\alpha}\middle|^{(\alpha,\alpha)}_{(\frac{n}{2},1),(1,1)}\right]$$

is the fundamental solution of the n-dimensional fractional diffusion equation with the Riemann-Liouville derivative in terms of Fox's $H-$function.

Further for $n = 1$ we use the following estimates [22]:

$$\left|\frac{\partial^m}{\partial x^m} G_\alpha(x,t)\right| \leq C t^{-\frac{\alpha(1+m)}{2}-1+\alpha} \exp\left\{-\mu_m t^{-\frac{\alpha}{2-\alpha}} x^{\frac{2}{2-\alpha}}\right\},\ m = 0, 1, 2, 3, \quad (6.50)$$

for $x^2 \geq t^\alpha$; where $\mu_0 := (2-\alpha)\alpha^{\alpha/(2-\alpha)} \cdot 2^{-2/(2-\alpha)}$, and as μ_m is any positive number less than μ_0;

$$\left|\frac{\partial^m}{\partial x^m} G_\alpha(x,t)\right| \leq C t^{-\frac{\alpha(m-1)}{2}-1},\ m = 0, 1, 2, 3, \quad (6.51)$$

for $x^2 \leq t^\alpha$, $(x,t) \in R_T$. In (6.50), (6.51) the letter C denotes various positive constants.

We also note the equality [22]:

$$\int_{\mathbb{R}} G_\alpha(x,t)dx = C_0 t^{\alpha-1},\ t \in [0,T], \quad (6.52)$$

where C_0 depends only on α.

Remark 6.2 When the estimates (6.50), (6.51) for the function $G_\alpha(x,t)$ are fulfilled and the functional classes for given functions are chosen appropriately, the integrals in (6.49) converge.

In (6.40), introducing the notations $\int\limits_0^t k(\tau)v(x,t-\tau)d\tau + k(t)\varphi''(x) + f_{txx}(x,t) =: F(x,t)$, $\varphi''''(x) + f_{xx}(x,0) =: \Phi(x)$ and applying the formula (6.49) to direct problem (6.40), (6.41) for $n = 1$, we obtain the integral equation for determining $v(x,t)$:

$$v(x,t) = v_0(x,t) - \int_0^t \int_{\mathbb{R}} G_\alpha(x-\xi, t-\tau) \int\limits_0^\tau k(\zeta)v(\xi, \tau-\zeta)d\zeta d\xi d\tau, \quad (6.53)$$

where

$$v_0(x,t) := \int_{\mathbb{R}} G_\alpha(x-\xi,t)\left[\varphi''''(\xi) + f_{\xi\xi}(\xi,0)\right]d\xi$$

$$+ \int_0^t \int_{\mathbb{R}} G_\alpha(x-\xi,t-\tau)\left[k(\tau)\varphi''(\xi) + f_{\tau\xi\xi}(\xi,\tau)\right]d\xi d\tau. \quad (6.54)$$

Following assertion holds true:

Lemma 6.5 *If* $k(t) \in C[0, T]$, $f(x,t) \in C^1\left([0, T], H^{2+l}(\mathbb{R})\right)$, $l \in (0, 1)$, $\varphi(x) \in C^4(\mathbb{R})$, *then there exists a unique solution of the integral equation (6.53)* $t^{\alpha-1}v(x,t) \in C^{1-\alpha, 2}(R_T)$.

Proof For proof of Lemma 6.5 we use the method of successive approximations and consider the sequence of functions:

$$v_n(x,t) = -\int_0^t \int_{\mathbb{R}} G_\alpha(x-\xi, t-\tau) \int_0^\tau k(\zeta) v_{n-1}(\xi, \tau-\zeta) d\zeta d\xi d\tau, n = 1, 2, \cdots, \tag{6.55}$$

where $v_0(x,t)$ is determined by the equality (6.54).

Introduce the notations

$$\varphi_0 := \|\varphi\|_{C^4(\mathbb{R})} = \sum_{i=0}^{4} \sup_{x\in\mathbb{R}} \left| \frac{d^i \varphi(x)}{dx^i} \right|, \quad f_0 := \max\left\{\|f\|_\alpha^{2+l}, \|f_t\|_\alpha^{2+l}\right\}$$

and in what follows $\|k\|_\alpha$ means the norm of k in $C_\alpha[0, T]$.

Here we also use the following relation:

$$I_{0+}^\alpha t^\mu = \frac{\Gamma(\mu+1)}{\Gamma(\alpha+\mu+1)} t^{\alpha+\mu}, \ \mu \in \mathbb{R}. \tag{6.56}$$

Now using (6.54), (6.55), (6.52), (6.56), we estimate the modulus of $v_0(x,t)$ in the domain R_T as

$$|v_0(x,t)| \le C_0 \left[t^{\alpha-1}(\varphi_0 + f_0) + (\varphi_0\|k\|_\alpha + f_0)\Gamma(\alpha) I_{0+}^\alpha t^{\alpha-1}\right]$$

$$= t^{\alpha-1} C_0 \left[\varphi_0 + f_0 + (\varphi_0\|k\|_\alpha + f_0)\frac{\Gamma^2(\alpha)}{\Gamma(2\alpha)} T^\alpha\right] =: t^{\alpha-1}\varphi_{00}.$$

Similarly, estimate $v_1(x,t)$

$$|v_1(x,t)| \le \varphi_{00} C_0 \|k\|_\alpha \Gamma^2(\alpha) I_{0+}^\alpha \left(I_{0+}^\alpha t^{\alpha-1}\right) = \varphi_{00} C_0 \|k\|_\alpha \Gamma^3(\alpha) \frac{t^{3\alpha-1}}{\Gamma(3\alpha)}.$$

In this way, for arbitrary $n = 1, 2, \ldots$, we have

$$|v_n(x,t)| \le \varphi_{00}\Gamma(\alpha)\left(C_0\|k\|_\alpha \Gamma^2(\alpha)\right)^n \frac{t^{(2n+1)\alpha-1}}{\Gamma((2n+1)\alpha)}.$$

Multiplying the last inequality by $t^{1-\alpha}$, we find estimates

$$t^{1-\alpha}\left|v_n(x,t)\right| \le \varphi_{00}\Gamma(\alpha)\left(C_0\|k\|_\alpha\Gamma^2(\alpha)\right)^n \frac{t^{2\alpha n}}{\Gamma\left(2\alpha n+\alpha\right)},$$

which are fulfilled for all $n = 0, 1, 2, \ldots$ in $\overline{R}_T$.

It follows from the above estimates that the series

$$t^{1-\alpha}v(x,t) = \sum_{n=0}^{\infty} t^{1-\alpha}v_n(x,t)$$

converges uniformly in $\overline{R}_T$, since it can be majorized in $\overline{R}_T$ by the convergent numerical series

$$\varphi_{00}\Gamma\left(\alpha\right)\sum_{n=0}^{\infty}\left(C_0\|k\|_\alpha\Gamma^2(\alpha)\right)^n \frac{T^{2\alpha n}}{\Gamma\left(2\alpha n+\alpha\right)}.$$

This means the following estimate for the solution of the integral equation (6.53) holds

$$t^{1-\alpha}\left|v(x,t)\right| \le \varphi_{00}\Gamma\left(\alpha\right)\sum_{n=0}^{\infty}\left(C_0\|k\|_\alpha\Gamma^2(\alpha)\right)^n \frac{T^{2\alpha n}}{\Gamma\left(2\alpha n+\alpha\right)}$$

$$= \varphi_{00}\Gamma(\alpha)E_{2\alpha,\alpha}\left(C_0\|k\|_\alpha\Gamma^2(\alpha)T^{2\alpha}\right),\ (x,t)\in\overline{R}_T, \tag{6.57}$$

where $E_{2\alpha,\alpha}(\cdot)$ is the Mittag-Leffler function of a nonnegative real argument defined in Sect. 1.1.3, formula (1.26).

Under the assumptions of Lemma 6.5 and based on the estimates (6.50), (6.51) (see also [18]) one has $t^{1-\alpha}v_n(x,t) \in C^{1-\alpha,\,2}\left(R_T\right)$. According to the general theory of integral equations, this implies that the same property will be possessed the function $t^{1-\alpha}v(x,t)$ in R_T. As usual, this function is a solution of integral equation (6.53).

In what follows, we need an a priori estimate for the norm of the difference between the solution of the original integral equation (6.53) and the solution of this equation with perturbed functions $\tilde{k}$, $\tilde{f}$ and $\tilde{\varphi}$. Let $\tilde{\vartheta}(x,t)$ be a solution of the integral equation (6.53) corresponding to the functions $\tilde{k}$, $\tilde{f}_{xx}$, $\tilde{f}_{txx}$, $\tilde{\varphi}''$ and $\tilde{\varphi}''''$, i.e.,

$$\tilde{v}(x,t) = \tilde{v}_0(x,t) - \int_0^t\int_{\mathbb{R}} G_\alpha(x-\xi,t-\tau)\int_0^\tau \tilde{k}(\zeta)\tilde{v}(\xi,\tau-\zeta)d\zeta d\xi d\tau, \tag{6.58}$$

where $\tilde{v}_0$, according to (6.54), is determined as

$$\tilde{v}_0(x,t) := \int_{\mathbb{R}} G_\alpha(x-\xi,t)\left[\tilde{\varphi}''''(\xi)+\tilde{f}_{\xi\xi}(\xi,0)\right]d\xi$$

$$+\int_0^t\int_{\mathbb{R}} G_\alpha(x-\xi,t-\tau)\left[\tilde{k}(\tau)\tilde{\varphi}''(\xi)+\tilde{f}_{\tau\xi\xi}(\xi,\tau)\right]d\xi d\tau. \tag{6.59}$$

Composing the difference $v - \tilde{v}$ with the help of Eqs. (6.53), (6.58) and introducing the notations $v - \tilde{v} = \overline{v}$, $v_0 - \tilde{v}_0 = \overline{v}_0$, $k - \tilde{k} = \overline{k}$, we obtain the integral equation

$$\overline{v}(x,t) = \overline{v}_0(x,t) - \int_0^t \int_{\mathbb{R}} G_\alpha(x-\xi, t-\tau) \int_0^\tau \overline{k}(\zeta)\tilde{v}(\xi, \tau-\zeta)d\zeta d\xi d\tau$$

$$- \int_0^t \int_{\mathbb{R}} G_\alpha(x-\xi, t-\tau) \int_0^\tau k(\zeta)\overline{v}(\xi, \tau-\zeta)d\zeta d\xi d\tau, \qquad (6.60)$$

from which, based on (6.57), it is derived the following linear integral inequality for $\left|\overline{\vartheta}(x,t)\right|$:

$$|\overline{v}(x,t)| \le |\overline{v}_0(x,t)| + \varphi_{00} C_0 \Gamma^4(\alpha) E_{2\alpha,\alpha}\left(C_0 \|\tilde{k}\|_\alpha \Gamma^2(\alpha) T^{2\alpha}\right) \frac{T^{3\alpha-1}}{\Gamma(3\alpha)} \|\overline{k}\|_\alpha$$

$$+ \|k\|_\alpha \int_0^t \int_{\mathbb{R}} G_\alpha(x-\xi, t-\tau) \int_0^\tau (\tau-\zeta)^{\alpha-1} |\overline{v}(\xi,\zeta)| d\zeta d\xi d\tau. \qquad (6.61)$$

It follows from the equalities (6.54) and (6.59) the estimate

$$|\overline{v}_0(x,t)| \le C_0 t^{\alpha-1} \Bigg[\left(1 + \|\tilde{k}\|_\alpha \frac{\Gamma^2(\alpha)}{\Gamma(2\alpha)} T^\alpha\right) \|\overline{\varphi}\|_{C^4(\mathbb{R})}$$
$$+ \left(1 + \frac{\Gamma^2(\alpha)}{\Gamma(2\alpha)} T^\alpha\right) \|\overline{f}\|_\alpha^{2+l} + \varphi_0 \frac{\Gamma^2(\alpha)}{\Gamma(2\alpha)} T^\alpha \|\overline{k}\|_\alpha \Bigg],$$

where $\overline{\varphi} = \varphi - \tilde{\varphi}$, $\overline{f} = f - \tilde{f}$.

Set

$$\lambda = \lambda\left(\alpha,\ T,\ \|\tilde{k}\|_\alpha,\ \tilde{\varphi}_0,\ f_0\right)$$

$$= \max\left\{ C_0 \left(1 + \|\tilde{k}\|_\alpha \frac{\Gamma^2(\alpha)}{\Gamma(2\alpha)} T^\alpha\right), C_0 \left(1 + \frac{\Gamma^2(\alpha)}{\Gamma(2\alpha)} T^\alpha\right), \right.$$

$$\left. \frac{\Gamma^2(\alpha)}{\Gamma(2\alpha)} T^\alpha \varphi_0,\ \varphi_{00} C_0 \Gamma^4(\alpha) E_{2\alpha,\alpha}\left(C_0 \|\tilde{k}\|_\alpha \Gamma^2(\alpha) T^{2\alpha}\right) \frac{T^{2\alpha}}{\Gamma(3\alpha)} \right\}.$$

Applying the successive approximation method to inequality (6.61) with the help of the scheme

$$|\overline{v}(x,t)|_0 \le \lambda\left(\|\overline{\varphi}\|_{C^4(\mathbb{R})} + \|\overline{f}\|_\alpha^{2+l} + \|\overline{k}\|_\alpha\right),$$

$$|\overline{v}(x,t)|_n \le \|k\|_\alpha \int_0^t \int_{\mathbb{R}} G_\alpha(x-\xi, t-\tau) \int_0^\tau (\tau-\zeta)^{\alpha-1} |\overline{v}(\xi,\zeta)| d\zeta d\xi d\tau,\ n = 1, 2, ...,$$

we arrive at the estimate

$$t^{1-\alpha}\left|\overline{v}(x,t)\right| \\ \leq \lambda\varphi_{00}\Gamma(\alpha)E_{2\alpha,\alpha}\left(C_0\|k\|_\alpha\Gamma^2(\alpha)T^{2\alpha}\right)\left(\|\overline{\varphi}\|_{C^4(\mathbb{R})}+\left\|\overline{f}\right\|_\alpha^{2+l}+\left\|\overline{k}\right\|_\alpha\right), \tag{6.62}$$

which will be used in the next section. Indeed the expression (6.62) is the stability estimate for the solution of the Cauchy-type problem (6.40) and (6.41). The uniqueness for this solution follows from (6.62). □

6.2.3 Investigation of the Inverse Problem Solvability

We set in (6.53) $x = 0$ and use the overdetermination condition (6.42). Taking into account the function $v_0(x,t)$ in the form (6.54), after simple conversion, we get the following integral equation for determining $k(t)$:

$$k(t) = k_0(t) - \frac{1}{\varphi(0)}\int_0^t k(\tau)\left[g'(t-\tau) + \int_{\mathbb{R}} G_\alpha(\xi, t-\tau)\varphi''(\xi)d\xi\right]d\tau \\ + \frac{1}{\varphi(0)}\int_0^t\int_{\mathbb{R}} G_\alpha(\xi,t-\tau)\int_0^\tau k(\zeta)v(\xi,\tau-\zeta)d\zeta d\xi d\tau, \tag{6.63}$$

where

$$k_0(t) := \frac{1}{\varphi(0)}\left[f_t(0,t) - \frac{\partial^\alpha}{\partial t^\alpha}g'(t)\right. \\ \left. - \int_{\mathbb{R}} G_\alpha(\xi,t)\left[\varphi''''(\xi) + f_{\xi\xi}(\xi,0)\right]d\xi - \int_0^t\int_{\mathbb{R}} G_\alpha(\xi,t-\tau)f_{\tau\xi\xi}(\xi,\tau)d\xi d\tau\right].$$

Note that the solution of integral equation (6.53) depends on k, i.e., $v = v(x,t;k)$. We introduce the operator

$$G[k](t) = k_0(t) - \frac{1}{\varphi(0)}\int_0^t k(\tau)\left[g'(t-\tau) + \int_{\mathbb{R}} G_\alpha(\xi, t-\tau)\varphi''(\xi)d\xi\right]d\tau \\ + \frac{1}{\varphi(0)}\int_0^t\int_{\mathbb{R}} G_\alpha(\xi,t-\tau)\int_0^\tau k(\zeta)v(\xi,\tau-\zeta)d\zeta d\xi d\tau. \tag{6.64}$$

Then Eq. (6.63) can be written in a more convenient form as

$$k(t) = G[k](t), \ t \in [0, T]. \tag{6.65}$$

Fix a number $\rho > 0$ and consider the ball

$$\Phi_\alpha^T (k_0, \rho) := \{k(t) : k(t) \in C_\alpha[0, T], \ \|k - k_0\|_\alpha \leq \rho\}.$$

The main result of this section is the following assertion:

Theorem 6.3 *If* $\left(t^{1-\alpha} f, \, t^{1-\alpha} f_t\right) \in C\left([0, \, T], \ H^{2+l}(\mathbb{R})\right)$, $\varphi(x) \in C^4(\mathbb{R})$, $\left(g(t), \, \frac{\partial^\alpha}{\partial t^\alpha} g(t)\right) \in C_\alpha^1[0, \, T]$, $|\varphi(0)| \neq 0$, $\varphi(0) = g(0)$, $\varphi''(0) = \partial_t^\alpha g(0) - f(0, 0)$, *then there exists a number* $T^* \in (0, \, T)$, *such that there exists a unique solution* $k(t) \in C_\alpha[0, \, T^*]$ *of the inverse problem (6.33)–(6.35).*

Proof At first we show that for an enough small $T > 0$ the operator G maps the ball $\Phi_\alpha^T (k_0, \rho)$ into itself; i.e., the condition $k(t) \in \Phi_\alpha^T (k_0, \rho)$ implies that $G[k](t) \in \Phi_\alpha^T (k_0, \rho)$. Indeed, for any continuous function $k(t)$, the function $G[k(t)]$ calculated using formula (6.65) will be continuous. Moreover, based on (6.52), (6.56), (6.57), estimating the norm of the differences, we find

$$\begin{aligned}
&|G[k](t) - k_0(t)| \\
&\leq \frac{\|k\|_\alpha}{|\varphi(0)|}\Big[\left(\|g'\|_\alpha + C_0\varphi_0\right)\Gamma(\alpha) I_{0+}^\alpha t^{\alpha-1} + C_0\varphi_{00}\Gamma^3(\alpha) \\
&\times E_{2\alpha,\alpha}\left(C_0\|k\|_\alpha \times \Gamma^2(\alpha)T^{2\alpha}\right) I_{0+}^\alpha\left(I_{0+}^\alpha t^{\alpha-1}\right)\Big] \leq \frac{t^{\alpha-1}\|k\|_\alpha}{|\varphi(0)|} \\
&\times\left[\left(\|g'\|_\alpha + C_0\varphi_0\right)\frac{\Gamma^2(\alpha)}{\Gamma(2\alpha)} t^\alpha + C_0\varphi_{00} E_{2\alpha,\alpha}\left(C_0\|k\|_\alpha\Gamma^2(\alpha)T^{2\alpha}\right)\frac{\Gamma^4(\alpha)}{\Gamma(3\alpha)} t^{2\alpha}\right].
\end{aligned}$$

Note that the expression on the right side of this inequality is a monotonically increasing function of T, $\|k\|$, and from $q(t) \in \Phi_\alpha^T (k_0, \rho)$ it follows the inequality

$$\|k\|_\alpha \leq \rho + \|k_0\|_\alpha. \tag{6.66}$$

In view of these facts, we obtain for $t \in [0, T]$

$$\begin{aligned}
\|G[k](t) - k_0(t)\|_\alpha &= \max_{t\in[0,T]} t^{1-\alpha}|G[k](t) - k_0(t)| \\
\leq \frac{\|k_0\|_\alpha + \rho}{|\varphi(0)|}&\Big[\left(\|g'\|_\alpha + C_0\varphi_0\right)\frac{\Gamma^2(\alpha)}{\Gamma(2\alpha)} T^\alpha + C_0\varphi_{00} \\
&\times E_{2\alpha,\alpha}\left(C_0\left(\|k_0\|_\alpha + \rho\right)\Gamma^2(\alpha)T^{2\alpha}\right)\frac{\Gamma^4(\alpha)}{\Gamma(3\alpha)} T^{2\alpha}\Big].
\end{aligned}$$

Let T_1 be the smallest positive root of the equation

$$m_1(T) = \frac{(\|k_0\|_\alpha + \rho)}{|\varphi(0)|}\Big[\left(\|g'\|_\alpha + C_0\varphi_0\right)\frac{\Gamma^2(\alpha)}{\Gamma(2\alpha)}T^\alpha$$
$$+ C_0\varphi_{00}E_{2\alpha,\alpha}\left(C_0\left(\|k_0\|_\alpha + \rho\right)\Gamma^2(\alpha)T^{2\alpha}\right)\frac{\Gamma^4(\alpha)}{\Gamma(3\alpha)}T^{2\alpha}\Big] = \rho.$$

Note that the unique such root exists, since the function $m_1(T)$ is monotonically increasing and its range is positive real numbers. Then, G maps $\Phi_\alpha^T(k_0, \rho)$ into itself for each fixed $T \in [0, T_1]$.

Now consider functions $\left\{k(t), \tilde{k}(t)\right\} \in \Phi_\alpha^T(k_0, \rho)$ and estimate the distance between $G[k](t)$ and $G[\tilde{k}](t)$ in the space $C_\alpha[0, T]$. The function $\tilde{k}(t)$ corresponds to $\tilde{v}(x, t)$, which satisfies Eq. (6.58) with $\tilde{k}(t)$. We have

$$|G[k](t) - G[k_0](t)| \le \frac{t^{\alpha-1}}{|\varphi(0)|}\Big[\left(\|g'\|_\alpha + C_0\varphi_0\right)\frac{\Gamma^2(\alpha)}{\Gamma(2\alpha)}t^\alpha$$
$$+C_0\varphi_{00}E_{2\alpha,\alpha}\left(C_0\|\tilde{k}\|_\alpha\Gamma^2(\alpha)T^{2\alpha}\right)\frac{\Gamma^4(\alpha)}{\Gamma(3\alpha)}t^{2\alpha}\left(1 + \|k\|_\alpha\right)\Big]\|\bar{k}\|_\alpha.$$

Taking into account relation (6.66) for functions k and $\tilde{k}$, from the previous inequality we have

$$\left\|G[k](t) - G[\tilde{k}](t)\right\|_\alpha$$
$$= \max_{t\in[0,T]} t^{1-\alpha}\left|G[k](t) - G[\tilde{k}](t)\right| \le \frac{1}{|\varphi(0)|}\Big[\left(\|g'\|_\alpha + C_0\varphi_0\right)\frac{\Gamma^2(\alpha)}{\Gamma(2\alpha)}T^\alpha$$
$$+ C_0\varphi_{00}E_{2\alpha,\alpha}\left(C_0\left(\|\tilde{k_0}\|_\alpha + \rho\right)\Gamma^2(\alpha)T^{2\alpha}\right)\frac{\Gamma^4(\alpha)}{\Gamma(3\alpha)}T^{2\alpha}\left(1 + \lambda\left(\|k\|_\alpha + \rho\right)\right)\Big]\|\bar{k}\|_\alpha.$$

Here we also have replaced in definitions of constants λ and φ_{00} functions $\|k\|_\alpha$ and $\|\tilde{k}\|_\alpha$ by right-hand sides of (6.66).

Now let T_2 be the smallest positive root of the equation

$$m_2(T) = \frac{1}{|\varphi(0)|}\Big[\left(\|g'\|_\alpha + C_0\varphi_0\right)\frac{\Gamma^2(\alpha)}{\Gamma(2\alpha)}T^\alpha$$
$$+C_0\varphi_{00}E_{2\alpha,\alpha}\left(C_0\left(\|\tilde{k_0}\|_\alpha + \rho\right)\Gamma^2(\alpha)T^{2\alpha}\right)\frac{\Gamma^4(\alpha)}{\Gamma(3\alpha)}T^{2\alpha}\left(1 + \lambda\left(\|k\|_\alpha + \rho\right)\right)\Big] = 1.$$

Then, for $T \in [0, T_2)$ it follows that the distance between the functions $G[k](t)$ and $G[\tilde{k}](t)$ in the functional space $C_\alpha[0, T]$ is less than the distance between the functions $k(t)$ and $\tilde{k}(t)$ in $C_\alpha[0, T]$. Consequently, if we choose $T^* < \min(T_1, T_2)$, then the operator G is a contraction in the ball $\Phi_\alpha^{T^*}(k_0, \rho)$. Then, the operator G has a unique fixed point in the ball $\Phi_\alpha^{T^*}(k_0, \rho)$; i.e., there exists a unique solution of Eq. (6.65). Theorem 6.3 is proven. □

Let T be a positive fixed number. Consider the set $\Omega(\omega_0)$ ($\omega_0 > 0$ is some fixed number) of the given functions $(f,\ \varphi,\ g)$ for which all conditions from Theorem 6.3 are fulfilled and $\max\left\{\|f\|_\alpha^{2+l},\ \|f_t\|_\alpha^{2+l},\ \|\varphi\|_{C^4(\mathbb{R})},\ \|g'\|_\alpha,\ \left\|\frac{\partial^\alpha}{\partial t^\alpha}g'\right\|_\alpha\right\} \le \omega_0$. By $Q(\omega_1)$ we denote the class of functions $k(t) \in C_\alpha[0,\ T]$ satisfying the inequality $\|k\|_\alpha \le \omega_1$ with some fixed positive number ω_1.

Theorem 6.4 *Let* $(f,\ \varphi,\ g) \in \Omega(\omega_0)$, $\left(\tilde{f},\ \tilde{\varphi},\ \tilde{g}\right) \in \Omega(\omega_0)$ *and* $\left(k,\ \tilde{k}\right) \in Q(\omega_1)$. *Then If* $\alpha \in (\frac{1}{2}, 1)$, *for the solution of the inverse problem (6.33)–(6.35) the following stability estimate holds:*

$$\|k - \tilde{k}\|_\alpha \le d\kappa, \tag{6.67}$$

where

$$\kappa := \Bigg(\|f - \tilde{f}\|_\alpha^{2+l} + \|f_t - \tilde{f}_t\|_\alpha^{2+l} + \|\varphi - \tilde{\varphi}\|_{C^4(\mathbb{R})} + \|g' - \tilde{g}'\|_\alpha + \left\| \frac{\partial^\alpha}{\partial t^\alpha} g' - \frac{\partial^\alpha}{\partial t^\alpha} \tilde{g}' \right\| \Bigg),$$

the constant d depends only on T, α, g_0, ω_0, ω_1.

Proof To prove this theorem, using (6.63) we write down the equations for $\tilde{k}(t)$ and then estimate the difference $k(t) - \tilde{k}(t)$

$$\left|k(t) - \tilde{k}(t)\right| \le d_0\kappa + d_1 \int_0^t \tau^{\gamma-1}(t-\tau)^{\alpha-1} \left|k(\tau) - \tilde{k}(\tau)\right| d\tau,\ t \in [0,\ T], \tag{6.68}$$

where d_0 and d_1 depend on the same constants as d.

Here we use the following assertion as a generalized analog of Gronwall's inequality ([23], p. 189):

Proposition 6.3 *Suppose* $b \ge 0$, $\alpha > 0$, $\gamma > 0$, $\alpha + \gamma > 1$ *and* $a(t)$ *is nonnegative function locally integrable on* $0 \le t < T$ *and suppose* $t^{\gamma-1}\omega(t)$ *is nonnegative and locally integrable on* $0 \le t < T$ *with*

$$\omega(t) \le a(t) + b \int_0^t (t-s)^{\alpha-1} s^{\gamma-1} \omega(s) ds,$$

then, there holds

$$\omega(t) \le a(t) W_{\alpha,\gamma}\left((b\Gamma(\alpha))^{\frac{1}{\alpha+\gamma-1}}\, t \right),$$

where $W_{\alpha,\gamma}(t) = \sum\limits_{m=0}^{\infty} c_m t^{m(\alpha+\gamma-1)}$, $c_0 = 1$, $\frac{c_{m+1}}{c_m} = \frac{\Gamma(m(\alpha+\gamma-1)+\gamma)}{\Gamma(m(\alpha+\gamma-1)+\alpha+\gamma)}$ *for* $m \ge 0$. *As* $t \to +\infty$ $W_{\alpha,\gamma}(t) = O\left(t^{\frac{1}{2}\left(\frac{\alpha+\gamma-1}{\alpha} - \gamma\right)} \exp\left(\frac{\alpha}{\alpha+\gamma-1} t^{\frac{\alpha+\gamma-1}{\alpha}} \right) \right)$.

From Proposition 6.3, we get the estimate

$$\left|k(t)-\tilde{k}(t)\right| \leq \kappa d_0 W_{\alpha,\alpha}\left((d_1\Gamma(\alpha))^{\frac{1}{2\alpha-1}}\, t\right),\ t\in[0,\ T].$$

Setting

$$d = d_0 W_{\alpha,\alpha}\left((d_1\Gamma(\alpha))^{\frac{1}{2\alpha-1}}\, T\right),$$

we obtain (6.67). □

From Theorem 6.4 readily follows the uniqueness theorem for any $T > 0$:

Theorem 6.5 *Let the functions* $k,\ f,\ \varphi,\ g$ *and* $\tilde{k},\ \tilde{f},\ \tilde{\varphi},\ \tilde{g}$ *have the same meaning as in Theorem 6.3. Moreover, if* $f=\tilde{f},\ \varphi=\tilde{\varphi},\ g=\tilde{g}$ *for* $t\in[0,\ T]$, *and* $\alpha\in(\frac{1}{2},1)$, *then* $k(t)=\tilde{k}(t)$ *for* $t\in C_\alpha[0,\ T]$.

Remark 6.3 By found the function $k(t)$ the solution to direct problem (6.40), (6.41) is determined as solution of integral equation (6.53). Then, the solution to direct problem (6.33), (6.34) is found by virtue of the formula

$$u(x,t)=\varphi(x)+\int_0^t\int_{-\infty}^x (x-\xi)v(\xi,\tau)d\xi d\tau,$$

provided that $(u,u_x)\to 0$ as $|x|\to\infty$, and $(u,\ u_t)\in C^{1-\alpha,4}(R_T)$.

6.3 Conclusion

In this chapter, we have proved the local existence on the segment $[0,\ T^*]$, the uniqueness and obtained the stability estimate of solutions to the inverse problems (6.1)–(6.4) (Sect. 6.1) and (6.33)–(6.35) (Sect. 6.2). The proofs of Theorems 6.2 and 6.3 are constructive, i.e., in fact the algorithm for finding T^* was given. Hence, in inverse problem of Sect. 6.1 solving the system of the Volterra-type integral equations (6.17) and (6.28), for example, by the method of successive approximations, we uniquely find the functions $v_n(t),\ k(t)$ for $T\in[0,\ T^*]$. Then $v(x,t)$ is defined by formula (6.14) in R_T. Further solving the system of integral equations (6.53) and (6.63) for inverse problem of Sect. 6.2 by the same method of successive approximations, we uniquely determine the functions $v(x,t),\ k(t)$ for $T\in[0,\ T^*]$ in the domain R_T, thus defining the solution to the inverse problem (6.33)–(6.35), i.e., the kernel function $k(t)$ on the segment $[0,\ T^*]$.

References

1. Lesnic, D.: Inverse Problems with Applications in Science and Engineering. Leeds University, United Kingdom (2022)
2. Karuppiah, K., Kim, J.K., Balachandran, K.: Parameter identification of an integro-differential equation. Nonlinear Funct. Anal. Appl. **20**(2), 169–185 (2015)
3. Ivanchov, M., Vlasov, V.: Inverse problem for a two dimensional strongly degenerate heat equation. Electron. J. Differ. Equ. **77**, 1–17 (2018)
4. Huntul, M.J., Lesnic, D., Hussein, M.S.: Reconstruction of time-dependent coefficients from heat moments. Appl. Math. Comput. **301**, 233–253 (2017)
5. Avdonin, S., Ivanov, S., Wang, J.: Inverse problems for the heat equation with memory. Inverse Probl. Imaging **13**(1), 31–38 (2019)
6. Wu, B., Gao, Y., Yan, L., et al.: Existence and uniqueness of an inverse memory kernel for an integro-differential parabolic equation with free boundary. J. Dyn. Control Syst. **24**, 237–252 (2018)
7. Durdiev, D.K., Zhumaev, Z.Z.: Problem of determining a multidimensional thermal memory in a heat conductivity equation. Methods Funct. Anal. Topol. **25**(3), 219–226 (2019)
8. Durdiev, D.K., Zhumaev, Z.Z.: Problem of determining the thermal memory of a conducting medium differential equations. Differ. Equ. **56**(6), 785–796 (2020)
9. Durdiev, D.K., Rashidov, A.S.: Inverse problem of determining the kernel in an integro-differential equation of parabolic type. Differ. Equ. **50**(1), 110–116 (2014). (in Russian)
10. Durdiev, D.K., Shishkina, E.L., Sitnik, S.M.: The explicit formula for solution of anomalous diffusion equation in the multi-dimensional space. Lobachevskii J. Math. **42**(6), 1264–1273 (2021)
11. Schumer, R., Benson, D.A., Meerschaert, M.M., Baeumer, B.: Fractal mobile/immobile solute transport. Water Resour. Res. **39**(10), 1–12 (2023)
12. Durdiev, D.K., Jumaev, J.J.: Inverse problem of determining the kernel of integro-differential fractional diffusion equation in bounded domain. Russ Math. **67**, 1–13 (2023)
13. Kilbas, A.A., Srivastava, H.M., Trujillo, J.J.: Theory and applications of fractional differential equations. Elsevier, Amsterdam (2006)
14. Kian, Y., Yamamoto, M.: Global uniqueness in an inverse problem for time fractional diffusion equations. J. Differ. Equ. **264**(2), 1146–1170 (2018)
15. Durdiev, D.K.: Inverse coefficient problem for the time-fractional diffusion equation. Euras. J. Math. Comput. Appl. **9**(1), 44–54 (2021)
16. Ma, W., Sun, L.: Inverse potential problem for a semilinear generalized fractional diffusion equation with spatio-temporal dependent coefficients. Inverse Prob. **39**, 015005 (2023)
17. Durdiev, D.K., Rahmonov, A.A., Bozorov, Z.R.: A two-dimensional diffusion coefficient determination problem for the time-fractional equation. Math. Methods Appl. Sci. **44**(13), 10753–10761 (2021)
18. Durdiev, D.K., Rahmonov, A.A.: A multidimensional diffusion coefficient determination problem for the time-fractional equation. Turk. J. Math. **46**(6), 2250–2263 (2022)
19. Wang, H., Wu, B.: On the well-posedness of determination of two coefficient in a fractional integro differential equation. Chin. Ann. Math. **35**(3), 447–468 (2014)
20. Durdiev, D.K.: Convolution kernel determination problem for the time-fractional diffusion equation. Physica D **457**, 133959 (2024)
21. Ladyzhenskaya, A., Solonnikov, V.A., Uraltseva, N.N.: Linear and quasilinear equations of parabolic type. Nauka, Moscow, (in Russian); English translation: American Mathematical Society, Providence, Rhode Island, 1968 (1967)
22. Eidelman, S.D., Kochubei, A.N.: Cauchy problem for fractional diffusion equations. J. Differ. Equ. **199**, 211–255 (2004)
23. Henry, D.: Geometric Theory of Semy Linear Equations. Berlin, Germany (1981)
24. Durdiev, U.D.: Problem of determining the reaction coefficient in a fractional diffusion equation. Differ. Equ. **57**(9), 1195–1204 (2021)

Chapter 7
Determining of Two Unknown Functions

In this chapter, the method of integral equations, as applied in Chap. 1, is adapted to the case where two unknown functions must be determined simultaneously. Here, the dimensionality of the information corresponds to the number of unknowns to be identified. Nonlinear inverse problems are considered, specifically, the simultaneous definition of the time-dependent zeroth-order coefficient and the source function in Sect. 7.1 and the definition of the source function and the convolution kernel in Sect. 7.2 in the time-fractional diffusion equation. In Sect. 7.1, we study the initial-boundary value problem with the usual Cauchy initial condition and Dirichlet boundary conditions, and in Sect. 7.2, the initial-boundary value problem with the Cauchy initial condition and Neumann boundary conditions as a direct problem. To study inverse problems, two overdetermination conditions of the integral type are used. To establish the existence of a solution, Schauder's and Banach's fixed point theorems are applied. The local existence and global uniqueness theorems are proven.

The presentation of this chapter corresponds to the articles [1, 2], published by the author of this book jointly with his student.

7.1 Inverse Problem of Determining Time-Dependent Coefficient and Source Function

In this section, a nonlinear inverse problem involving the determination of a time-dependent zero-order coefficient and source function for the time-fractional diffusion equation with overdetermination conditions is investigated. Firstly, the direct problem is studied. Using the Fourier method, the direct problem is reduced to an equivalent system of integral equations. Then, by employing estimates of the Mittag-Leffler function and generalized singular Gronwall inequalities, an estimate for the solution of the direct problem is obtained in terms of the norms of the unknown functions. The inverse problem is further reduced to an equivalent system of integral equations.

D. K. Durdiev, *Inverse Problems for Fractional Diffusion Equations*,
Industrial and Applied Mathematics, https://doi.org/10.1007/978-981-96-5338-6_7

To establish the existence of a solution, Schauder's theorem is applied. The results include proofs of local existence and global uniqueness theorems.

7.1.1 The Potential and Source Recovering Problem Formulation

Let $0 < T < +\infty$ and $0 < l < +\infty$ be fixed numbers, and let D_T denote a rectangular domain in the (x, t)-plane defined by $0 < x < l$ and $0 < t \leq T$. Consider the nonhomogeneous time-fractional diffusion equation with a reaction term:

$$\frac{\partial^\alpha}{\partial t^\alpha}(u(x,t) - u(x,0)) - u_{xx}(x,t) + q(t)u(x,t) = p(t)f(x,t), (x,t) \in D_T, \tag{7.1}$$

with the initial condition

$$u(x,0) = \varphi(x), \quad x \in [0, l], \tag{7.2}$$

the boundary conditions

$$u|_{x=0} = u|_{x=l} = 0, \quad \varphi(0) = \varphi(l) = 0, \quad t \in [0, T], \tag{7.3}$$

where $\frac{\partial^\alpha}{\partial t^\alpha}$ is the Riemann-Liouville fractional derivative of order $0 < \alpha \leq 1$ with respect to the time variable (see (1.6)), $q(t), t > 0$ is the source control term, $p(t)f(x,t)$ is a source term, and $\varphi(x)$ is the initial temperature.

If the functions $q(t)$, $p(t)$, $\varphi(x)$, $f(x,t)$ are known, the problem of finding $u(x,t)$ satisfying (7.1)–(7.3) is called direct problem. However, the problem here is that the source functions $p(t), q(t)$ are unknown, which needs to be determined by energy conditions

$$\int_0^l \omega_i(x)u(x,t)dx = h_i(t), \; i = 1, 2, \; t \in [0, T], \tag{7.4}$$

here $\omega_i(x)$ and $h_i(t)$, $i = 1, 2$ are given functions of $x \in [0, 1]$ and $t \in [0, T]$, respectively.

A precedent model for a similar setup in microwave heating process is used in various applications in industry. The external energy is supplied to a target at a controlled level by the microwave generating equipment. However, the dielectric constant of the target material varies in space and time, resulting in spatially heterogeneous conversion of electromagnetic energy to heat. This can correspond to source term $p(t)f(x,t)$ in (7.1), where $p(t)$ is proportional to the power of the external energy source and $f(x,t)$ is the local conversion rate of microwave energy. It is needed to notice that this spatial variation of absorbing material does not greatly affect the thermal diffusivity, which is due to another material at a higher concentration(see [3]). The problem of finding a coefficient $q(t)$ together with the solution $u(x,t)$ of the heat equation with the integral overdetermination condition and different nonlocal boundary conditions are studied in [4–6].

Let $C^{2,\alpha}(D_T)$ be the class of continuous functions $u(x,t)$ together with the derivatives of the second order of the spatial variable x and fractional derivative ∂_t^α in t on D_T.

Definition 7.1 A triple of functions $(u(x,t), q(t), p(t))$ is said to be a classical solution of the inverse problem (7.1)-(7.4), if the functions $u(x,t) \in C^{2,\alpha}(D_T) \cap C(\overline{D}_T)$, $q(t) \in C[0,T]$ and $p(t) \in C[0,T]$ satisfy the equation (7.1), the condition (7.2) on $[0,l]$, and the statements (7.3), (7.4) on the interval $[0,T]$.

Suppose that the data of the problem (1)–(4) the functions φ, f and $h_i (i=1,2)$ satisfy the following assumptions:

(K1) $\varphi(x) \in \Phi^3[0,l]$ *(for definition of* $\Phi^k[0,l]$ *see Sect. 1.1.3);*

(K2) $f(x,t) \in C^{3,1}_{x,t}(\overline{D}_T)$, $f^{(4)}_{xxxx}(x,t) \in C\left([0,T],\ L_2(D_T)\right)$, $f(0,t) = f(l,t) = f''(0,t) = f''(l,t) = 0$;

(K3) $h_i(t) \in C^2[0,T]$, $\int_0^l \omega_i(x)\varphi(x)dx = h_i(0), i = 1,2$;

(K4) $\omega_i(x) \in C^2[0,l]$, $\omega_i(0) = \omega_i(l) = 0, i = 1,2$.

7.1.2 Initial-Boundary Value Problem for Nonhomogeneous Equation

In this section, we study the existence and uniqueness of the solution to the direct problem. Direct problem is to find $u(x,t)$ belonging to the class of functions $C^{2,\alpha}(D_T) \cap C(\overline{D}_T)$ and satisfying relations (7.1)–(7.3).

The solution to problem (7.1)–(7.3) can be expressed in the form:

$$u(x,t) = \sum_{n=1}^{\infty} u_n(t) \sin(\lambda_n x), \tag{7.5}$$

where

$$\lambda_n = \frac{\pi n}{l}, \quad u_n(t) = \sqrt{\frac{2}{l}} \int_0^l u(x,t) \sin(\lambda_n x) dx.$$

To determine of the desired functions $u_n(t)$ using separation of variables from (7.1), (7.5), we obtain

$$\left(\partial_t^\alpha u_n\right)(t) + \lambda_n^2 u_n(t) = F_n(t; u, p, q, f), \tag{7.6}$$

where

$$F_n(t; u, q, f) := p(t) f_n(t) - q(t) u_n(t),$$

$$f_n(t) = \sqrt{\frac{2}{l}} \int_0^l f(x, t) \sin(\lambda_n x) dx.$$

The initial condition (7.2) give

$$u_n(0) = \varphi_n, \quad \varphi_n = \sqrt{\frac{2}{l}} \int_0^l \varphi(x) \sin(\lambda_n x) dx. \tag{7.7}$$

Based on Theorem 1.1 (see Sect. 1.1.3) we conclude that the initial-value problem (7.6), (7.7) is equivalent in the space $C[0, T]$ to the Volterra integral equation of the second kind for determining $u_n(t)$ (for each $n \in \mathbb{N}$)

$$u_n(t) = \varphi_n E_{\alpha,1}(-\lambda_n^2 t^\alpha) + \int_0^t (t-\tau)^{\alpha-1} E_{\alpha,\alpha}\left(-\lambda_n^2 (t-\tau)^\alpha\right) p(\tau) f_n(\tau) d\tau +$$
$$- \int_0^t (t-\tau)^{\alpha-1} E_{\alpha,\alpha}\left(-\lambda_n^2 (t-\tau)^\alpha\right) q(\tau) u_n(\tau) d\tau. \tag{7.8}$$

Estimating the function $u_n(t)$ for $t \in [0, T]$, we obtain the following integral inequality

$$|u_n(t)| \le |\varphi_n| + \frac{t^\alpha \|f_n\| \|p\|}{\Gamma(\alpha+1)} + \frac{\|q\|}{\Gamma(\alpha)} \int_0^t (t-\tau)^{\alpha-1} |u_n(t)| d\tau,$$

where $\|q\| = \max_{t\in[0,T]} |q(t)|, \ \|p\| = \max_{t\in[0,T]} |p(t)|, \ \|f_n\| = \max_{t\in[0,T]} |f_n(t)|$.

Applying Lemma 1.4 (see Sect. 1.1.3) to the above inequality, we arrive at the following estimate

$$|u_n(t)| \le \left(|\varphi_n| + \frac{t^\alpha \|f_n\| \|p\|}{\Gamma(\alpha+1)} \right) E_{\alpha,1}(\|q\| t^\alpha). \tag{7.9}$$

Using equality (7.6), we have an estimate for $\partial^\alpha u_n(t)$:

$$|\partial^\alpha u_n(t)| \le (\lambda_n^2 + \|q\|) \left(|\varphi_n| + \frac{t^\alpha \|f_n\| \|p\|}{\Gamma(\alpha+1)} \right) E_{\alpha,1}(\|q\| t^\alpha) + \|p\| \|f_n\|.$$

Thus we have proved the following lemma:

Lemma 7.1 *For any $t \in [0, T]$ and $n \in \mathbb{N}$, the following estimates are valid:*

$$|u_n(t)| \le \left(|\varphi_n| + \frac{t^\alpha \|f_n\| \|p\|}{\Gamma(\alpha+1)} \right) E_{\alpha,1}(\|q\| t^\alpha),$$

$$|\partial^{\alpha} u_n(t)| \le (\lambda_n^2 + \|q\|)\left(|\varphi_n| + \frac{t^{\alpha}\|f_n\|\|p\|}{\Gamma(\alpha+1)}\right)E_{\alpha,1}(\|q\|t^{\alpha}) + \|p\|\|f_n\|.$$

Further formally, from (7.5), by term-by-term differentiation, we obtain the series:

$$\partial_t^{\alpha} u(x,t) = \sum_{n=1}^{\infty} \partial^{\alpha} u_n(t)\sin(\lambda_n x), \tag{7.10}$$

$$u_{xx}(x,t) = \sum_{n=1}^{\infty} \lambda_n^2 u_n(t)\sin(\lambda_n x), \tag{7.11}$$

In view of Lemma 7.1, the sums of the series (7.5), (7.10), and (7.11) for any $(x,t) \in D_T$, in absolute value do not exceed the numerical series

$$C_0 \sum_{n=1}^{\infty} \left(n^2|\varphi_n| + n^2\|f_n\|\right),$$

where $C_0 > 0$ is constant independent on n, φ and f.

We present the following auxiliary lemma.

Lemma 7.2 *If the conditions (K1), (K2) hold, then the following equalities are valid:*

$$\varphi_n = \frac{1}{\lambda_n^3}\varphi_n^{(3)}, \quad f_n(t) = \frac{1}{\lambda_n^3} f_n^{(3)}(t), \tag{7.12}$$

where

$$\varphi_n^{(3)} = \sqrt{\frac{2}{l}}\int_0^l \varphi^{(3)}(x)\cos(\lambda_n x)dx, \; f_n^{(3)}(t) = \sqrt{\frac{2}{l}}\int_0^l f_{xxx}^{(3)}(x,t)\cos(\lambda_n x)dx,$$

with the estimates:

$$\sum_{n=1}^{\infty} |\varphi_n^{(3)}|^2 \le \|\varphi^{(3)}\|_{L_2(0,l)}, \quad \sum_{n=1}^{\infty} |f_n^{(3)}(t)|^2 \le \|f^{(3)}\|_{L_2(0,l)\times C[0,T]}. \tag{7.13}$$

To prove this lemma, we note that if conditions (K1), (K2) are satisfied, the equalities (7.12) can easily be obtained by integrating by parts the integrals for the Fourier coefficients of the functions φ, f. The relations (7.13) are Bessel's inequalities.

If the functions $\varphi(x)$ and $f(x,t)$ satisfy the conditions of Lemma 7.2, then due to representations (7.12) and (7.13) series (7.5), (7.10) and (7.11) converge uniformly in the domain D_T, therefore, function $u(x,t)$ satisfies relations (7.1)–(7.3).

Using the above results, we obtain the following assertion.

Lemma 7.3 *Let* $(q(t), p(t)) \in C[0, T]$ *and assume that (K1), (K2) are satisfied. Then, there exists a unique solution* $u(x,t) \in C^{2,\alpha}(D_T) \cap C\left(\overline{D_T}\right)$ *to the direct problem* (7.1)–(7.3).

Remark 7.1 In the proof of Lemma 7.3, there were used Proposition 1.2 and the estimate $\left|E_{\alpha,\beta}(z)\right| \le C$, following from Proposition 1.3 (see Sect. 1.1). In fact, this ensures the inclusion $u(x,t) \in C^{2,\alpha}\left(\overline{D_T}\right)$ holds.

To study the inverse problem we need an estimate for the norm of the difference between the solution of the original integral equation (7.8) and the solution of this equation with perturbed functions $\widetilde{p}, \widetilde{q}, \widetilde{\varphi}_n, \widetilde{f}_n$. Let $\widetilde{u}_n(t)$ be solution of the integral equation (7.8) corresponding to the functions $\widetilde{p}, \widetilde{q}, \widetilde{\varphi}_n, \widetilde{f}_n$; i.e.,

$$\widetilde{u}_n(t) = \widetilde{\varphi}_n E_{\alpha,1}(-\lambda_n^2 t^\alpha) + \int_0^t (t-\tau)^{\alpha-1} E_{\alpha,\alpha}\left(-\lambda_n^2 (t-\tau)^\alpha\right) \widetilde{p}(\tau) \widetilde{f}_n(\tau) d\tau -$$

$$- \int_0^t (t-\tau)^{\alpha-1} E_{\alpha,\alpha}\left(-\lambda_n^2 (t-\tau)^\alpha\right) \widetilde{q}(\tau) \widetilde{u}_n(\tau) d\tau. \tag{7.14}$$

By considering the difference $u - \widetilde{u}$ with the help of the equations (7.8), (7.14) and introducing the notations $u - \widetilde{u} = \overline{u}_n$, $p - \widetilde{p} = \overline{p}$, $q - \widetilde{q} = \overline{q}$, $f_n - \widetilde{f_n} = \overline{f}_n$, we obtain the integral equation

$$\overline{u}_n(t) = \overline{\varphi}_n E_{\alpha,1}(-\lambda_n^2 t^\alpha) + \int_0^t (t-\tau)^{\alpha-1} E_{\alpha,\alpha}\left(-\lambda_n^2 (t-\tau)^\alpha\right) \overline{p}(\tau) f_n(\tau) d\tau +$$

$$+ \int_0^t (t-\tau)^{\alpha-1} E_{\alpha,\alpha}\left(-\lambda_n^2 (t-\tau)^\alpha\right) \widetilde{p}(\tau) \overline{f}_n(\tau) d\tau -$$

$$- \int_0^t (t-\tau)^{\alpha-1} E_{\alpha,\alpha}\left(-\lambda_n^2 (t-\tau)^\alpha\right) \overline{q}(\tau) u_n(\tau) d\tau -$$

$$- \int_0^t (t-\tau)^{\alpha-1} E_{\alpha,\alpha}\left(-\lambda_n^2 (t-\tau)^\alpha\right) \widetilde{q}(\tau) \overline{u}_n(\tau) d\tau.$$

From this, the following linear integral inequality for $|\overline{u}_n(t)|$ is derived:

$$|\overline{u}_n(t)| \le |\overline{\varphi}_n| + \frac{t^\alpha \|\overline{f}_n\| \|\widetilde{p}\|}{\Gamma(\alpha+1)} + \frac{t^\alpha \|f_n\| \|\overline{p}\|}{\Gamma(\alpha+1)} +$$

$$+ \frac{\|\overline{q}\| t^\alpha}{\Gamma(\alpha+1)} \left(|\varphi_n| + \frac{t^\alpha \|f_n\| \|p\|}{\Gamma(\alpha+1)} \right) E_\alpha(\|q\| t^\alpha) + \frac{\|\widetilde{q}\|}{\Gamma(\alpha)} \int_0^t (t-\tau)^{\alpha-1} |\overline{u}_n(\tau)| d\tau.$$

Using Lemma 7.3 from last inequality, we arrive at the estimate:

$$|\overline{u}_n(t)| \le \Big\{ |\overline{\varphi}_n| + \frac{t^\alpha \|\overline{f}_n\| \|\widetilde{p}\|}{\Gamma(\alpha+1)} + \frac{t^\alpha \|f_n\| \|\overline{p}\|}{\Gamma(\alpha+1)} +$$

$$+ \frac{\|\overline{q}\| t^\alpha}{\Gamma(\alpha+1)} \Big(|\varphi_n| + \frac{t^\alpha \|f_n\| \|p\|}{\Gamma(\alpha+1)} \Big) E_{\alpha,1}(|q(t)| t^\alpha) \Big\} E_{\alpha,1}(\|\widetilde{q}\| t^\alpha). \tag{7.15}$$

The estimate (7.15) will be used in the next subsection.

7.1.3 Solvability of the Inverse Problem: Schauder's Fixed Point Theorem

In this subsection, the inverse problem is studied as the problem of determining the functions $q(t)$, $p(t)$ from relations (7.1)–(7.4), using Schauder's fixed point theorem.

Let us multiply alternately (7.1) by $\omega_i(x)$, $(i = 1, 2)$ and integrate over x from 0 to l:

$$\int_0^l \omega_i(x) \partial_t^\alpha u(x,t) dx - \int_0^l \omega_i(x) u_{xx} dx + q(t) \int_0^l \omega_i(x) u(x,t) dx$$

$$= p(t) \int_0^l \omega_i(x) f(x,t) dx, \quad i = 1, 2, \quad x \in (0, l), \quad t \in (0, T]. \tag{7.16}$$

By Remark 7.1, all integrals in this relation are well defined.

After integrating by parts in all integrals of (7.16), in view of conditions (7.2)–(7.4) and (K4), we obtain the equality

$$\partial_t^\alpha h_i(t) - \int_0^l \omega_i''(x) u(x,t) dx + q(t) h_i(t) = p(t) \int_0^l \omega_i(x) f(x,t) dx, \quad i = 1, 2. \tag{7.17}$$

Solving the system (7.17) with respect to the unknown functions $p(t)$ and $q(t)$, we obtain the following integral equations:

$$p(t) = \frac{1}{\Delta(t)} \sum_{\substack{i,j=1 \\ i \ne j}}^{2} (-1)^j h_j(t) \Big[\partial^\alpha h_i(t) - \sum_{n=1}^{\infty} u_n(t; p; q) \lambda_n^2 \omega_{in}^{(2)} \Big], \tag{7.18}$$

$$q(t) = \frac{1}{\Delta(t)} \sum_{\substack{i,j=1 \\ i \ne j}}^{2} (-1)^j \Big[\sum_{n=1}^{\infty} f_n(t) \omega_{jn} \Big(\partial^\alpha h_i(t) - \sum_{n=1}^{\infty} u_n(t; p; q) \lambda_n^2 \omega_{in}^{(2)} \Big) \Big], \tag{7.19}$$

where

$$\Delta(t) = h_1(t) \sum_{n=1}^{\infty} f_n(t)\omega_{2n} - h_2(t) \sum_{n=1}^{\infty} f_n(t)\omega_{1n};$$

$$\omega_{in} = \sqrt{\frac{2}{l}} \int_0^l \omega_i(x) \sin(\lambda_n x) dx; \quad \omega_{in}^{(2)} = \frac{1}{\lambda_n^2} \sqrt{\frac{2}{l}} \int_0^l \omega_i''(x) \sin(\lambda_n x) dx.$$

The main result of this section is presented as follows:

Theorem 7.1 *Let (K1)-(K4) and* $|\Delta(t)| > \Delta_0 > 0$ *be satisfied. Then there exists a number* $T^* \in (0, T)$*, such that there exists a unique solution* $p(t),\ q(t) \in C[0, T^*]$ *of the inverse problem* (7.1)–(7.4).

Proof. We consider the operator equation

$$v = \Lambda v \tag{7.20}$$

where $v = (v_1, v_2) = (p, q)$ is the vector function. $\Lambda = (\Lambda_1, \Lambda_2)$ is defined by the right-hand sides of equations (7.18), (7.19):

$$\Lambda v_1(t) = v_{01}(t) - \frac{1}{\Delta(t)} \sum_{\substack{i,j=1 \\ i \neq j}}^{2} (-1)^j h_j(t) \sum_{n=1}^{\infty} u_n(t; v_1; v_2) \lambda_n^2 \omega_{in}^{(2)},$$

$$\Lambda v_2(t) = v_{02}(t) - \frac{1}{\Delta(t)} \sum_{\substack{i,j=1 \\ i \neq j}}^{2} (-1)^j \sum_{n=1}^{\infty} f_n(t)\omega_{jn} \sum_{n=1}^{\infty} u_n(t; v_1; v_2) \lambda_n^2 \omega_{in}^{(2)}, \tag{7.21}$$

Let $v_0 := (v_{01}, v_{02})$, where

$$v_{01}(t) = \frac{1}{\Delta(t)} \sum_{\substack{i,j=1 \\ i \neq j}}^{2} (-1)^j h_j(t) \partial^\alpha h_i(t),$$

$$v_{02}(t) = \frac{1}{\Delta(t)} \sum_{\substack{i,j=1 \\ i \neq j}}^{2} (-1)^j \sum_{n=1}^{\infty} f_n(t)\omega_{jn} \partial^\alpha h_i(t).$$

Consider the functional space of functions $v \in C(D_T)$ with the norm given by

$$\|v\| = \max \left\{ \max_{t \in [0,T]} |v_1(t)|, \max_{t \in [0,T]} |v_2(t)| \right\}.$$

Let us recall some concepts:

A family $\chi(x)$ of functions defined on a closed interval is said to be *uniformly bounded* if there exists a number M such that $|\chi(x)| < M$ for all x and for all χ belonging to the given family;

A family of functions is said to be *equicontinuous* if for every $\epsilon > 0$ there is a $\delta > 0$ such that $|\chi(x_1) - \chi(x_2)| < 0$ for all x_1, x_2 such that $|x_1 - x_2| < \delta$ and for all χ in the given family.

Now we present the following well-known theorem (see [7], p. 718; [8], p. 54):

Arzela's Theorem. A necessary and sufficient condition that a family of continuous functions defined on the closed interval $[a, b]$ be compact (completely continuous) in $C[a, b]$ is that this family be uniformly bounded and equicontinuous.

In this subsection, we use the following theorem (see [9]):

Theorem 7.2 *(Generalized mean value theorem). Suppose that $g \in C[a, b]$ and ${}_a\partial_t^\alpha g \in C[a, b]$ for Let $0 < \alpha \le 1$, then we have*

$$g(t) = g(a) + \frac{1}{\Gamma(\alpha)} {}_a\partial_t^\alpha g(\tau)(t-a)^\alpha$$

with $a \le \tau \le t$ for all $t \in (a, b]$. Here ${}_a\partial_t^\alpha$ is the Caputo derivative with the beginning at point a defined by

$${}_a\partial_t^\alpha g(t) = \frac{1}{\Gamma(1-\alpha)} \int_a^t (t-\tau)^{-\alpha} \frac{d}{d\tau} g(\tau) d\tau.$$

Let $S_d(0) = \{v : \|v\| \le d\}$, where d is a positive and sufficiently large number.

To show the existence of the solutions $q(t)$, $p(t)$ to the equation (7.20), we use Shauder's fixed point theorem (see Theorem (Schauder principle) of Sect. 1.2). To apply this theorem, we first determine the conditions under which the operator Λ maps the set $S_d(0)$ into itself.

First, we establish the uniform boundedness of the operator Λ. Specifically, we demonstrate that there exist $T^* \in (0, T]$ and $\rho \in (0, d]$ such that the inequality $\|\Lambda v\| \le \rho$ is hold, where

$$\|\Lambda v\| = \max\left\{ \max_{t\in[0,T]} |\Lambda v_1(t)|, \max_{t\in[0,T]} |\Lambda v_2(t)| \right\}.$$

For $a \in S_d(0)$ and $t \in [0, T]$, by Lemma 7.1, we have

$$\|\Lambda v_1\| \le v_{00} + \frac{2h_0}{\Delta_0} \sum_{n=1}^{\infty} |\omega_n| \lambda_n^2 \left(|\varphi_n| + \frac{T^\alpha \|f_n\| d}{\Gamma(\alpha+1)} \right) E_{\alpha,1}(dT^\alpha) =: \gamma_1(T),$$

$$\|\Lambda v_2\| \le v_{00} + \frac{2}{\Delta_0} \sum_{n=1}^{\infty} \|f_n\| |\omega_n| \sum_{n=1}^{\infty} \lambda_n^2 |\omega_n^{(2)}| \left(|\varphi_n| + \frac{T^\alpha \|f_n\| d}{\Gamma(\alpha+1)} \right) \times E_{\alpha,1}(dT^\alpha) =: \gamma_2(T),$$

where $v_{00} = \max\left\{\max_{t\in[0,T]}|v_{01}(t)|, \max_{t\in[0,T]}|v_{02}(t)|\right\}$,
$h_0 = \max\left\{\max_{t\in[0,T]}|h_1(t)|, \max_{t\in[0,T]}|h_2(t)|\right\}, 0 < \Delta_0 < |\Delta(t)|$.

Note that $\gamma_1(T), \gamma_2(T)$ are the positive and increasing functions of T, and

$$\gamma_1(0) := v_{00} + \frac{2h_0}{\Delta_0}\sum_{n=1}^{\infty}|\omega_n|\lambda_n^2|\varphi_n|,$$

$$\gamma_2(0) := v_{00} + \frac{2}{\Delta_0}\sum_{n=1}^{\infty}\|f_n\||\omega_n|\sum_{n=1}^{\infty}\lambda_n^2|\omega_n^{(2)}||\varphi_n|.$$

Therefore the equations $\gamma_i(T) = \rho, i = 1, 2$ have unique positive roots if $\rho > \gamma_0 = \max\{\gamma_1(0), \gamma_2(0)\}$. Then for the sufficiently small T, the operator Λ is uniformly bounded.

The following statement holds.

Lemma 7.4 *Suppose (K1)-(K3) conditions are satisfied. Then, for all T, the operator Λ is uniformly bounded and equicontinuous.*

Proof Let $(t_1, t_2) \in (0, T)$, $t_1 \neq t_2$. Using Theorem 7.2 estimate the expressions $|\Lambda v_1(t_1) - \Lambda v_1(t_2)|$, $|\Lambda v_2(t_1) - \Lambda v_2(t_2)|$ as follows

$$|\Lambda v_1(t_1) - \Lambda v_1(t_2)| \leq \frac{1}{\Gamma(\alpha+1)}\left|{}_{t_1}\partial_t^{\alpha} v_{01}(t)\right|\left|t_1 - t_2\right|^{\alpha}$$

$$+\sum_{\substack{i,j=1\\ i\neq j}}^{2}\sum_{n=1}^{\infty}\lambda_n^2|\omega_{in}^{(2)}|\Big[\frac{|u_n(t_1, v_1, v_2)|}{|\Delta(t_1)|}|h_j(t_1) - h_j(t_2)|$$

$$+\left|\sum_{\substack{i,j=1\\ i\neq j}}^{2}(-1)^j\sum_{n=1}^{\infty}\lambda_n^2\omega_{in}^{(2)}\left(\frac{h_j(t_1)}{\Delta(t_1)}u_n(t_1, v_1, v_2) - \frac{h_j(t_2)}{\Delta(t_2)}u_n(t_2, v_1, v_2)\right)\right|$$

$$+\frac{|h_j(t_2)|}{|\Delta(t_1)|}|u_n(t_1, v_1, v_2) - u_n(t_2, v_1, v_2)| + \frac{|u_n(t_2)|}{|\Delta(t_1)\Delta(t_2)|}|\Delta(t_1) - \Delta(t_2)|\Big]$$

$$\leq \frac{1}{\Gamma(\alpha+1)}\left|{}_{t_1}\partial_t^{\alpha} v_{01}(t)\right|\,|t_1 - t_2|^{\alpha}$$

$$+\frac{1}{\Gamma(\alpha+1)}\sum_{\substack{i,j=1\\ i\neq j}}^{2}\sum_{n=1}^{\infty}\lambda_n^2|\omega_{in}^{(2)}|\frac{|u_n(t_1, v_1, v_2)|}{|\Delta(t_1)|}\left|{}_{t_1}\partial_t^{\alpha} h_j(t)\right|\,|t_1 - t_2|^{\alpha}$$

$$+\frac{1}{\Gamma(\alpha+1)}\sum_{\substack{i,j=1\\ i\neq j}}^{2}\sum_{n=1}^{\infty}\lambda_n^2|\omega_{in}^{(2)}|\frac{|h_j(t_2)|}{|\Delta(t_1)|}\left|{}_{t_1}\partial_t^{\alpha} u_n(t)\right|\,|t_1 - t_2|^{\alpha}$$

$$+\frac{1}{\Gamma(\alpha+1)}\sum_{\substack{i,j=1\\i\neq j}}^{2}\sum_{n=1}^{\infty}\lambda_n^2|\omega_{in}^{(2)}|\frac{|u_n(t_2)|}{|\Delta(t_1)\Delta(t_2)|}\left|{}_{t_1}\partial_t^\alpha\Delta(t)\right||t_1-t_2|^\alpha.$$

The definition of the Caputo derivative yields

$$\begin{aligned}{}_{t_1}\partial_t^\alpha u_n(t;v_1,v_2)&=\frac{1}{\Gamma(1-\alpha)}\int_{t_1}^{t}(t-\tau)^{-\alpha}\frac{\partial}{\partial\tau}u_n(\tau;v_1,v_2)d\tau\\&=\frac{1}{\Gamma(1-\alpha)}\int_{0}^{t}(t-\tau)^{-\alpha}\frac{\partial}{\partial\tau}u_n(\tau;v_1,v_2)d\tau\\&-\frac{1}{\Gamma(1-\alpha)}\int_{0}^{t_1}(t-\tau)^{-\alpha}\frac{\partial}{\partial\tau}u_n(\tau;v_1,v_2)d\tau\\&=\partial_t^\alpha u_n(\tau;v_1,v_2)-\partial_{t_1}^\alpha u_n(\tau;v_1,v_2).\end{aligned}$$

Now, in view of (7.6), we express $\partial_t^\alpha u_n(t;p,q)$ and $\partial_{t_1}^\alpha u_n(t;p,q)$ for $t_1>0$ as follows

$$\partial_t^\alpha u_n(t;p,q)+(\lambda_n^2+q(t))u_n(t;p,q)=p(t)f_n(t)\quad for\ \ t\in(t_1,t_2),$$

$$\partial_{t_1}^\alpha u_n(t;p,q)+(\lambda_n^2+q(t_1))u_n(t_1;p,q)=p(t)f_n(t_1)\quad for\ \ t_1\in(0,T].$$

Subtracting these equations from each other, we get

$$\begin{aligned}{}_{t_1}\partial_t^\alpha u_n(t;p,q)&=p(t)f_n(t)-(\lambda_n^2+q(t))u_n(t;p,q)\\&-(p(t)f_n(t_1)-(\lambda_n^2+q(t_1))u_n(t_1;p,q)).\end{aligned}$$

Then, in view of (7.9), we have for all $t_1>0$ and $t\in(t_1,\ t_2)$ that

$$\begin{aligned}\left|{}_{t_1}\partial_t^\alpha u_n(t;p,q)\right|&\le 2\|p\|\|f_n\|\\&+2(\lambda_n^2+\|q\|)\Big(|\varphi_n|+\frac{T^\alpha}{\Gamma(\alpha+1)}\|f_n\|\|p\|\Big)E_\alpha(\|q\|T^\alpha).\end{aligned}$$

Using the above estimate, we get

$$\begin{aligned}|\Lambda v_1(t_1)-\Lambda v_1(t_2)|&\le\frac{1}{\Gamma(\alpha+1)}\Big[|{}_{t_1}\partial_t^\alpha v_{01}(t)|\\&+\frac{2h_0}{\Delta_0}\sum_{\substack{i,j=1\\i\neq j}}^{2}\sum_{n=1}^{\infty}\lambda_n^2|\omega_{in}^{(2)}|\Big(|\varphi_n|+\frac{T^\alpha}{\Gamma(\alpha+1)}\|f_n\|d\Big)E_\alpha(dT^\alpha))\\&+\frac{2h_0}{\Delta_0}\sum_{\substack{i,j=1\\i\neq j}}^{2}\sum_{n=1}^{\infty}\lambda_n^2|\omega_{in}^{(2)}|\Big(2d\|f_n\|\end{aligned}$$

$$+ 2(\lambda_n^2 + d)\Big(|\varphi_n| + \frac{T^\alpha}{\Gamma(\alpha+1)}\|f_n\|d\Big)E_\alpha(dT^\alpha)\Big)$$

$$+ \frac{1}{\Delta_0}\sum_{\substack{i,j=1\\ i\neq j}}^{2}\sum_{n=1}^{\infty}\lambda_n^2|\omega_{in}^{(2)}|\left(|\varphi_n| + \frac{T^\alpha}{\Gamma(\alpha+1)}\|f_n\|d\right)E_\alpha(dT^\alpha))\Big]|t_1-t_2|^\alpha .$$

$$\|\Lambda v_2(t_1) - \Lambda v_2(t_2)\| \le \frac{1}{\Gamma(\alpha+1)}\Big[\ \big\|{}_{t_1}\partial_t^\alpha v_{02}(t)\big\|_{C[0,T]}$$

$$+ \sum_{\substack{i,j=1\\ i\neq j}}^{2}\sum_{n=1}^{\infty}\sum_{m=1}^{\infty}\lambda_n^2|\omega_{jm}||\omega_{in}^{(2)}|\Big(\frac{|f_m(t_1)|}{|\Delta(t_1)|}|{}_{t_1}\partial_t^\alpha v_n(t)| + \frac{|u_n(t_1)|}{|\Delta(t_1)|}|{}_{t_1}\partial_t^\alpha f_m(t)|$$

$$+ \frac{|f_m(t_2)|}{|\Delta(t_1)||\Delta(t_2)|}|{}_{t_1}\partial_t^\alpha\Delta(t)|\Big)\Big]|t_1-t_2|^\alpha \le \frac{1}{\Gamma(\alpha+1)}\Big[\ \big\|{}_{t_1}\partial_t^\alpha v_{02}(t)\big\|_{C[0,T]}$$

$$+ \sum_{\substack{i,j=1\\ i\neq j}}^{2}\sum_{n=1}^{\infty}\sum_{m=1}^{\infty}\lambda_n^2|\omega_{jm}||\omega_{in}^{(2)}|\Big(\frac{|f_m|}{|\Delta_0|}\Big(2d\|f_n\|$$

$$+ 2(\lambda_n^2 + d)\Big(|\varphi_n| + \frac{T^\alpha}{\Gamma(\alpha+1)}\|f_n\|d\Big)E_\alpha(dT^\alpha)\Big)$$

$$+ \frac{|{}_{t_1}\partial_t^\alpha f_m(t)|}{|\Delta_0|}\left(|\varphi_n| + \frac{T^\alpha}{\Gamma(\alpha+1)}\|f_n\|d\right)E_\alpha(dT^\alpha)$$

$$+ \frac{|f_m(t_2)|}{|\Delta_0|^2}|{}_{t_1}\partial_t^\alpha\Delta(t)|\Big)\Big]|t_1-t_2|^\alpha.$$

Due to the fulfillment of the conditions of Theorem 7.1, the series on the right sides of these estimates are uniformly convergent.

Fix an arbitrary $\epsilon > 0$. From the above estimates, it follows that there exists $\delta = \delta(\epsilon)$ such that for any $t_1,\ t_2 \in [0, T]$ satisfying condition $|t_1 - t_2| < \delta$, the inequality

$$|\Lambda q(t_1) - \Lambda q(t_2)| \le \epsilon$$

holds.

Therefore, the set $\Lambda\ (S_d(0))$ is equicontinuous. □

Let us prove the uniqueness of this solution.

Lemma 7.5 *If the operator equation (7.20) has a solution for all* $v = (p, q) \in S_d(0)$*, then it is unique in the class* $C[0, T]$.

Proof Let there be two solutions (p, q) and $(\tilde{p}, \tilde{q})$ of the equation (7.20) with the same data $\varphi,\ f,\ h$. We denote their differences by $\overline{q}(t) = q - \tilde{q},\ \ \overline{p}(t) = p - \tilde{p}$. For the difference $\overline{q}(t), \overline{p}(t)$ we obtain the integral equations

$$\overline{q}(t) = -\frac{1}{\Delta(t)}\sum_{\substack{i,j=1\\ i\neq j}}^{2}(-1)^j h_j(t)\sum_{n=1}^{\infty}\overline{u}_n\,(t; q,\, p)\,\lambda_n^2\omega_{in}^{(2)}, \tag{7.22}$$

$$\overline{p}(t) = -\frac{1}{\Delta(t)} \sum_{\substack{i,j=1 \\ i \neq j}}^{2} (-1)^j \sum_{n=1}^{\infty} f_n(t)\omega_{jn} \sum_{n=1}^{\infty} \overline{u}_n\left(t; q, p\right) \lambda_n^2 \omega_{in}^{(2)}, \tag{7.23}$$

where $\overline{u}_n\left(t; q, p\right) = u_n\left(t; q, p\right) - \tilde{u}_n\left(t; \tilde{q}, \tilde{p}\right)$. For $\overline{u}_n(t)$ from (7.15) it follows estimate

$$|\overline{u}_n(t)| \leq \left\{ |\overline{\varphi}_n| + \frac{t^{\alpha} \|\overline{f}_n\| \|\tilde{p}\|}{\Gamma(\alpha+1)} + \frac{t^{\alpha} \| f_n \| \|\overline{p}\|}{\Gamma(\alpha+1)} + \right.$$

$$\left. + \frac{\|\overline{q}\| t^{\alpha}}{\Gamma(\alpha+1)} \left(|\varphi_n| + \frac{t^{\alpha} \| f_n \| \| p \|}{\Gamma(\alpha+1)} \right) E_{\alpha,1}(|q(t)| t^{\alpha}) \right\} E_{\alpha,1}(\|\overline{q}\| t^{\alpha}). \tag{7.24}$$

From this and taking into account (7.22), (7.23), we find $\overline{q}(t) = 0, \overline{p}(t) = 0$ i.e., $q(t) = \tilde{q}(t)$, $p(t) = \tilde{p}(t)$ for $t \in [0, T]$. □

Lemmas (7.4) and (7.5) complete the proof of Theorem 7.1.

7.2 On Determining the Source and Convolution Kernel

This section studies an inverse problem for the diffusion equation on a rectangular domain with Neumann boundary conditions and integral-type overdetermination conditions. Here, the source function and the convolution kernel are defined. The existence and uniqueness of the solution to the inverse problem are proved using the contraction mapping principle.

The partial differential equations of fractional order have begun to play an important role, particularly in the modeling of so-called anomalous phenomena, over the last few decades. The works [10–15] are entirely devoted to various applications of fractional differential equations in physics, chemistry, engineering, astrophysics, and other fields. In this context, the so-called time-fractional diffusion equation, which is derived from the classical diffusion equation by replacing the first-order time derivative with a fractional derivative of order α with $0 < \alpha < 1$, deserves special mention. Differential equations of fractional order enable the representation of long-memory effects and nonlocal dependencies in many anomalous processes.

Due to both applied and mathematical interests, the inverse problem for the diffusion equation is one of the most extensively studied problems. In [16, 17], the authors examined the uniqueness of simultaneously recovering the space-dependent two kernels for an integro-differential parabolic equation. The results of global uniqueness were provided therein.

Consider *the inverse problem* of finding functions $(u(x,t), a(t), k(t))$, which satisfy the equation

$$\partial_t^\alpha u - u_{xx} = \int_0^t k(\tau)u(x,t-\tau) + a(t)f(x,t), \quad (x,t) \in D_T, \tag{7.25}$$

with initial (7.2), boundary conditions

$$u_x(0,t) = u_x(l,t) = 0, \quad t \in [0,T] \tag{7.26}$$

and the overdetermination conditions (7.4).

As before by $C^{2,\alpha}(D_T)$, we denote the class of functions that are twice continuously differentiable with respect to x and continuously differentiable of order α in the Caputo derivative sense with respect to t in the domain D_T.

Definition 7.2 A triple of functions $(u(x,t), a(t), k(t))$ is said to be a classical solution of the inverse problem (7.25), (7.2), (7.26), (7.4), if the functions $u(x,t) \in C^{2,\alpha}(D_T) \cap C^{1,0}(\overline{D}_T)$, $a(t) \in C^1(0,T]$ and $k(t) \in C(0,T]$ satisfy the equation (7.25), the condition (7.2) on $[0,l]$, and the statements (7.26), (7.4) on the interval $[0,T]$.

Now we introduce some functional spaces.

Let $[a,b]$ be a finite interval $(-\infty < a < b < \infty)$ and $\gamma \in \mathbb{C}$ $(0 \leq \Re(\gamma) < 1)$. We introduce the weighted space $C_\gamma[a,b]$ of functions f given on $(a,b]$, such that the function $(x-a)^{1-\gamma} f(x) \in C[a,b]$, and

$$\|f\|_\gamma = \|(x-a)^{1-\gamma} f(x)\|, \quad C_0[a,b] = C[a,b].$$

For $n \in \mathbb{N}$ we denote by $C_\gamma^n[a,b]$ the Banach space of functions $f(x)$ which are continuously differentiable on $[a,b]$ up to order $n-1$ and have the derivative $f^{(n)}(x)$ of order n on $(a,b]$ such that $f^{(n)}(x) \in C_\gamma[a,b]$ with norm

$$\|f\|_{C_\gamma^n[a,b]} = \sum_{k=0}^{n-1} \|f^{(k)}\|_{C[a,b]} + \|f^{(n)}\|_{C_\gamma[a,b]},$$

and $C_\gamma^0[a,b] = C_\gamma[a,b]$.

In the next subsection, we construct Green's function for the fractional diffusion operator with the Neumann conditions.

7.2.1 Green's Function with the Neumann Boundary Conditions

Consider the problem of determining a function $w(x,t)$ hat satisfies the equations

$$\partial_t^\alpha w(x,t) - w_{xx} = F(x,t), \quad (x,t) \in D_T, \tag{7.27}$$

$$w(x,0) = \psi(x), \quad x \in [0,l], \tag{7.28}$$

$$w_x(0,t) = w_x(l,t) = 0, \quad t \in [0,T], \tag{7.29}$$

where ∂_t^α is the Caputo fractional derivative of the order $\alpha \in (0,1]$ (for definition see formula (1.4)).

We seek a solution to the problem (7.27)–(7.29) in the form of an expansion in Fourier series using the eigenfunctions of the second boundary value problem:

$$w(x,t) = \frac{1}{\sqrt{l}} w_0(t) + \sqrt{\frac{2}{l}} \sum_{n=1}^{\infty} w_n(t) cos\left(\frac{\pi n x}{l}\right), \tag{7.30}$$

every component of which must satisfies the equation (7.27) and the initial condition (7.28), i.e.,

$$\partial_t^\alpha w_n(t) + \left(\frac{\pi n x}{l}\right)^2 w_n(t) = F_n(t), \qquad w_n(0) = \psi_n, \tag{7.31}$$

where

$$F_0(t) = \frac{1}{\sqrt{l}} \int_0^l F(\xi,t) d\xi, \quad F_n(t) = \sqrt{\frac{2}{l}} \int_0^l F(\xi,t) cos\left(\frac{\pi n x}{l}\right) d\xi; \tag{7.32}$$

$$\psi_0 = \frac{1}{\sqrt{l}} \int_0^l \psi(\xi) d\xi, \quad \psi_n = \sqrt{\frac{2}{l}} \int_0^l \psi(\xi) cos\left(\frac{\pi n x}{l}\right) d\xi, \ n \in \mathbb{N}. \tag{7.33}$$

Solving the Cauchy problem (7.31) (see Lemma 1.4, formula (1.25)), we obtain

$$w_0(t) = \psi_0 + \frac{1}{\Gamma(\alpha)} \int_0^t (t-\tau)^{\alpha-1} F_0(\tau) d\tau, \tag{7.34}$$

$$w_n(t) = \psi_n E_{\alpha,1}\left(-\left(\frac{\pi n x}{l}\right)^2 t^\alpha\right)$$
$$+ \int_0^t (t-\tau)^{\alpha-1} E_{\alpha,\alpha}\left(-\left(\frac{\pi n x}{l}\right)^2 (t-\tau)^\alpha\right) F_n(\tau) d\tau, \ n \in \mathbb{N}. \tag{7.35}$$

Substituting (7.34) and (7.35) into (7.30), and taking into account the formulas (7.32), (7.33), we obtain a solution to the initial-boundary problem (7.27)-(7.29):

$$w(x,t) = \int_0^l G_{\alpha,1}(x,\xi,t)\psi(\xi)d\xi$$
$$+\int_0^t \int_0^l G_{\alpha,\alpha}(x,\xi,t-\tau)F(\xi,\tau)d\xi d\tau, \quad (x,t) \in \overline{D}_T, \tag{7.36}$$

where it was introduced notation

$$G_{\alpha,\beta}(x,\xi,t) := \sqrt{\frac{2}{l}}\Bigg\{\frac{t^{\beta-1}}{\sqrt{2}\Gamma(\beta)}$$
$$+\sum_{n=1}^{\infty} t^{\beta-1} E_{\alpha,\beta}\Big[-\Big(\frac{\pi n}{l}\Big)^2 t^\alpha\Big] \cos\frac{\pi n}{l}x \cos\frac{\pi n}{l}\xi\Bigg\}. \tag{7.37}$$

The function $G_{\alpha,\beta}(x,\xi,t)$ will be called (by analogy with the classical theory of such problems) Green's function of the second initial-boundary value problem for the fractional diffusion equation. For $\beta \neq 1$, this function contains a weak singularity at the point $t = 0$.

Since

$$\frac{\partial^{1-\alpha}}{\partial t^{1-\alpha}}(1) = \frac{t^{\alpha-1}}{\Gamma(\alpha)}, \quad \frac{\partial^{1-\alpha}}{\partial t^{1-\alpha}} E_{\alpha,1}\Big[-\Big(\frac{a\pi n}{l}\Big)^2 t^\alpha\Big] = t^{\alpha-1} E_{\alpha,\alpha}\Big[-\Big(\frac{a\pi n}{l}\Big)^2 t^\alpha\Big],$$

where $\frac{\partial^{1-\alpha}}{\partial t^{1-\alpha}}$ is the Riemann-Liouville fractional derivative of the order $1-\alpha$ with respect to t (see (1.6)), then $G_{\alpha,\alpha}$ is the Riemann-Liouville derivative of the order $1-\alpha$ with respect to t of $G_{\alpha,1}$, i.e., $G_{\alpha,\alpha}(x,\xi,t) = \frac{\partial^{1-\alpha}}{\partial t^{1-\alpha}} G_{\alpha,1}(x,\xi,t)$.

The function $G_{\alpha,\beta}(x,\xi,t-\tau)$ at $\beta = 1$, $\tau = 0$ satisfies (in the generalized sense) the following equations:

$$\frac{\partial^\alpha G_{\alpha,1}}{\partial t^\alpha} = \frac{\partial^2 G_{\alpha,1}}{\partial x^2},$$

$$\lim_{t\to 0} G_{\alpha,1}(x,\xi,t) = \delta(x-\xi), \quad \frac{\partial G_{\alpha,1}}{\partial x}(0,\xi,t) = \frac{\partial G_{\alpha,1}}{\partial x}(l,\xi,t) = 0,$$

where $\delta(\cdot)$ is Dirac delta function, and at $\beta = \alpha$, the following equations hold:

$$\frac{\partial^\alpha G_{\alpha,\alpha}}{\partial t^\alpha} = \frac{\partial^2 G_{\alpha,\alpha}}{\partial x^2} + \delta(t-\tau)\delta(x-\xi),$$

$$\lim_{t\to 0} G_{\alpha,\alpha}(x,\xi,t) = 0, \quad \frac{\partial G_{\alpha,\alpha}}{\partial x}(0,\xi,t) = \frac{\partial G_{\alpha,\alpha}}{\partial x}(l,\xi,t) = 0.$$

7.2.2 Solvability of the Initial-Boundary Value Problem

In this subsection, we consider problem (7.25), (7.2), (7.26) of determining $u(x,t)$ for given functions $k(t)$, $a(t)$ and we prove the existence and uniqueness of the solution for this problem.

We will seek the solution of problem (7.25), (7.2), (7.26) in the form (7.5), with

$$u_0(t) = \frac{1}{\sqrt{l}} \int_0^l u(\xi, t) d\xi, \quad u_n(t) = \sqrt{\frac{2}{l}} \int_0^l u(x, t) cos \lambda_n x d\xi, \ n \in \mathbb{N}.$$

In view of (7.31), we obtain following equations from (7.25):

$$\left(\partial_t^\alpha u_0\right)(t) = F_0(t; u, a, k, f), \tag{7.38}$$

$$\left(\partial_t^\alpha u_n\right)(t) + \lambda_n^2 u_n(t) = F_n(t; u, a, k, f), \tag{7.39}$$

where

$$F_0(t; u, a, k) := a(t) f_0(t) - \int_0^t k(\tau) u_0(t-\tau) d\tau,$$

$$F_n(t; u, a, k) := a(t) f_n(t) - \int_0^t k(\tau) u_n(t-\tau) d\tau.$$

The initial condition (7.2) gives

$$u_0(0) = \frac{1}{\sqrt{l}} \int_0^l \vartheta(x, 0) dx = \frac{1}{\sqrt{l}} \int_0^l \varphi(x) dx = \varphi_0, \tag{7.40}$$

$$u_n(0) = \sqrt{\frac{2}{l}} \int_0^l \vartheta(x, 0) cos \lambda_n x dx = \sqrt{\frac{2}{l}} \int_0^l \varphi(x) cos \lambda_n x dx = \varphi_n, \tag{7.41}$$

where

$$\varphi_n = \sqrt{\frac{2}{l}} \int_0^l \varphi(x) cos \lambda_n x dx,$$

$$f_n(t) = \sqrt{\frac{2}{l}} \int_0^l f(x, t) cos \lambda_n x dx, \ \lambda_n = \frac{\pi n}{l}, \ n \in \mathbb{N}.$$

Due to (7.34) and (7.35) we have that the initial-value problems (7.38)–(7.40) and (7.39)–(7.41) are equivalent in the space $C[0, T]$ to the Volterra-type integral equations of the second kind

$$u_0(t) = \varphi_0 + \frac{1}{\Gamma(\alpha)} \int_0^t (t-\tau)^{\alpha-1} a(\tau) f_0(\tau) d\tau$$
$$+ \frac{1}{\Gamma(\alpha)} \int_0^t (t-\tau)^{\alpha-1} \int_0^\tau k(s) u_0(\tau - s) ds d\tau;$$
$$u_n(t) = \varphi_n E_\alpha(-\lambda_n^2 t^\alpha) + \int_0^t (t-\tau)^{\alpha-1} E_{\alpha,\alpha}(-\lambda_n^2 (t-\tau)^\alpha) a(\tau) f_n(\tau) d\tau$$
$$+ \int_0^t (t-\tau)^{\alpha-1} E_{\alpha,\alpha}(-\lambda_n^2 (t-\tau)^\alpha) \int_0^\tau k(s) u_n(\tau - s) ds d\tau.$$

We will prove existence and uniqueness of a solution to the integral equation for $t \in [0, T]$. We estimate the modulus of $u_n(t)$ on $[0, T]$ as

$$|u_0(t)| \le |\varphi_0|$$
$$+ \frac{t^\alpha}{\Gamma(\alpha+1)} \|a\| \|f_0\| + \frac{1}{\Gamma(\alpha)} t \|k\| \int_0^t (t-\tau)^{\alpha-1} |u_0(\tau)| d\tau,$$

$$|u_n(t)| \le |\varphi_n|$$
$$+ \frac{t^\alpha}{\Gamma(\alpha+1)} \|a\| \|f_n\| + \frac{1}{\Gamma(\alpha)} t \|k\| \int_0^t (t-\tau)^{\alpha-1} |u_n(\tau)| d\tau.$$

Applying Lemma 1.4 to the last relation, we obtain the following estimate

$$|u_n(t)| \le \left(|\varphi_n| + \frac{t^\alpha}{\Gamma(\alpha+1)} \|a\| \|f_n\| \right) E_\alpha(t^{\alpha+1} \|k\|), n = 0, 1, 2, \ldots.$$

Using this, from equalities (7.38) and (7.39), we have an estimate for $\partial_t^\alpha u_n(t)$:

$$|\partial_t^\alpha u_n(t)| \le (\lambda_n^2 + \|k\| T)$$
$$\times \left(|\varphi_n| + \frac{t^\alpha}{\Gamma(\alpha+1)} \|a\| \|f_n\| \right) E_\alpha(t^{\alpha+1} \|k\|) + \|a\| \|f_n\|, \quad n = 0, 1, 2, \ldots.$$

The above calculations prove the validity of the following lemma.

Lemma 7.6 *For any $t \in [0, T]$, the estimates*

$$|u_n(t)| \le \overline{C}_1 \left(|\varphi_n| + \|f_n\| \right), \quad |\partial_t^\alpha u_n(t)| \le \overline{C}_2 \left(\lambda_n^2 |\varphi_n| + \lambda_n^2 \|f_n\| \right), n = 0, 1, 2, \ldots,$$

are valid, where $\overline{C}_i$, $i = 1, 2$, are positive constants independent on functions φ and f.

The following lemma expresses sufficient conditions for the existence of a solution to the direct problem:

Lemma 7.7 *If the conditions*
(1) $\varphi(x) \in \Phi^2[0,l]$ *(for definition of* $\Phi^k[0,l]$ *see Sect. 1.1.3),*
(2)

$$f(x,t) \in C^{2,1}(\overline{D}_T), f^{(3)}_{xxx}(.,t) \in C\left([0,T], L_2(0,l)\right),$$
$$f(0,t) = f(l,t) = f''(0,t) = f''(l,t) = 0$$

are fulfilled, then the following equalities are true

$$\varphi_n = \frac{1}{\lambda_n^3}\varphi_n^{(3)}, \quad f_n(t) = \frac{1}{\lambda_n^3} f_n^{(3)}(t),$$

where

$$\varphi_n^{(3)} = \sqrt{\frac{2}{l}}\int_0^l \varphi^{(3)}(x)\sin\lambda_n x dx, \quad f_n^{(3)}(t) = \sqrt{\frac{2}{l}}\int_0^l f^{(3)}_{xxx}(x,t)\sin\lambda_n x dx$$

with the following estimates

$$\sum_{n=1}^{\infty}|\varphi_n^{(3)}|^2 \le \|\varphi^{(3)}\|_{L_2(0,l)}, \quad \sum_{n=1}^{\infty}|f_n^{(3)}(t)|^2 \le \|f^{(3)}\|_{L_2(0,l)\times C[0,T]}. \tag{7.42}$$

Lemma 7.7 can be proved similarly to Lemma 7.2 in Sect. 7.1.
Using the above calculations, we have the following main result.

Lemma 7.8 *Let* $a(t), k(t) \in C[0,T]$. *If conditions of Lemmas 7.6 and 7.7 are satisfied, then there exists a unique solution* $u(x,t) \in C^{2,\alpha}(D_T) \cap C^{1,0}(\overline{D}_T)$ *for the direct problem (7.25), (7.2), (7.26).*

Next we will reduce the inverse problem to a form convenient for research.
Auxiliary problem.

Lemma 7.9 *Problem (7.25), (7.2), (7.26), (7.4) is equivalent to the problem of finding the functions* $(\vartheta(x,t), a(t), k(t))$ *from the equations*

$$\frac{\partial^\alpha}{\partial t^\alpha}\vartheta - \vartheta_{xx} = k(t)\varphi(x)$$
$$+\int_0^t k(\tau)\vartheta(x,t-\tau)d\tau + a'(t)f(x,t) + a(t)f_t(x,t), \quad (x,t) \in D_T, \tag{7.43}$$

$$I_t^{1-\alpha}\vartheta(x,t)\Big|_{t=0} = \varphi''(x) - a(0)f(x,0), \quad x \in [0,l], \tag{7.44}$$

$$\vartheta_x(x,t)\Big|_{x=0} = \vartheta_x(x,t)\Big|_{x=l} = 0, \quad t \in [0,T], \tag{7.45}$$

$$\int_0^l \omega_i(x)\vartheta(x,t)dx = h_i'(t), \quad i = 1,2, \quad t \in [0,T], \tag{7.46}$$

where $\frac{\partial^\alpha}{\partial t^\alpha}$ *is the Riemann-Liouville time-fractional derivative,* $\vartheta(x,t) = u_t(x,t)$.

Proof Introducing new function $\vartheta(x,t) = u_t(x,t)$, we differentiate the equalities (7.25), (7.4) and (7.26) with respect to t. For this, we use the following equalities

$$\frac{\partial}{\partial t}\partial_t^\alpha u(x,t) = \frac{\partial}{\partial t}\frac{1}{\Gamma(1-\alpha)}\int_0^t \frac{u_\tau(x,\tau)}{(t-\tau)^\alpha}d\tau$$
$$= \frac{1}{\Gamma(1-\alpha)}\frac{\partial}{\partial t}\int_0^t \frac{\vartheta(x,\tau)}{(t-\tau)^\alpha}d\tau =: \frac{\partial^\alpha}{\partial t^\alpha}\vartheta(x,t).$$

As a result, one has the problem of finding functions $\vartheta(x,t)$, $k(t)$, $a(t)$. The initial condition (7.43) was obtained from equality (7.25) by setting $t = 0$. It is not difficult to show that the reverse changes occur. □

7.2.3 *Local Solvability of the Kernel and Source Recovering Inverse Problem*

During this subsection, let us suppose that $\omega_i(x)$, $h_i(t)$ and $\varphi(x)$ satisfy the following assumptions:

(Q1) $\varphi(x) \in \Phi^2[0,l]$,

(Q2) $f(x,t) \in C^{2,1}(\overline{D}_T)$, $f^{(3)}_{xxx}(.,t) \in L_2[0,l]$, $f(0,t) = f(l,t) = f''_{xx}(0,t) = f''_{xx}(l,t) = 0$, $\omega_i(x) \in C^2[0,l]$, $\omega_i(0) = \omega_i(l) = 0$,

(Q3) $h_i(t) \in C^1[0,T]$ and $\int_0^l \omega_i(x)\varphi(x)dx = h_i(0)$, $i = 1,2$.

Similarly to Sect. 7.2.2, we will seek the solution of problem (7.43)–(7.46) in the form

$$\vartheta(x,t) = \frac{1}{\sqrt{l}}\vartheta_0(t) + \sqrt{\frac{2}{l}}\sum_{n=1}^{\infty}\vartheta_n(t)cos\left(\frac{\pi n x}{l}\right),$$

where

$$\vartheta_0(t) = \frac{1}{\sqrt{l}}\int_0^l \vartheta(x,t)dx,\quad \vartheta_n(t) = \sqrt{\frac{2}{l}}\int_0^l \vartheta(x,t)cos\left(\frac{\pi n x}{l}\right)dx.$$

In view of formula 3.1.34 in [18], which gives a solution to the Cauchy-type problem for a time-fractional equation with the Riemann-Liouville derivative, and using the technique for deriving formula (7.36) in Sect. 7.2.2, it is straightforward to obtain the following Volterra-type integral equation for the solution to the problem (7.43)–(7.46):

$$\begin{aligned}\vartheta(x,t) &= t^{\alpha-1}\int_0^l G_{\alpha,\alpha}(x,\xi,t)(\varphi''(\xi)-a(0)f(\xi,0))d\xi \\ &+ \int_0^l\int_0^t (t-\tau)^{\alpha-1}G_{\alpha,\alpha}(x,\xi,t-\tau)k(\tau)\varphi(\xi)d\tau d\xi \\ &+ \int_0^l\int_0^t (t-\tau)^{\alpha-1}G_{\alpha,\alpha}(x,\xi,t-\tau)\int_0^\tau k(s)\vartheta(x,\tau-s)dsd\tau d\xi \\ &+ \int_0^l\int_0^t (t-\tau)^{\alpha-1}G_{\alpha,\alpha}(x,\xi,t-\tau)a'(\tau)f(\xi,\tau)d\tau d\xi \\ &+ \int_0^l\int_0^t (t-\tau)^{\alpha-1}G_{\alpha,\alpha}(x,\xi,t-\tau)a(\tau)f_\tau(\xi,\tau)d\tau d\xi, \end{aligned} \tag{7.47}$$

where $G_{\alpha,\beta}(x,\xi,t)$ is defined by (7.37).

We also note the equality

$$\int_0^l G_{\alpha,\alpha}(x,\xi,t)dx = C_0, \quad t\in[0,T],$$

where C_0 depends on α, l.

The inverse problem is thereby reduced to the problem of finding the functions $a(t), k(t)$ from (7.43)–(7.46). Starting from these relations, we construct a system of integral equations for the functions a and k.

By multiplying the function $\omega_i(x)$ with the equation (7.43), then integrating from 0 until l and taking into account the conditions (7.46) we obtain the equations for the unknown functions $a'(t)$ and $k(t)$.

$$\begin{aligned}k(t) &= \frac{1}{\Delta(t)}\sum_{\substack{i,j=1\\ i\neq j}}^{2}(-1)^j\rho_j(t)\Big(\frac{\partial^\alpha}{\partial t^\alpha}h_i'(t) - \int_0^t k(\tau)h_i'(t-\tau)d\tau \\ &\quad - a(t)\int_0^l \omega_i(x)f_t(x,t)dx - \int_0^l \omega_i''(x)\vartheta(x,t)dx\Big),\end{aligned} \tag{7.48}$$

$$\begin{aligned}a'(t) &= \frac{1}{\Delta(t)}\sum_{\substack{i,j=1\\ i\neq j}}^{2}(-1)^j\phi_j\Big(\frac{\partial^\alpha}{\partial t^\alpha}h_i'(t) - \int_0^t k(\tau)h_i'(t-\tau)d\tau \\ &\quad - a(t)\int_0^l \omega_i(x)f_t(x,t)dx - \int_0^l \omega_i''(x)\vartheta(x,t)dx\Big),\end{aligned} \tag{7.49}$$

where $\Delta(t) = \rho_2(t)\phi_1 - \rho_1(t)\phi_2$,

$$\rho_j(t) = \int_0^l \omega_j(x)f(x,t)dx, \quad \phi_j = \int_0^l \omega_j(x)\varphi(x)dx, \quad i = 1,2,$$

$$a(t) = a(0) + \int_0^t a'(\tau)d\tau. \tag{7.50}$$

Suppose $a(0)$ is a given number.

We write the system of equations (7.47)–(7.50) as an operator equation

$$\psi = A\psi, \tag{7.51}$$

where

$$\psi = (\psi_1, \psi_2, \psi_3, \psi_4) := \big(\vartheta(x,t),\ k(t) + a(t)\beta(t),\ a'(t) + \gamma(t)a(t),\ a(t)\big),$$

$$\beta(t) = \frac{1}{\Delta(t)}\left(\rho_2(t)\int_0^l \omega_1(x) f_t(x,t)dx - \rho_1(t)\int_0^l \omega_2(x) f_t(x,t)dx\right),$$

$$\gamma(t) = \frac{1}{\Delta(t)}\left(\int_0^l \omega_1(x) f_t(x,t)dx\phi_2 - \int_0^l \omega_2(x) f_t(x,t)dx\phi_1\right),$$

is the vector function(we denote the linear space of such vector functions by $C_4\left(\overline{D}_T\right)$), while the operator A acts in the space $C_4\left(\overline{D}_T\right)$ and has the form $A = (A_1, A_2, A_3, A_4)^T$. Unknown functions are represented by functions $\psi_1, \psi_2, \psi_3, \psi_4$ as follows

$$\vartheta(x,t) = \psi_1; \quad k(t) = \psi_2 - \psi_4\beta(t); \quad a'(t) = \psi_3 - \psi_4\gamma(t); \quad a(t) = \psi_4.$$

If we introduce the vector function

$$\begin{aligned}
&\psi_0(t) = (\psi_{01}, \psi_{02}, \psi_{03}, \psi_{04}) \\
&\quad := \Big[t^{\alpha-1}\int_0^l G_\alpha(x,\xi,t)(\varphi''(\xi) - a(0)f(\xi,0))dx, \\
&\quad \frac{1}{\Delta(t)}\sum_{\substack{i,j=1\\ i\neq j}}^{2}(-1)^j \rho_j(t)\frac{\partial^\alpha}{\partial t^\alpha}h_i'(t),\ \frac{1}{\Delta(t)}\sum_{\substack{i,j=1\\ i\neq j}}^{2}(-1)^j \phi_j\frac{\partial^\alpha}{\partial t^\alpha}h_i'(t),\ \ a(0)\Big],
\end{aligned}$$

then, according to equations (7.47)–(7.50), the components of the operator A in equation (7.51) become

$$\begin{aligned}
A_1\psi &= \psi_{01} \\
&- \int_0^l\int_0^t (t-\tau)^{\alpha-1}G_{\alpha,\alpha}(x,\xi,t-\tau)\Big(\psi_2(\tau) - \psi_4(\tau)\beta(\tau)\varphi(\xi)\Big)d\tau d\xi \\
&+ \int_0^l\int_0^t (t-\tau)^{\alpha-1}G_{\alpha,\alpha}(x,\xi,t-\tau) \\
&\times \int_0^\tau \Big(\psi_2(s) - \psi_4(s)\beta(s)\Big)\psi_1(\xi,\tau-s)dsd\tau d\xi \\
&+ \int_0^l\int_0^t (t-\tau)^{\alpha-1}G_{\alpha,\alpha}(x,\xi,t-\tau)\Big(\psi_2(\tau) - \psi_3(\tau)\gamma(\tau)\Big)f_\tau(\xi,\tau)d\tau d\xi
\end{aligned}$$

$$+\int_0^l\int_0^t (t-\tau)^{\alpha-1}G_{\alpha,\alpha}(x,\xi,t-\tau)\psi_4(\tau)f_\tau(\xi,\tau)d\tau d\xi.$$

$$A_2\psi = \frac{1}{\Delta(t)}\sum_{\substack{i,j=1\\ i\neq j}}^{2}(-1)^j\rho_j(t)$$
$$\times\Big(\int_0^t(\psi_2(\tau)-\psi_4(\tau)\beta(\tau))h_i'(t-\tau)d\tau+\int_0^l\omega_i''(x)\psi_1(x,t)dx\Big),$$

$$A_3\psi(t) = \frac{1}{\Delta(t)}\sum_{\substack{i,j=1\\ i\neq j}}^{2}(-1)^j\phi_j\Big(\int_0^t(\psi_2(\tau)-\psi_4(\tau)\beta(\tau))h_i'(t-\tau)d\tau$$
$$+\int_0^l\omega_i''(x)\psi_1(x,t)dx\Big),$$
$$A_4\psi(t) = a(0)+\int_0^t(\psi_3(\tau)-\psi_4(\tau)\gamma(\tau))d\tau.$$

We define the class of functions $C_\alpha^{2,\alpha}(D_T)$:

$$C_\alpha^{2,\alpha}(D_T) = \Big\{v(x,t):\ v(\cdot,t)\in C^2(0,l),$$
$$\text{and } v(x,\cdot),\ \partial_t^\alpha v(x,\cdot)\in C_\alpha(0,T],\ (x,t)\in D_T\Big\}$$

and norm

$$\|v\|_\alpha = \|t^{1-\alpha}v(x,t)\|_{C(D_T)}.$$

Theorem 7.3 *Let (Q1)–(Q3) and* $\|\Delta(t)\|_{C[0,T]}>\delta>0$ *be satisfied. Then, there exists a number* $T^*\in(0,T)$, *such that there is a unique solution* $\vartheta(x,t)\in C_\alpha^{2,\alpha}(D_{T^*})$, $a(t)\in C_\alpha^1[0,T^*]$, $k(t)\in C_\alpha[0,T^*]$ *of the inverse problem (7.43)–(7.46).*

Proof Let us consider the previously introduced linear space C_4 of vector functions ψ. In this space, we introduced the weighted norm by the formula $\|\psi\|_\alpha = \max_{1\le i\le 4}\{\|\psi_i\|_\alpha\}$, where $\|\psi_1\|_\alpha = \max_{(x,t)\in D_T} t^{1-\alpha}|\psi_1(x,t)|$, $\|\psi_i\|_\alpha = \max_{t\in[0,T]} t^{1-\alpha}|\psi_i(t)|$, $i=2,3,4$.

By $W(\psi_0,\|\psi_0\|)$, we denote the set of vector functions $\psi(x,t)$ that satisfy the inequality $\|\psi-\psi_0\|_\alpha\le\|\psi_0\|_\alpha$. Obviously, $\|\psi\|_\alpha\le 2\|\psi_0\|_\alpha$ for $\psi(x,t)\in W(\psi_0,\|\psi_0\|)$.

Now, we are going to prove that A is a contraction operator in the complete metric space $W(\psi_0,\|\psi_0\|)$ provided the number T is sufficiently small. Let us verify the first condition of the contraction operator. To simplify calculations, we denote

$$h_0 := \|h_i\|_{C^2[0,T]},\, i = 1, 2;\; \beta_0 = \max_{t\in[0,T]} |\beta(t)|\,;\; \gamma_0 = \max_{(t)\in[0,T]} |\gamma(t)|\,,$$

$$\varphi_0 := \|\varphi\|_{C^2[0,1]};\; \rho_0 = \|\rho_i\|_{C[0,T]},\;\; \omega_0 := \|\omega\|_{C^2[0,1]},$$

$$\phi_0 = \max\{\phi_i\},\, i = 1, 2,\;\; f_0 := \max_{(x,t)\in\overline{D}_T} |f(x,t)|.$$

Let $\psi \in W(\psi_0, \|\psi_0\|)$. Then, for $(x, t) \in D_T$, we have the estimates

$$\begin{aligned}
\|(A\psi)_1 - \psi_{01}\|_\alpha &= \max_{(x,t)\in\overline{D}_T} t^{1-\alpha}|(A\psi)_1 - \psi_{01}|\\
&\le t^\alpha C_0 B(\alpha,\alpha)\varphi_0(\|\psi_2\|_\alpha + \beta_0\|\psi_4\|_\alpha)\\
&+ t^{2\alpha} C_0 \alpha^{-1} B(\alpha, 2\alpha) B(\alpha,\alpha)(\|\psi_2\|_\alpha + \|\psi_4\|_\alpha \beta_0)\|\psi_1\|_\alpha\\
&+ t^\alpha C_0 B(\alpha,\alpha)(\|\psi_2\|_\alpha + \|\psi_4\|_\alpha + \|\psi_3\|_\alpha \gamma_0)\| f_0\\
&\le 4C_0\Big(T^\alpha B(\alpha,\alpha)\varphi_0(1+\beta_0) + T^\alpha B(\alpha,\alpha)B(\alpha,2\alpha)(1+\beta_0)\|\psi_0\|_\alpha\\
&+ T^\alpha B(\alpha,\alpha)(2+\gamma_0) f_0\Big)\|\psi_0\|_\alpha := \Lambda_1(T)\|\psi_0\|_\alpha;
\end{aligned}$$

$$\begin{aligned}
\|(A\psi)_2 - \psi_{02}\|_\alpha &= \max_{t\in[0,T]} t^{1-\alpha}|(A\psi)_2 - \psi_{02}|\\
&\le \frac{2\rho_0}{\delta}\Big(t\alpha^{-1} h_0(\|\psi_2\|_\alpha + \beta_0\|\psi_4\|_\alpha) + \omega_0\|\psi_1\|_\alpha\Big)\\
&\le \frac{4\rho_0}{\delta}\Big(2\alpha^{-1}T(1+\beta_0)h_0 + \omega_0\Big)\|\psi_0\|_\alpha := \Lambda_2(T)\|\psi_0\|_\alpha;
\end{aligned}$$

$$\begin{aligned}
\|(A\psi)_3 - \psi_{03}\|_\alpha &= \max_{t\in[0,T]} t^{1-\alpha}|(A\psi)_3 - \psi_{03}|\\
&\le \frac{2\phi_0}{\delta}\Big(t\alpha^{-1} h_0(\|\psi_2\|_\alpha + \gamma_0\|\psi_4\|_\alpha) + \omega_0\|\psi_1\|_\alpha\Big)\\
&\le \frac{4\phi_0}{\delta}\Big(2\alpha^{-1}T(1+\gamma_0)h_0 + \omega_0\Big)\|\psi_0\|_\alpha := \Lambda_3(T)\|\psi_0\|_\alpha;
\end{aligned}$$

$$\begin{aligned}
\|(A\psi)_4 - \psi_{04}\|_\alpha &= \max_{t\in[0,T]} t^{1-\alpha}|(A\psi)_4 - \psi_{04}|\\
&\le t^\alpha \alpha^{-1}(\|\psi_3\|_\alpha + \gamma_0\|\psi_4\|_\alpha) \le 2T^\alpha\alpha^{-1}(1+\gamma_0)\|\psi_0\|_\alpha\\
&:= \Lambda_4(T)\|\psi_0\|_\alpha,
\end{aligned}$$

where $B(\cdot,\cdot)$ is Euler's beta function (see [18, p. 26]).

It follows from the above estimates that the operator A maps the set $W(\psi_0, \|\psi_0\|)$ to itself if the number T satisfies the condition

$$\max_{1\le i\le 4} \Lambda_i(T) < 1. \tag{7.52}$$

Now we verify the second condition of the contraction operator. Let $\psi^k := (\psi_1^k, \psi_2^k, \psi_3^k, \psi_4^k)$, and suppose $\psi^k \in W(\psi_0, \|\psi_0\|)$ for $k = 1, 2, 3, 4$. We use the following obvious inequalities:

$$|\psi_k^1\psi_s^1 - \psi_k^2\psi_s^2|_\alpha \le |\psi_k^1 - \psi_k^2|_\alpha |\psi_s^1|_\alpha + |\psi_s^1 - \psi_s^2|_\alpha |\psi_k^2|_\alpha \le 4\|\psi_0\|_\alpha \|\psi^1 - \psi^2\|_\alpha$$

to obtain

$$\begin{aligned}\|(A\psi^1 - A\psi^2)_1\|_\alpha &= \max_{(x,t)\in\overline{D}_T} t^{1-\alpha}|(A\psi^1 - A\psi^2)_1| \\ &\le C_0\Big(T^\alpha B(\alpha,\alpha)\varphi_0(2+\beta_0+\gamma_0) \\ &+ 4T^{2\alpha}\alpha^{-1}B(\alpha,\alpha)B(\alpha,2\alpha)(1+\beta_0)\|\psi_0\|_\alpha \\ &+ T^\alpha B(\alpha,\alpha)(2+\gamma_0)f_0\Big)\|\psi^1 - \psi^2\|_\alpha := \Upsilon_1(T)\|\psi^1 - \psi^2\|_\alpha;\end{aligned}$$

$$\begin{aligned}\|(A\psi^1 - A\psi^2)_2\|_\alpha &= \max_{t\in[0,T]} t^{1-\alpha}|(A\psi^1 - A\psi^2)_2| \\ &\le \frac{2\rho_0}{\delta}\Big(\alpha^{-1}T(1+\beta_0)h_0 + \omega_0\Big)\|\psi^1 - \psi^2\|_\alpha := \Upsilon_2(T)\|\psi^1 - \psi^2\|_\alpha;\end{aligned}$$

$$\begin{aligned}\|(A\psi^1 - A\psi^2)_3\|_\alpha &= \max_{t\in[0,T]} t^{1-\alpha}|(A\psi^1 - A\psi^2)_3| \\ &\le \frac{2\phi_0}{\delta}\Big(\alpha^{-1}T(1+\gamma_0)h_0 + \omega_0\Big)\|\psi^1 - \psi^2\|_\alpha := \Upsilon_3(T)\|\psi^1 - \psi^2\|_\alpha; \\ \|(A\psi^1 - A\psi^2)_4\|_\alpha &= \max_{t\in[0,T]} t^{1-\alpha}|(A\psi^1 - A\psi^2)_4| \\ &\le T^\alpha\alpha^{-1}(1+\gamma_0)\|\psi^1 - \psi^2\|_\alpha := \Upsilon_4(T)\|\psi^1 - \psi^2\|_\alpha.\end{aligned}$$

As follows from the above estimates, the positive number T is chosen from the inequality

$$\max_{1\le i\le 4} \Upsilon_i(T) < 1, \tag{7.53}$$

then the operator A is contraction on $W(\psi_0, \|\psi_0\|)$.

If the number T is small enough for conditions (7.52), (7.53) to be satisfied, then A is a contraction operator in the subspace $W(\psi_0, \|\psi_0\|)$ of the complete metric space $C_4\left(\overline{D}_T\right)$. Then, according to the Banach principle, there exists a unique solution of equation (7.52) in $W(\psi_0, \|\psi_0\|)$. □

Remark The triple of functions $\vartheta(x, t), a(t), k(t)$ constructed above is the classical solution of the problem (7.43)–(7.46) in the domain D_T, the triple of functions $u(x, t), a(t), k(t)$ forms the classical solution to the inverse problem (7.25), (7.2), (7.26), (7.4), where

$$u(x, t) = \varphi(x) + \int_0^t \vartheta(x, \tau)d\tau$$

and $u(x, t) \in C^{2,\alpha}(D_T) \cap C^{1,0}(\overline{D}_T)$.

7.3 Conclusion

In this chapter, inverse problems involving the simultaneous determination of two unknown functions that depend on a time variable are investigated. The dimension of the information corresponds to the number of unknowns to be determined. Nonlinear inverse problems involving the simultaneous determination of the time-dependent zero coefficient and the source function in the fractional time diffusion equation, as well as the source and kernel functions in the same equation with a convolution integral, are considered. The direct problem is formulated as an initial-boundary value problem with Dirichlet boundary conditions in Sect. 7.1 and Neumann boundary conditions in Sect. 7.2, with the overdetermination conditions specified in the form of an integral. Theorems guaranteeing the unique solvability of these problems are proved.

References

1. Durdiev, D.K., Zhumaev, Zh.Zh., Turdiev, H.H.: Inverse problem for determining time dependent coefficient and source functions in a time-fractional diffusion equation. J. Math. Sci. (2024). https://doi.org/10.1007/s10958-024-07204-y
2. Durdiev, D.K., Zhumaev, Z.Z.: Recovering source function and kernel for a time-fractional diffusion equation in the bounded domain. Lobachevskii J. Math. **45**(4), 1691–1702 (2024)
3. Ismailov, M., Çiçek, M.: Inverse source problem for a time-fractional diffusion equation with nonlocal boundary conditions. Appl. Math. Model. **40**, 4891–4899 (2016)
4. Ivanchov, M.I., Pabyrivska, N.V.: Simultaneous determination of two coefficients of a parabolic equation in the case of nonlocal and integral conditions. Ukrain. Math. J. **53**(5), 674–684 (2001)
5. Cannon, J.R., Lin, Y., Wang, S.: Determination of source parameter in parabolic equation. Meccanica **27**, 85–94 (1992)
6. Kerimov, N.B., Ismailov, M.I.: An inverse coefficient problem for the heat equation in the case of nonlocal boundary conditions. J. Math. Anal. Appl. **369**, 546–554 (2012)
7. Evans, L.: Partial Differential Equations. Grad. Stud. Math. vol. 19. American Mathematical Society, Providence (2002)
8. Kolmogorov A.N., Fomin S.V.: Theory of Functions and Functional Analysis, Volume 1: Metric and Normed Spaces. Graylock Press, Rochester, N.Y (1957)
9. Odibat, Z.M., Shawagfeh, N.T.: Generalized Taylor's formula. Appl. Math. Comput. **186**, 286–293 (2007)

10. Diethelm, K., Freed, A.D.: Reaction Engineering and Molecular Properties. Springer, Heidelberg (1999)
11. Gaul, L., Klein, P., Kempfle, S.: Damping description involving fractional operators. Mech. Syst. Signal Process. **5**, 81–881 (1991)
12. Glockle, W.G., Nonnenmacher, T.F.: A fractional calculus approach to self-similar protein dynamics. Biophys. J. **68**, 46–53 (1995)
13. Mainardi, F.: Fractals and Fractional Calculus in Continuum Mechanics. Springer, Wien (1997)
14. Metzler, R., Schick, W., Kilian, H.G., Nonnenmacher, T.F.: Relaxation in filled polymers: a fractional calculus approach. J. Chem. Phys. **103**, 7180–7186 (1995)
15. Podlubny, I.: Fractional Differential Equations. Academic, San Diego (1999)
16. Durdiev, D.K., Zhumaev, Z.Z.: Memory kernel reconstruction problems in the integrodifferential equation of rigid heat conductor. Math. Meth. Appl. Sci. **45**(14), 8374–8388 (2022)
17. Durdiev, D.K., Jumaev, J.J., Atoev, D.D.: Inverse problem on determining two kernels in integro-differential equation of heat flow. Ufa Math. J. **15**(2), 119–134 (2023)
18. Kilbas, A.A., Srivastava, H.M., Trujillo, J.J.: Theory and Application of Fractional Differential Equations. North-Holland Mathematical Studies, Elsevier, Amsterdam (2006)

GPSR Compliance

The European Union's (EU) General Product Safety Regulation (GPSR) is a set of rules that requires consumer products to be safe and our obligations to ensure this.

If you have any concerns about our products, you can contact us on ProductSafety@springernature.com

In case Publisher is established outside the EU, the EU authorized representative is:

Springer Nature Customer Service Center GmbH
Europaplatz 3
69115 Heidelberg, Germany

Batch number: 10584501

Printed by Printforce, the Netherlands